CHEMISTRY AND PHYSICS OF COMPLEX MATERIALS

Concepts and Applications

CHEMISTRY AND PHYSICS OF COMPLEX MATERIALS

Concepts and Applications

Edited by

**Maria Rajkiewicz, DSc, Wiktor Tyszkiewicz, PhD,
and Zbigniew Wertejuk, PhD**

Apple Academic Press

TORONTO   NEW JERSEY

Apple Academic Press Inc. | Apple Academic Press Inc.
3333 Mistwell Crescent | 9 Spinnaker Way
Oakville, ON L6L 0A2 | Waretown, NJ 08758
Canada | USA

©2014 by Apple Academic Press, Inc.

First issued in paperback 2021

Exclusive worldwide distribution by CRC Press, a member of Taylor & Francis Group

No claim to original U.S. Government works

ISBN 13: 978-1-77463-289-5 (pbk)
ISBN 13: 978-1-926895-60-4 (hbk)

Library of Congress Control Number: 2013951881

Library and Archives Canada Cataloguing in Publication

Chemistry and physics of complex materials: concepts and applications/edited by Maria Rajkiewicz, DSc, Wiktor Tyszkiewicz, PhD, and Zbigniew Wertejuk, PhD.

Includes bibliographical references and index.
ISBN 978-1-926895-60-4
1. Materials. 2. Materials--Analysis. 3. Chemistry, Technical. 4. Materials science.
I. Rajkiewicz, Maria, editor of compilation II. Tyszkiewicz, Wiktor, editor of compilation
III. Wertejuk, Zbigniew, editor of compilation

TA403.6.C54 2013   620.1'1   C2013-906848-1

Apple Academic Press also publishes its books in a variety of electronic formats. Some content that appears in print may not be available in electronic format. For information about Apple Academic Press products, visit our website at **www.appleacademicpress.com** and the CRC Press website at **www.crcpress.com**

ABOUT THE EDITORS

Maria Rajkiewicz, DSc

Professor Maria Rajkiewicz is Head of the Division of the Institute for Engineering of Polymer Materials and Dyes, in Warsaw, Poland. She is a well-known specialist in the field of synthesis, investigation of properties and applications of low molecular compounds, oligomers, polymers, composites, and nanocomposites. She is a contributor or co-contributor to several monographs and the author of about 100 original papers.

Wiktor Tyszkiewicz, PhD

Wiktor Tyszkiewicz, PhD, is Professor at the Military Institute of Chemistry and Radiometry in Warsaw, Poland. He has a long and varied career, including being a member of the Polish Delegation to ICCS in Vietnam; he has participated in many NATO workshops on environmental problems emanating from military installations and activities. He has headed the Biological Laboratory at the Military Institute of Chemistry and Radiometry, among many other roles.

Zbigniew Wertejuk, PhD

ZbigniewWertejuk, PhD, is Professor at the Military Institute of Chemistry and Radiometry in Warsaw, Poland. He has worked as a researcher and was also Research and Development Deputy Director and Director of the Institute. He has served as the Head of the Organization for the Prohibition of Chemical Weapons Expert Team as well as a NATO Long-Term Scientific Study national representative.

CONTENTS

LIST OF CONTRIBUTORS

M. Anachkov
Institute of Catalysis, Bulgarian Academy of Sciences, "11 Acad. G. Bonchev" str., 1113 Sofia Bulgaria

J. Aneli
Iv. Javakhishvili Tbilisi State University, Institute of Macromolecular Chemistry and Polymeric Materials, I. Chavchavadze Ave., 13, Tbilisi 0179, Georgia, E-mail: JimAneli@yahoo.com

V. A. Babkin
403343 SF VolgSABU, c. Mikhailovka, Region Volgograd, Michurina 21, Russia. E-mail: sfi@reg. avtlg.ru

M. Belitski
Institute of Catalysis, Bulgarian Academy of Sciences, 11 Bonchev Str., 1113 Sofia, Bulgaria

Ana Beltrán
Analytical Chemistry, Nutrition and Food Sciences Department, University of Alicante, P.O. Box 99, 03080. Alicante, Spain

A. Berlin
N. N. Semenov Institute of Chemical Physics, Russian Academy of Sciences, 4 Kosygin str., Moscow 119991 Russia, email: Berlin@chph.ras.ru

V. I. Binyukov
N M Emanuel Institute of Biochemical Physics, Russian Academy of Sciences, ul. Kosygina 4, 119334 Moscow, Russian Federation. Tel. (7-495) 939 71 40, Fax (7-495) 137 41 01

C. Bueno-Ferrer
Analytical Chemistry, Nutrition and Food Sciences Department, University of Alicante, P.O. Box 99, E-03080, Alicante, Spain

N. Burgos
Analytical Chemistry, Nutrition and Food Sciences Department, University of Alicante, P.O. Box 99, E-03080, Alicante, Spain

E. B. Burlakova
Emanuel Institute of Biochemical Physics of the Russian Academy of Sciences, Moscow, Russia

L. B. Dudnik
Emanuel Institute of Biochemical Physics of Russian Academy of Sciences, Moscow, Russia

N. M. Evteeva
Establishment of the Russian Academy of Sciences Institute of biochemical physics of N.M. Emanuelja, Kosygina, 4, 119991 Moscow, Russia, Fax: (095 1374101)

C. Garrigós, María
Analytical Chemistry, Nutrition and Food Sciences Department, University of Alicante, P.O. Box 99, 03080. Alicante, Spain

V. Georgiev
Institute of Catalysis, Bulgarian Academy of Sciences, 11 Bonchev Str., 1113 Sofia, Bulgaria

A. K. Haghi
University of Guilan, Rsaht, Iran, E-mail: **Haghi@Guilan.ac.ir**

Alfonso Jiménez
Analytical Chemistry, Nutrition and Food Sciences Department, University of Alicante, P.O. Box 99, E-03080, Alicante, Spain; Telephone: +34-965909660; E-mail: alfjimenez@ua.es

N. Koiava
Iv. Javakhishvili Tbilisi State University, Institute of Macromolecular Chemistry and Polymeric Materials, I. Chavchavadze Ave., 13, Tbilisi 0179, Georgia

S. M. Lomakin
Establishment of the Russian Academy of Sciences Institute of biochemical physics of N.M. Emanuelja, Kosygina, 4, 119991 Moscow, Russia, Fax: (095 1374101)

E. Markarashvili
Iv. Javakhishvili Tbilisi State University, Institute of Chemistry, I. Chavchavadze 3, Tbilisi 0128, Georgia

L. I. Matienko
N M Emanuel Institute of Biochemical Physics, Russian Academy of Sciences, ul. Kosygina 4, 119334 Moscow, Russian Federation. Tel. (7-495) 939 71 40, Fax (7-495) 137 41 01, E-mail: matienko@sky.chph.ras.ru

L. I. Mazaletskaya
Emanuel Institute of Biochemical Physics of Russian Academy of Sciences, Moscow, Russia, E-mail: lim@sky.chph.ras.ru

L. A. Mosolova
N M Emanuel Institute of Biochemical Physics, Russian Academy of Sciences, ul. Kosygina 4, 119334 Moscow, Russian Federation. Tel. (7-495) 939 71 40, Fax (7-495) 137 41 01

O. Mukbaniani
Iv. Javakhishvili Tbilisi State University, Institute of Chemistry, I. Chavchavadze 3, Tbilisi 0128, Georgia, E-mail: Omarimui@yahoo.com

Elena Pekhtasheva
G. V. Plekhanov Russian Economic University, 36, Stremyannyi way, 117997 Moscow, Russia, E-mail: pekhtashevael@mail.ru

Mercedes A. Peltzer
Analytical Chemistry, Nutrition and Food Sciences Department, University of Alicante, P.O. Box 99, 03080. Alicante, Spain

S. Rakovsky
Institute of Catalysis, Bulgarian Academy of Sciences, "11 Acad. G. Bonchev" str., 1113 Sofia Bulgaria, E-mail: rakovsky@ic.bas.bg

Marina Ramos
Analytical Chemistry, Nutrition and Food Sciences Department, University of Alicante, P.O. Box 99, 03080. Alicante, Spain; E-mail: marina.ramos@ua.es, Tel: +34965903400 Ext. 1187, Fax: +34965903527

N. I. Sheludchenko
Emanuel Institute of Biochemical Physics of Russian Academy of Sciences, Moscow, Russia

L. N. Shishkina
Emanuel Institute of Biochemical Physics of Russian Academy of Sciences, Moscow, Russia

Anamika Singh
Division of Reproductive and Child Health, Indian Council of Medical Research, New Delhi

Rajeev Singh
Division of Reproductive and Child Health, Indian Council of Medical Research, New Delhi, E-mail: 10rsingh@gmail.com

T. Tatrishvili
Iv. Javakhishvili Tbilisi State University, Institute of Chemistry, I. Chavchavadze 3, Tbilisi 0128, Georgia

Arancha Valdés
Analytical Chemistry, Nutrition and Food Sciences Department, University of Alicante, P.O. Box 99, 03080. Alicante, Spain

A. A. Volodkin
Establishment of the Russian Academy of Sciences Institute of biochemical physics of N.M. Emanuelja, Kosygina, 4, 119991 Moscow, Russia, Fax: (095 1374101)

G. E. Zaikov
Institute of Biochemical Physics, Russian Academy of Sciences, 4 Kosygin Street, 117334 Moscow, Russia. E-mail: GEZaikov@Yahoo.com

Gennady Zaikov
N.M. Emanuel Institute of Biochemical Physics, Russian Academy of Sciences, 4, Kosygin str., 119334 Moscow, Russia, E-mail: chembio@sky.chph.ras.ru

V. Zhigacheva
Emanuel Institute of Biochemical Physics of the Russian Academy of Sciences, Moscow, Russia, E-mail: zhigacheva@mail.ru

LIST OF ABBREVIATIONS

AcP	acetophenone
AHH	acute hypobaric hypoxia
AO	antioxidants
BBIC	bioluminescent bioreporter integrated circuit
CEF	compensation effect
CHP	cumylhydroperoxide
CNF	carbon nanofiber
DADPM	diaminodiphenylmethane
DADPS	diaminodiphenylsulphone
DATPO	diaminotriphenyloxide
DCHP	dicumylhydroperoxide
DMA	dynamic mechanical analysis
DMPC	dimethylphenylcarbinol
DSC	differential scanning calorimetry
DTA	differential thermal analysis
DTBP	2,6-di-*tert*-butylphenol
ECH	epichlorohydrin
EMAP	equilibrium modified atmosphere packaging
ERM	estimation of reaction mechanism
FCCP	carbonylcyanide-*p*-trifluoromethoxyphenylhydrazone
GFP	green fluorescent protein
HAART	highly active retroviral therapy
HMTA	Hexamethylenetetramine
HPLC	high performance liquid chromatography
IW	insufficient watering
LPO	lipidperoxidation
MF	melaphen
MMD	molecular mass distribution
MPDA	methylphenyldiamine
MTHPA	methyltetrahydrophthalic anhydride

NSP	nanoscale particles
OIT	oxidation induction time
ONRL	Oak Ridge National Laboratory
P	pyraphen
PEPA	polyethylene polyamine
PGR	plant growth regulators
PPSSO	polyphenylsilsesquioxanes
PVC	polyvinylcyclohexane
ROS	reactive oxygen species
SPME	solid phase microextraction
TGA	thermogravimetric analysis
TP	terpenephenols

LIST OF SYMBOLS

ΔH_m (J g^{-1})	latent heat of fusion of the sample
ΔH_m^o	theoretical latent heat of fusion for 100% crystalline PP
C_v	the axis of symmetry
$d_{H...O}$	the bond length in the complex
D_{O3}	diffusion coefficient of ozone in the solution
e	the base of natural logarithm
g	the earth acceleration
h	Planck's constant
I	the current intensity
I_o	the intensity of the chemiluminiscent signal
j	the number of axes of symmetry
k_B	Boltzmann's constant
K_d	the dissociation constant of the complex
k_L	coefficient of mass transfer in the liquid phase
m	the mass of the molecule
n	the number of carbon atoms in the cycle
N_a	Avogadro's number
r	the bond order
r_A and r_B	van der Waals radii of the reagents
s	time interval of renovation
SE	steric energy
T	absolute temperature
U	the height of tunneling occurrence
v	rate of the gas flow
V	volume of the liquid phase
W	PP weight fraction in the sample
W'_{DMPC}	the rate of DMPC formation at square termination
W_{O3}	the rate of ozone absorption
Z_o	collision factor

Greek Symbols

α	Henry's coefficient
δ	thickness of the bound layer in hydrodynamic model
η	viscosity of the solvent
ν	kinematics viscosity of the solvent
ξ	empiric coefficient
ρ	solvent density
ω	relative rate

PREFACE

This book provides innovative chapters on the growth of educational, scientific, and industrial research activities among chemists, biologists and polymer and chemical engineers and provides a medium for mutual communication between international academia and the industry. This book publishes significant research and reviews reporting new methodologies and important applications in the fields of complex materials. This book also provides a comprehensive presentation of the concepts, properties, and applications of complex materials. It also provides the first unified treatment for the broad subject of materials. The authors of each chapter use a fundamental approach to define the structure and properties of a wide range of solids on the basis of the local chemical bonding and atomic order present in the material. Emphasizing the physical and chemical origins of different material properties, this important volume focuses on the most technologically important materials being utilized and developed by scientists and engineers.

In Chapter 1, active packaging based on the release of Carvacrol and Thymol for fresh food is presented. In Chapter 2, the importance of vegetable oils as platform chemicals for synthesis of thermoplastic bio-based polyurethanes is reviewed in detail. In Chapter 3, a new concept on genetically engineered microbial biosensors is reported. In Chapter 4, degradation and stabilization of fur and leather are well described. In Chapter 5, new developments in materials chemistry and physics are presented. In Chapter 6, the quantum-chemical calculation in chemical reaction is well described. In Chapter 7, energy of a homolytic cleavage of communication OH in 4-Replaced 2,6-di-*tert*.butyl phenols is investigated. Results of quantum-chemical calculations of phenols and phenoxyls radicals a method of Hartrii-Focks with parameters UHF, RHF are used for calculations of energy of homolytic decomposition of communication OH bonds. Results of the analysis of dependences of calculations $D_{(OH)}$ from k_7 allow to recommend approach PM6 with parameter RHF for calculation $D_{(OH)}$ of sterically hindered phenols.

In Chapter 8, a study on reaction of ozone with some oxygen containing organic compounds is presented. This chapter, based on 92 references, is focused on degradation of organics by ozonation and it comprises various classes of oxygen-containing organic compounds – alcohols, ketones, ethers and hydroxybenzenes. The mechanisms of a multitude of ozone reactions with these compounds in organic solvents are discussed in detail, presenting the respective reaction schemes, and the corresponding kinetic parameters are given and some thermodynamic parameters are also listed. The dependences of the kinetics and the mechanism of the ozonation reactions on the structure of the compounds on the medium and on the reaction conditions are revealed. The various possible applications of ozonolysis are specified and discussed. All these reactions have practical importance for the protection of the environment. In Chapter 9, the kinetics and mechanism of the selective ethylbenzene oxidation is studied. In this chapter, AFM method in the analytical purposes to research the possibility of the formation of supramolecular structures on basis of heterobinuclear heteroligand triple complexes $Ni(II)(acac)_2 \cdot NaSt(or\ LiSt) \cdot PhOH$ were applied.

It is also shown what the self-assembly-driven growth seems to be due to H-bonding of triple complexes $Ni(II)(acac)_2 \cdot NaSt(or\ LiSt) \cdot PhOH$ with a surface of modified silicone, and further formation supramolecular nanostructures $\{Ni(II)(acac)_2 \cdot NaSt(or\ LiSt) \cdot PhOH\}_n$ due to directional intermolecular (phenol–carboxylate) H-bonds, and, possibly, other non-covalent interactions (van Der Waals-attractions and π-bonding).

In Chapter 10, a detailed study of composite materials on the basis of epoxy containing organosilicon compounds is presented. The synthesis of organosiloxane monomers and polymers of linear and cyclic structure with epoxy groups via catalytic oxidation reactions of compounds with vinyl group and condensation reactions of compounds with hydroxyl or ethoxyl groups with epichlorohydrin have been summarized in this chapter. The hydrosilylation reaction of linear polyhydromethylsiloxane with allyl glycidyl ether is also considered. Some ring opening reactions of epoxy groups have been studied.

The composite materials on the basis of epoxy containing organosiloxane compounds and their physical-chemical properties have been considered as well.

In Chapter 11, the reactivity of terpenephenols and their phenoxyl radicals in reactions of oxidation is investigated. The stoichiometric factors of inhibition and the rate constants of the terpenephenols (TP) with isobornyl and isocamphyl substituents were determined by the reaction with peroxy radicals of ethylbenzene. The reactivity was found to decrease for *o*-alkoxy compared with *o*-alkyl substituent caused by the intramolecular hydrogen bond formation that is conformed by FTIR-spectroscopy. The inhibitory activity for mixtures of terpenephenols with 2,6-di-*tert*-butyl phenols in the initiated oxidation of ethylbenzene was also studied in this chapter. In spite of the similar antiradical activities of terpenephenols with isobornyl and isocamphyl sunstituents, the reactivity of phenoxyl radicals formed from them are substantially different that is resulted from the kinetic data for mixtures of terpenephenols with sterically hindered phenols. In Chapter 12, reactions of ozone with hydrocarbons are explained. In Chapter 13, which is the last chapter of this book, generation of reactive oxygen species by mitochondria is introduced. In this chapter the influence of spatially-hindered phenols and plant growth regulators (PGR) on the intensity of lipid peroxidation (LPO) in membranes of mitochondria were studied. It is shown that under stress conditions, the intensity of fluorescence of LPO products in membranes of these organelles increases three to four times. Spatially-hindered phenols and PGRs decrease the LPO intensity to the control values and maintain thereby a high functional activity of mitochondria. Prevention of dysfunction of mitochondria is associated with enhancement of the resistance of plant and animal organisms to the action of stressors.

This book provides a platform for innovative research, new concepts, and novel developments in the application of complex materials to solve related issues. The book contains advances on both fundamental and applied research in various branches of chemical technologies and allied areas. The book opens with a presentation of classical models, moving on to increasingly more complex quantum mechanical and dynamical theories. Coverage and examples are drawn from complex materials in polymers. This new book:

- is a collection of chapters that highlight some important areas of current interest in polymer products and chemical processes.
- focuses on topics with more advanced methods.

- emphasizes on precise mathematical development and actual experimental details.
- analyzes theories to formulate and prove the physicochemical principles.
- gives an up-to-date and thorough exposition of the present state-of-the-art complex materials.
- familiarizes the reader with new aspects of the techniques used in the examination of polymers, including chemical, physico-chemical and purely physical methods of examination.
- describes the types of techniques now available to the chemist and technician, and discusses their capabilities, limitations and applications.

**— Maria Rajkiewicz, DSc, Wiktor Tyszkiewicz, PhD,
and Zbigniew Wertejuk, PhD**

CHAPTER 1

ACTIVE PACKAGING BASED ON THE RELEASE OF CARVACROL AND THYMOL FOR FRESH FOOD

MARINA RAMOS, ANA BELTRÁN, ARANCHA VALDÉS, MERCEDES A. PELTZER, ALFONSO JIMÉNEZ, and MARÍA C. GARRIGÓS

CONTENTS

1.1   INTRODUCTION

Active packaging is becoming an emerging food technology to improve quality and safety of food products. One of the most common approaches is based on the release of antioxidant/antimicrobial compounds from the packaging material. In this work an antifungal active packaging system based on the release of carvacrol and thymol was optimized to increase the post-harvest shelf life of fresh strawberries and bread during storage. Thermal properties of the developed packaging material were determined by differential scanning calorimetry (DSC) and thermogravimetric analysis (TGA). Volatile compounds in food samples contained in active packaging systems were monitored by using headspace solid phase microextraction followed by gas chromatography analysis (HS-SPME-GC-MS) at controlled conditions. The obtained results provided evidences that exposure to carvacrol and thymol is an effective way to enlarge the quality of strawberries and bread samples during distribution and sale.

Active packaging is the one of most relevant approaches to increase the protection and shelf-life of fresh food [12, 40]. Active packaging is broad concept that can be defined as a system in which the product, the packaging material and the environment interact in a positive way to extend food shelf-life [9, 27]. Many different active agents can be incorporated into the packaging material to improve its functionality. The migration of the active compound may be achieved by direct contact between food and the packaging material or through gas phase diffusion from the inner packaging layer to the food surface [7, 8, 14, 27].

Food can be subjected to microbial contamination that is mainly caused by bacteria, yeasts and fungi. Many of these microorganisms can cause undesirable reactions and can deteriorate organoleptic and nutritional properties of foods [21]. Different antimicrobial (AM) agents (most of them synthetic) are commonly incorporated directly into the food to diminish food spoilage by microorganisms. But this strategy has several disadvantages such as: the rising consumer refusal for food with synthetic additives; the useless addition of these agents to the food bulk when spoilage occurs primarily on the surface and the undesirable modification of food flavor. Therefore, antimicrobial packaging is a promising method to protect food from microbial contamination with no need of using synthetic additives in food composition [22].

Natural AM agents (AMs) have attracted much attention from food and packaging industries by their potential action in food preservation. According to Davidson and Zivanovic, natural AM agents can be classified by their sources in vegetal (herbs and essential oils (EOs)), animals (lysozyme, lactoferrin), microbials (nisin, natamycin) or antimicrobial polymers (chitosan) [10]. In these sense, many studies have focused on the AMs present in EOs extracted from plants (basil, thyme, oregano, cinnamon, clove, rosemary) consisting on complex mixtures of different compounds including terpenoids, esters, aldehydes, ketones, acids and alcohols [13]. These plant EOs are volatile liquids characterized by a strong odor [3]. Extracts derived from herbs and EOs contain many natural compounds such as thymol, linalool and carvacrol with a broad AM range against different pathogenic and spoilage microorganisms including Gram-negative [25, 41] and Gram-positive species [16, 24]; as well as against yeast [20] and molds [37]. In general, these additives are considered to be safe and they have been classified as GRAS (Generally Recognized As Safe) by the American Food and Drug Administration [21].

Rodriguez et al. studied the addition of EOs to a wax coating in order to develop an antimicrobial active packaging and they assessed their ability to preserve strawberries from microorganism contamination by the release of AMs from the coating [38]. During this study, there was no direct contact between the EOs and the food product. Therefore, the chemicals responsible for the inhibition of the pathogen growth must have been the natural volatile compounds (eugenol, carvacrol, thymol) present in the headspace packaging. In other study, carvacrol was added to chitosan-based films for active packaging and its antimicrobial efficiency against food spoilage microorganisms was demonstrated by using a headspace chromatographic technique [23]. Gutierrez et al. used a cinnamon-based active material to increase more than three times the shelf-life of a complex bakery product with minimal changes in the packaging and no additional manipulation steps [15].

In addition to microbial contamination, there are changes in the food macroscopic properties that also induced biochemical reactions and chemical alterations in tissues, such as changes in the volatile profile [6] and development of undesirable chemicals (i.e., ethanol or acetaldehyde) associated with changes in the respiratory paths [42]. Flavor is one of the main factors influencing consumer's food choice [32]. Volatile compounds are important contributors to flavor and odor of fruits, being aroma a very important quality attribute of strawberries. The flavor of strawberries is comprised of a complex mixture of esters, aldehydes, alcohols, furans and sulfur compounds. Esters are the main headspace volatiles. Methyl esters amount increases with the plant maturation, while ethyl esters amount do not change significantly during the fruit growth [36].

In the case of bread, aroma has been largely studied and many methods have been developed to identify the compounds responsible of flavor [5]. More than 540 different compounds have been described in the complex volatile fraction of bread [39], being alcohols, aldehydes, esters, ketones, acids, pyrazines and pyrrolines the most important quantitatively, but furans, hydrocarbons and lactones were also identified [30].

Solid Phase Microextraction (SPME) has become one of the preferred techniques in aroma analysis, offering solvent free, rapid sampling with low cost and easy preparation. Also, it is sensitive, selective and compatible with low detection limits [18]. Placed in the sample headspace, SPME is a non-destructive and non-invasive method to evaluate volatile and semi-volatile compounds. In this sense, the extraction of volatile compounds released from a great number of foods has been carried out by using HS-SPME technique [29, 33].

Besides the increasing concern in recent years about the use of synthetic polymers due to their poor biodegradability and high permanence in the environment after their use, these materials are still highly competitive by their many advantages including low cost, good processability and excellent mechanical and physical properties. Therefore, the development of antimicrobial packaging materials manufactured from synthetic polymers, such as low-density polyethylene (LDPE), high-density polyethylene (HDPE), polystyrene (PS), polyethylene terephthalate (PET) and polypropylene (PP) is important by their commercial benefits for food packaging [21].

This study focuses on the optimization of antifungal active systems based on the controlled release of carvacrol and thymol from polypropylene (PP) films. The effectiveness of the developed active films was evaluated by studying the headspace volatile composition of two food samples (bread and strawberries) stored at different conditions. This study was carried out by headspace Solid Phase Micro-Extraction (SPME) followed by gas chromatography analysis (HS-SPME-GC-MS). Results were correlated with the antimicrobial activity by visual observation of the fungal growth in the studied food (bread and strawberries). In addition, the effect of the studied additives on the thermal properties of the developed active films was also carried out.

1.2 EXPERIMENTAL

1.2.1 MATERIALS

The polymer used in this work was polypropylene (PP) ECOLEN HZ10K (Hellenic Petroleum, Greece), kindly supplied in pellets by Ashland Chemical Hispania (Barcelona, Spain). Melt flow index (MFI) was 3.2 g 10 min^{-1} determined according to ASTM-D1238 standard (230°C, 2.16 Kg), and density 0.9 g cm^{-3}. Carvacrol (98%) and thymol (99.5%) were obtained from Sigma-Aldrich (Madrid, Spain).

Strawberries and sliced bread were purchased from a local market. Damaged, non-uniform, unripe or overripe strawberries were removed and the selected fruits were stored for at least 2 h at 3°C to ensure their thermal equilibrium. Strawberries were selected for this study due to their rapid post-harvest deterioration, which constitutes a problem on their commercial distribution. Sliced bread was selected due to the increasing consumer demand for fresh bread with long shelf-life.

1.2.2 PREPARATION OF ACTIVE FILMS

Active films were prepared by melt-blending followed by compression molding by using a method previously reported [34]. A Haake Polylab QC mixer (ThermoFischer Scientific, Walham, USA) at 190°C for 6 min at rotation speed of 50 rpm was used. Both additives were introduced in the mixer once the polymer was already in the melt state to avoid unnecessary losses and to ensure their presence in the final materials. The active antimicrobial films were obtained at 190°C in a hot press (Carver Inc, Model 3850, USA) for 12 minutes. The average thickness of the films was around 200 μm measured with a Digimatic Micrometer Series 293 MDC-Lite (Mitutoyo, Japan) at five random positions around the film. The final appearance of the films was completely transparent and homogenous.

Two active formulations were prepared: PP containing 8 wt% of thymol (PPT8) and PP with 8 wt% of carvacrol (PPC8). An additional sample without any active compound was also prepared and used as control (PP0).

1.2.3 THERMAL CHARACTERIZATION OF ACTIVE FILMS

1.2.3.1 THERMOGRAVIMETRIC ANALYSIS (TGA)

TGA tests were performed in a TGA/SDTA 851 Mettler Toledo thermal analyzer (Schwarzenbach, Switzerland). Approximately 5 mg samples were weighed in alumina pans (70 μl) and were heated from 30°C to 700°C at a heating rate of 10°C min^{-1} under inert nitrogen atmosphere (flow rate 50 ml min^{-1}).

1.2.3.2 DIFFERENTIAL SCANNING CALORIMETRY (DSC)

1.2.3.2.1 DETERMINATION OF THERMAL PARAMETERS IN INERT ATMOSPHERE

DSC tests were conducted in a TA DSC Q-2000 instrument (New Castle, DE, USA) under inert nitrogen atmosphere. 3 mg samples were introduced in aluminium pans (40 μl) and were submitted to the following thermal program: heating from 0°C to 180°C at 10°C min^{-1} (3 min hold), cooling at 10°C min^{-1} to 0°C (3 min hold) and heating to 180°C at 10°C min^{-1}. The percentage of crystallinity (χ%) for each material was calculated according to the following equation:

$$\chi(\%) = \left(\frac{\Delta H_m}{W \Delta H_m^0} \right) . 100 \tag{1}$$

where ΔH_m (J g^{-1}) is the latent heat of fusion of the sample, W is the PP weight fraction in the sample, and ΔH_m° is the theoretical latent heat of fusion for 100% crystalline PP, 138 J g^{-1} [19].

1.2.3.2.2.   EVALUATION OF OXIDATION INDUCTION TIME

The antioxidant performance of carvacrol and thymol in the developed active films was also studied by DSC by determining their oxidation induction time, OIT (min) (Pospisil et al., 2003; Archodoulaki et al., 2006). The OIT value is defined as the time to the onset of an exothermic oxidation peak in oxidative atmosphere and it was determined by using oxygen and air, as the results obtained can be dependent on the type of atmosphere used for the analysis.

OIT tests were carried out by heating samples at 100°C min^{-1} under nitrogen (flow rate 50 ml min^{-1}) to the set temperature (200°C) according to ASTM-D3895-07 Standard. After 5 min, the atmosphere was switched to pure oxygen or air (50 ml min^{-1}). The heat flow was then recorded in isothermal conditions up to the detection of the exothermic peak indicating the beginning of the oxidation reaction. All tests were performed in triplicate for each formulation.

1.2.4.   STUDY OF THE EFFECTIVENESS OF THE ACTIVE FILMS TO PRESERVE PERISHABLE FOOD

1.2.4.1.   OBSERVATION OF FUNGAL GROWTH

The effectiveness of the developed active films was evaluated by putting them in contact with sliced bread and strawberries and further observing the occurrence of fungal growth on food samples with time. For this purpose, food samples were appropriately cut to be placed on the base of disposable polypropylene Petri dishes (inside dimensions: 88 mm diameter ×

12 mm high). An additional test was carried out with uncut strawberries, which were placed into a polyethylene suitable food container (250 ml, 4 cm high × 13 cm opening diameter) as shown in Fig. 1.

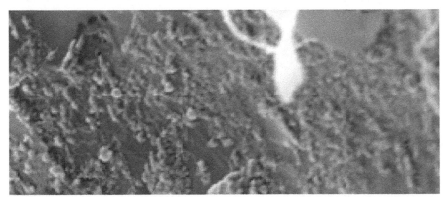

FIGURE 1 Experimental assembly used for headspace analysis of whole strawberries by HS-SPME.

Active films were cut with the appropriate dimensions to match the top of the lid of the used containers in order to release the antimicrobial studied agents (carvacrol and thymol) into the packaging headspace. The final containers were then sealed with "Parafilm" in order to avoid losses of volatile compounds and were incubated at 25°C and 50% RH in a CM 0/81 climatic chamber (Dycometal, Barcelona, Spain) during 15 days. Food samples stored with the control film (without active compounds) were also studied for comparison.

1.2.4.2.   HEADSPACE ANALYSIS BY HS-SPME-GC-MS

Food samples (whole strawberries and sliced bread) were placed into the polyethylene container described in the previous section in direct contact with the PP films and samples were extracted at selected times to determine the headspace composition. Containers were sealed and a PTFE/silicone septum was placed on their top part to allow the insertion of the SPME fiber for volatiles extraction (Fig. 1). Samples were then stored in a climatic chamber and tested at different temperatures and days of storage according to the conditions shown in Table 1. In this sense, 25°C and 4°C were selected in order to simulate ambient and refrigerated storage conditions, respectively. Three replicates were performed for each food sample and day of study.

TABLE 1 Storage and testing conditions in the headspace study of food by HS-SPME-GC-MS.

Food sample	Temperature (°C)	Days of study					
Slice bread	25	0	2	5	10	15	–
Whole strawberries	25	0	2	4	7	10	–
Whole strawberries	4	0	2	4	7	10	15

HS-SPME analysis of volatile compounds for food samples was performed according to a previously reported method applied to bread samples [29] with slight modifications. Similar conditions to those proposed by Blanda et al. [4] were used in the study with strawberries. The SPME fiber used was divinylbenzene/carboxen/polydimethylsiloxane (DVB/CAR/PDMS) 50/30 μm, StableFlex, 1 cm long mounted to an SPME manual holder assembly from Supelco (Bellefonte, PA) (Fig. 1). Prior to use, the fiber was conditioned by following the manufacturer's recommendations. The needle of the SPME device was inserted into the container through the septum and the fiber was exposed to the food sample headspace for 30 min at room temperature. The fiber was then retracted into the needle assembly, removed from the container, transferred to the injection port of the GC unit and immediately desorbed.

Analysis of volatiles produced in the headspace of bread and strawberries packed samples was performed by using a Perkin Elmer TurboMass Gold GC-MS (Boston, MA, USA) equipped with a split/splitless injector and a quadrupole mass spectrometer operating in electronic impact (EI) ionisation mode (70 eV). A SPB-5 capillary column (30 m × 0.25 mm × 0.25 μm; Supelco, Bellefonte, PA, USA) was used. The column temperature was programmed from 40°C (hold 10 min) to 120°C (hold 1 min) at 5°C min^{-1}, to 140°C at 2°C min^{-1} (hold 0 min) and to 230°C at 5°C min^{-1} (hold 8 min). Helium was used as carrier gas at a flow rate of 1 ml min^{-1}.

Ion source and GC/MS transfer line temperatures were 250 and 270°C, respectively. Injector temperature was 270°C and time for fiber desorption was fixed at 5 min in the splitless mode (1.5 min splitless-period). After every run, the SPME fiber was conditioned for 30 min at 270°C in the injector of the gas chromatograph followed by a blank analysis to avoid carryover of the fiber.

Identification of volatile compounds in strawberries and sliced bread headspace was performed in full scan mode (m/z 30–550). Carvacrol and thymol were identified by a combination of the NIST mass spectral library and gas chromatographic retention times of standard compounds. The rest of volatiles were tentatively identified by their GC/MS spectra. In this sense, the compounds having $\leq 90\%$ similarity with spectra in the NIST library were not taken into consideration. Chromatographic responses of detected volatile compounds (peak area counts) were monitored for comparative measurements of each compound in the studied samples.

1.3   RESULTS AND DISCUSSION

1.3.1.   THERMOGRAVIMETRIC ANALYSIS (TGA)

The effect of carvacrol and thymol on the thermal stability of PP films was studied by TGA under nitrogen atmosphere. The TGA patterns obtained for PP films showed a first degradation step at low temperatures (about 115°C) observed only for active films, which was associated to the thermal decomposition of carvacrol and/or thymol and a second step corresponding to the thermal degradation of the polymer matrix [34]. The first degradation step observed for active films was considered as an indirect confirmation of the presence of the active compounds (thymol and carvacrol) in the polymer matrix after processing and consequently their ability to act as active agents in these materials as it has been reported by other authors [28].

Table 2 summarizes the temperatures obtained for the main degradation step (T_{max}), ascribed to the PP thermal degradation. No significant differences were observed for T_{max} values in all samples. These results showed that the addition of carvacrol and thymol to the polymer matrix did not significantly affect its thermal degradation profile in inert nitrogen atmosphere. However, it would be expectable that a certain amount of carvacrol and thymol would be lost during processing, since materials are submitted to temperatures above the decomposition point of these additives. Therefore, the processing parameters, in particular temperature and time, should be optimized to avoid the excessive evaporation and consequent loss of these additives incorporated to PP [11].

TABLE 2 TGA, DSC and OIT parameters obtained for all samples.

Sample	T_{max} (°C)	T_c (°C)	T_m (°C)	ΔH_c (J g^{-1})	ΔH_m (J g^{-1})	χ (%)	OIT (min) Oxygen[a]	OIT (min) Air[a]
PP0	461	119	161	95.2	99.1	72	0.9 ± 0.3	1.3 ± 0.4
PPC8	462	118	161	89.2	48.1	38	8.5 ± 1.0	20.7 ± 2.8
PPT8	462	115	159	88.9	49.5	39	15.4 ± 1.7	38.8 ± 0.6

[a] **mean** ± SD ($n = 3$).

1.3.2. DIFFERENTIAL SCANNING CALORIMETRY (DSC)

1.3.2.1. DETERMINATION OF THERMAL PARAMETERS IN INERT ATMOSPHERE

Four parameters were determined for the thermal characterization of these materials by DSC (Table 2): cold-crystallization temperature, T_{cc} (°C); melting temperature, T_m (°C); crystallization enthalpy, ΔH_c (J g^{-1}) and melting enthalpy, ΔH_m (J g^{-1}). As can be seen in Table 2, melting and cold-crystallization temperatures as well as crystallization enthalpy did not show important differences for all the studied materials. Nevertheless, it should be highlighted that the melting enthalpy of PP0 sample was clearly higher than those obtained for the active materials. In this sense, crystallinity, χ(%), of samples was calculated according to Eq. (1) to evaluate if the addition of thymol and carvacrol could alter the crystallization behavior of PP. A higher value for χ(%) was determined for the PP0 sample. Therefore, it could be concluded that the PP crystallinity decreases significantly with the addition of thymol and carvacrol. This decrease in crystallinity could be due to the interactions between the polymer matrix and the additive molecules in the PP macromolecular network. A similar effect was reported for PP with the addition of some commercial synthetic antioxidants, such as Irgafos 168 and Irganox 1010 [2].

1.3.2.2.   EVALUATION OF THE OXIDATION INDUCTION TIME (OIT)

The determination of OIT is considered a simple, reliable and fast method for the evaluation of the antioxidants efficiency [31], corresponding to relative measurements of the materials stability against oxidation. The evaluation of the antioxidant performance of carvacrol and thymol in PP is important since they are supposed not only to play the role of active additives for food, but also to protect the polymer to oxidative degradation during processing and use.

The evaluation of OIT was carried out in two different atmospheres. Air was selected to get a similar situation to the real conditions during materials processing or food shelf-life, while the use of pure oxygen would represent the most aggressive conditions for oxidative degradation. Table 2 shows the results obtained for OIT in both atmospheres.

In both cases it was confirmed the higher efficiency of thymol as an antioxidant when compared to carvacrol. This behavior was also reported by other authors who demonstrated that the antioxidant efficiency of thymol was higher in sunflower oil samples [43]. In the case of air atmosphere, as expected, OIT values were higher than those obtained in pure oxygen atmosphere, since the experiment under air is less aggressive to materials [35].

In all cases, the increase in OIT values for PP with additives showed the existence of certain antioxidant effect after processing. These results are an additional confirmation that certain amounts of thymol and carvacrol are still remaining in all formulations after processing and they would be able to be released from the material to foodstuff as active additives.

1.3.3. STUDY OF THE EFFECTIVENESS OF THE ACTIVE FILMS TO PRESERVE PERISHABLE FOOD

1.3.3.1. OBSERVATION OF FUNGAL GROWTH

This study was conducted to evaluate the antimicrobial activity of the developed films and their ability to act in active packaging formulations to increase the fresh food shelf-life. It was based on the visual observation of the inhibition of the fungal growth on food samples by the action of the volatile active additives, carvacrol and thymol. In this sense, some studies by other authors showed the effectiveness of these compounds against different fungal strains of particular interest in the food industry [24].

Figure 2 shows the appearance of sliced strawberries and bread samples at the beginning of the experiment (day 0) and after the observation of microbial growth. Regarding strawberries, satisfactory results were obtained for samples in contact with the films with additives, since no fungal growth was observed until six days of storage. In the case of strawberries in contact with the pure PP film (PP0), a rapid growth of microorganisms was observed at the third day of treatment.

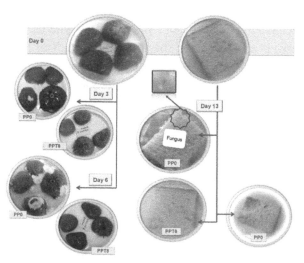

FIGURE 2 Study of the effectiveness of PP0 and active films containing 8 wt% of thymol (PPT8) to preserve cut bread and strawberries by observation of fungal growth.

On the other hand, the presence of microorganisms was observed in bread samples in contact with the PP0 film after 13 days of storage, in contrast with bread with the films with additives where no evidence of microbial contamination after 45 days of storage was observed.

However, it was noticed that strawberries lost their organoleptic properties in a few days, even before the visual evidence of fungal growth, when they were cut and stored. For this reason this study was also conducted for uncut strawberries (Fig. 3). For the PP0 film microbial growth was observed after 6 days of storage. However, strawberries in contact with the film containing 8 wt% of thymol (PPT8) remained unaltered after 13 days. At this storage time strawberries presented a physical deterioration due to the experimental storage conditions, but it is important to highlight that microbial growth was not observed until the end of the study (15 days).

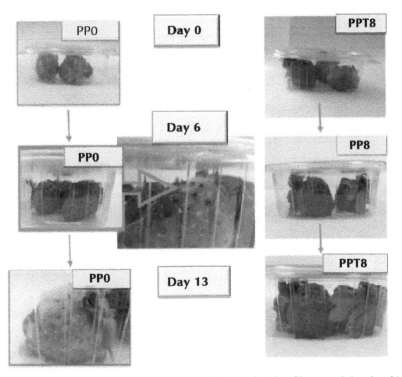

FIGURE 3   Evaluation of the effectiveness of PP0 and active film containing 8 wt% of thymol (PPT8) to preserve uncut strawberries by observation of fungal growth.

Similar studies were performed with uncut strawberries by other authors, getting satisfactory results for samples in contact with films treated with essential oils, such as cinnamon, oregano and thyme [38]. Regarding thyme and oregano essential oils, their antimicrobial activity is due to the high amount of carvacrol and thymol in their composition [17]. Other studies conducted in different fruits and vegetables also demonstrated the effectiveness of the constituents of different essential oils (eugenol, thymol, menthol or eucalyptol) to improve the organoleptic quality of food as well as to reduce the microbial growth, by using also a modified atmosphere [26].

In conclusion, results obtained from food samples in contact with PP films containing carvacrol and thymol evidenced the effectiveness of these compounds to improve the shelf-life of perishable food, such as strawberries and bread. Accordingly, these results also indicated the potential to use the developed films in active packaging systems to replace the direct addition of preservatives in food formulations.

1.3.3.2   HEADSPACE ANALYSIS BY HS-SPME-GC-MS

Figure 4 shows the levels of carvacrol, in terms of peak area counts, reached in the headspace of the containers with bread slices after 0, 2, 5, 10 and 15 days of storage at room temperature. As it can be seen, an increase in the amount of carvacrol released from the PP films was observed with time for the bread samples. A high release of carvacrol was observed at 2 days, being released more slowly after 5, 10 and 15 days of storage. This mechanism of controlled release could lead to shelf-life improvement of the stored samples retarding the post-harvest deterioration. This behavior was also observed for strawberries. Regarding the thymol release, a similar trend was shown for both test food samples.

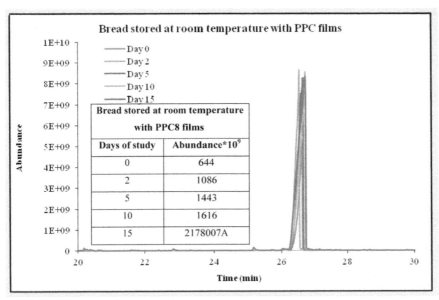

Bread stored at room temperature with PPC films

Bread stored at room temperature with PPC8 films	
Days of study	Abundance*10^9
0	644
2	1086
5	1443
10	1616
15	2178007A

FIGURE 4 Release of carvacrol in the headspace of bread slices after 0, 2, 5, 10 and 15 days of storage at room temperature.

Equilibrium modified atmosphere packaging (EMAP) is the most commonly used packaging technology to reduce the high respiration rate of strawberries. It is known that a suitable atmosphere composition can reduce the respiration rate of fruits and fungal growth with minimal alteration of organoleptic properties [36]. In this sense, Table 3 shows the compounds identified in the headspace of strawberries with PP0 films after 4 days. One of the most important processes occurring during fruit ripening is the increase in volatiles contributing to fruit aroma and flavor. The major volatiles identified for strawberries stored at room temperature include methyl-isopentanoate, 2-methyl-butylacetate, methyl-hexanoate and hexyl-acetate. The compounds methyl-butanoate and methyl-hexanoate were also found in the headspace composition of refrigerated strawberries at 4°C. These results are in accordance with those obtained by other authors when studying volatile compounds in the same food samples [4, 36].

TABLE 3 Identified compounds present in the headspace of food samples packed with PP0 films after 4 days.

Bread stored at room temperature		Strawberries stored at room temperature		Strawberries refrigerated at 4°C	
Time (min)	Compound	Time (min)	Compound	Time (min)	Compound
1.4	Ethanol	3.2	Methyl isopentano-ate	1.9	Hexane
5.7	2,4-dimethylhep-tane	5.8	2-methyl butylac-etate	3.1	Methylbutanoate
6.6	2,4-dimethyl hep-tene	7.4	Methyl hexanoate	3.9	Toluene
7.8	Isononane	10.6	Hexylacetate	6.8	Isopropyl butyr-ate
13.0	4-methyloctane			12.2	Methyl hexanoate

The addition of thymol and carvacrol to PP films significantly modified the initial atmosphere composition inside packages during the storage of food samples due to their release from the film. This fact could be related to the inhibition of the volatile identified compounds (Table 3) that did not were detected in samples in contact with PPT8 and PPC8 films after 4 days of storage.

On the other hand, ethanol was the main volatile found in the headspace of bread in contact with the PP0 film and stored at room temperature for 4 days. This compound was resulting from fermentation and/or lipid oxidation as it has been reported by other authors [30]. In this sense, commercial bread samples in contact with PPT8 or PPC8 films after 4 days of storage were characterized by significantly lower amounts of ethanol, suggesting a reduction on the lipid oxidation reactions by the presence of thymol and carvacrol. The improvement on the oxidative stability of bread could be attributed to the release of carvacrol and thymol increasing the shelf-life of bread.

From these results, it can be concluded that the release of both additives from active films to the headspace of the studied packaged foodstuff increased with the storage time, as expected. The volatiles profile obtained by HS-SPME-GC-MS was found to be different for samples in contact with PP0 and those with PPT8 and PPC8, due to the modification of the food headspace composition by the presence of these additives. Therefore, the release of thymol and carvacrol from the active PP films has shown to be effective in maintaining the quality of strawberries and bread during different storage conditions. Finally, it can be concluded that PP films with carvacrol and thymol could be a promising alternative to increase the shelf-life of different foodstuff.

1.4 CONCLUSIONS

Carvacrol and thymol have demonstrated their potential to be used as active additives in PP films for food packaging applications by their controlled antimicrobial release to foodstuff and also by the possibility to protect food from degradation processes. The addition of carvacrol and thymol did not significantly affect the thermal behavior of PP, but they modified the material cristallinity. PP films containing carvacrol and thymol showed a significant increase in OIT values, indicating that the polymer is well stabilized and a certain amount of these compounds remained in the polymer matrix after processing at relatively high temperatures. In this sense, these additives could be furthermore released from the material playing their role as antimicrobial additives. Therefore, it could be concluded that the addition of antimicrobial additives as carvacrol and thymol at 8 wt% to PP shows potential to improve the food quality and safety.

ACKNOWLEDGMENTS

Authors would like to thank Ashland Chemical Hispania for kindly supplying ECOLEN HZ10K PP (Hellenic Petroleum). Marina Ramos would like to thank University of Alicante (Spain) for UAFPU2011-48539721S predoctoral research grant.

KEYWORDS

- carvacrol
- eugenol
- parafilm
- thymol

REFERENCES

1. ASTM D 3895-07. Standard test method for oxidative-induction time of polyolefins by differential scanning calorimetry. In: Annual book of ASTM standards. *Amer. Soc. for Testing and Materials*, Philadelphia, PA, **2007**.
2. Alin, J.; Hakkarainen, M. Type of polypropylene material significantly influences the migration of antioxidants from polymer packaging to food simulants during microwave heating. *Journal of Applied Polymer Science*, **2010**, *118(2)*, 1084–1093.
3. Bakkali, F.; Averbeck, S.; Averbeck, D.; Idaomar, M. Biological effects of essential oils—A review. *Food and Chemical Toxicology*, **2008**, *46(2)*, 446–475.
4. Blanda, G.; Cerretani, L.; Cardinali, A.; Barbieri, S.; Bendini, A.; Lercker, G. Osmotic dehydrofreezing of strawberries: Polyphenolic content, volatile profile and consumer acceptance. *LWT—Food Science and Technology*, **2009**, *42(1)*, 30–36.
5. Cayot, N. Sensory quality of traditional foods. *Food Chemistry*, **2007**, *101(1)*, 154–162.
6. Chiralt, A.; Martínez-Navarrete, N.; Martínez-Monzó, J.; Talens, P.; Moraga, G.; Ayala, A.; Fito, P. Changes in mechanical properties throughout osmotic processes: Cryoprotectant effect. *J. Food Eng.*, **2001**, *49(2–3)*, 129–135.
7. Coma, V. Bioactive packaging technologies for extended shelf life of meat-based products. *Meat Science*, **2008**, *78(1–2)*, 90–103.
8. Conte, A.; Buonocore, G. G.; Bevilacqua, A.; Sinigaglia, M.; Del Nobile, M. A. Immobilization of Lysozyme on Polyvinylalcohol Films for Active Packaging Applications. *Journal of Food Protection*, **2006**, *69(4)*, 866–870.
9. Cutter, C. N. Microbial control by packaging: A review. *Critical Reviews in Food Science and Nutrition*, **2002**, *42(2)*, 151–161.
10. Davidson P. M. Z. S. The use of natural antimicrobials. *Food preservation techniques*. Boca Raton, Fla.: Woodhead Publishing Limited and CRC Press, **2003**.
11. Dobkowski, Z. Thermal analysis techniques for characterization of polymer materials. *Polymer Degradation and Stability*, **2006**, *91(3)*, 488–493.
12. Dong Sun Lee, K. L. Y.; Luciano Piergiovanni. *Food Packaging Science and Technology*. Boca Raton: CRC Press, **2008**.
13. Dorman, H. J. D.; Deans, S. G. Antimicrobial agents from plants: antibacterial activity of plant volatile oils. *Journal of Applied Microbiology*, **2000**, *88(2)*, 308–316.

14. Gemili, S.; Yemenicioğlu, A.; Altinkaya, S. A. Development of cellulose acetate based antimicrobial food packaging materials for controlled release of lysozyme. *J. Food Eng.,* **2009,** *90(4),* 453–462.
15. Gutieérrez, L.; Escudero, A.; Batlle, R. n.; Nerín, C. Effect of Mixed Antimicrobial Agents and Flavors in Active Packaging Films. *Journal of Agricultural and Food Chemistry,* **2009,** *57(18),* 8564–8571.
16. Gutiérrez, L.; Sánchez, C.; Batlle, R.; Nerín, C. New antimicrobial active package for bakery products. *Trends in Food Science & Technology,* **2009,** *20(2),* 92–99.
17. Hazzit, M.; Baaliouamer, A.; Faleiro, M. L.; Miguel, M. G. Composition of the essential oils of Thymus and Origanum species from Algeria and their antioxidant and antimicrobial activities. *Journal of Agricultural and Food Chemistry,* **2006,** *54(17),* 6314–6321.
18. Ho, C. W.; Wan Aida, W. M.; Maskat, M. Y.; Osman, H. Optimization of headspace solid phase microextraction (HS-SPME) for gas chromatography mass spectrometry (GC-MS) analysis of aroma compound in palm sugar (Arenga pinnata*). Journal of Food Composition and Analysis,* **2006,** *19(8),* 822–830.
19. Joseph, P. V.; Joseph, K.; Thomas, S.; Pillai, C. K. S.; Prasad, V. S.; Groeninckx, G.; Sarkissova, M. The thermal and crystallisation studies of short sisal fibre reinforced polypropylene composites. *Composites Part A: Applied Science and Manufacturing,* **2003,** *34(3),* 253–266.
20. Kuorwel, K. K.; Cran, M. J.; Sonneveld, K.; Miltz, J.; Bigger, S. W. Antimicrobial Activity of Natural Agents against Saccharomyces cerevisiae. *Packaging Technology and Science,* **2011a,** *24(5),* 299–307.
21. Kuorwel, K. K.; Cran, M. J.; Sonneveld, K.; Miltz, J.; Bigger, S. W. Essential Oils and Their Principal Constituents as Antimicrobial Agents for Synthetic Packaging Films. *Journal of Food Science,* **2011b,** *76(9),* R164–R177.
22. Kuorwel, K. K.; Cran, M. J.; Sonneveld, K.; Miltz, J.; Bigger, S. W. Migration of antimicrobial agents from starch-based films into a food simulant. *LWT — Food Science and Technology,* **2013,** *50(2),* 432–438.
23. Kurek, M.; Moundanga, S.; Favier, C.; Galić, K.; Debeaufort, F. Antimicrobial efficiency of carvacrol vapour related to mass partition coefficient when incorporated in chitosan based films aimed for active packaging. *Food Control,* **2013,** *32(1),* 168–175.
24. López, P.; Sánchez, C.; Batlle, R.; Nerín, C. Development of flexible antimicrobial films using essential oils as active agents. *Journal of Agricultural and Food Chemistry,* **2007a,** *55(21),* 8814–8824.
25. López, P.; Sánchez, C.; Batlle, R.; Nerín, C. Vapor-phase activities of cinnamon, thyme, and oregano essential oils and key constituents against foodborne microorganisms. *Journal of Agricultural and Food Chemistry,* **2007b,** *55(11),* 4348–4356.
26. Mastromatteo, M.; Conte, A.; Del Nobile, M. Combined use of modified atmosphere packaging and natural compounds for food preservation. *Food Engineering Reviews,* **2010a,** *2(1),* 28–38.
27. Mastromatteo, M.; Mastromatteo, M.; Conte, A.; Del Nobile, M. A. Advances in controlled release devices for food packaging applications. *Trends in Food Science & Technology,* **2010b,** *21(12),* 591–598.

28. Persico, P.; Ambrogi, V.; Carfagna, C.; Cerruti, P.; Ferrocino, I.; Mauriello, G. Nano-composite polymer films containing carvacrol for antimicrobial active packaging. *Polymer Engineering & Science*, **2009**, *49(7)*, 1447–1455.

29. Poinot, P.; Arvisenet, G.; Grua-Priol, J.; Colas, D.; Fillonneau, C.; Le Bail, A.; Prost, C. Influence of formulation and process on the aromatic profile and physical characteristics of bread. *Journal of Cereal Science*, **2008**, *48(3)*, 686–697.

30. Poinot, P.; Grua-Priol, J.; Arvisenet, G.; Rannou, C.; Semenou, M.; Bail, A. L.; Prost, C. Optimisation of HS-SPME to study representativeness of partially baked bread odorant extracts. *Food Research International*, **2007**, *40(9)*, 1170–1184.

31. Pomerantsev, A. L.; Rodionova, O. Y. Hard and soft methods for prediction of antioxidants' activity based on the DSC measurements. *Chemometrics and Intelligent Laboratory Systems*, **2005**, *79(1–2)*, 73–83.

32. Pozo-Bayón, M. A.; Guichard, E.; Cayot, N. Flavor Control in Baked Cereal Products. *Food Reviews International*, **2006**, *22(4)*, 335–379.

33. Quílez, J.; Ruiz, J. A.; Romero, M. P. Relationships Between Sensory Flavor Evaluation and Volatile and Nonvolatile Compounds in Commercial Wheat Bread Type Baguette. *Journal of Food Science*, **2006**, *71(6)*, S423–S427.

34. Ramos, M.; Jiménez, A.; Peltzer, M.; Garrigós, M. C. Characterization and antimicrobial activity studies of polypropylene films with carvacrol and thymol for active packaging. *Journal of Food Engineering*, **2012**, *109(3)*, 513–519.

35. Riga, A.; Collins, R.; Mlachak, G. Oxidative behavior of polymers by thermogravimetric analysis, differential thermal analysis and pressure differential scanning calorimetry. *Thermochimica Acta*, **1998**, *324*, 135–149.

36. Rizzolo, A.; Gerli, F.; Prinzivalli, C.; Buratti, S.; Torreggiani, D. Headspace volatile compounds during osmotic dehydration of strawberries (cv Camarosa): Influence of osmotic solution composition and processing time. *LWT — Food Science and Technology*, **2007**, *40(3)*, 529–535.

37. Rodriguez-Lafuente, A.; Nerin, C.; Batlle, R. Active Paraffin-Based Paper Packaging for Extending the Shelf Life of Cherry Tomatoes. *Journal of Agricultural and Food Chemistry*, **2010**, *58(11)*, 6780–6786.

38. Rodríguez, A.; Batlle, R.; Nerín, C. The use of natural essential oils as antimicrobial solutions in paper packaging. Part II. *Progress in Organic Coatings*, **2007**, *60(1)*, 33–38.

39. Ruiz, J. A.; Quilez, J.; Mestres, M.; Guasch, J. Solid-Phase Microextraction Method for Headspace Analysis of Volatile Compounds in Bread Crumb. *Cereal Chemistry Journal*, **2003**, *80(3)*, 255–259.

40. Singh, P.; Wani, A. A.; Saengerlaub, S. Active packaging of food products: Recent trends. *Nutrition and Food Science*, **2011**, *41(4)*, 249–260.

41. Suppakul, P.; Sonneveld, K.; Bigger, S. W.; Miltz, J. Diffusion of linalool and methylchavicol from polyethylene-based antimicrobial packaging films. *LWT — Food Science and Technology*, **2011**, *44(9)*, 1888–1893.

42. Tovar, B. Z.; García, H. S.; Mata, M. Physiology of pre-cut mango. I. ACC and ACC oxidase activity of slices subjected to osmotic dehydration. *Food Research International*, **2001**, *34(2–3)*, 207–215.

43. Yanishlieva, N. V.; Marinova, E. M.; Gordon, M. H.; Raneva, V. G. Antioxidant activity and mechanism of action of thymol and carvacrol in two lipid systems. *Food Chemistry*, **1999**, *64(1)*, 59–66.

CHAPTER 2

VEGETABLE OILS AS PLATFORM CHEMICALS FOR SYNTHESIS OF THERMOPLASTIC BIO-BASED POLYURETHANES

C. BUENO-FERRER, N. BURGOS, and A. JIMÉNEZ

CONTENTS

2.1 INTRODUCTION

During the last decade environmental concerns have risen in the modern society and have lead to the search of sustainable alternatives for the global development. Therefore, efforts in research for the new formulations in environmentally-friendly materials are currently increasing. In this sense, the substitution of materials obtained from petrochemical sources by others with renewable origin, such as vegetable oils, is one of the most studied possibilities to reduce the global dependency to fossil fuels. These materials are currently under development and the final goal of these new studies would be the reduction of the industrial carbon footprint and a significant improvement in the sustainability of the current production processes. The versatility and ability to chemical transformation of vegetable oils make them good candidates to be used as polymer matrices after some adequate modifications, giving rise to polymers such as thermoplastic polyurethanes (TPUs). In this context, the main goal of the present work is the proposal of thermoplastic formulations based on the use of vegetable oils as polymer matrices. The study of chemical structures, morphologies, thermal and mechanical behavior of different bio-based TPUs will be presented in this chapter.

The use of renewable raw materials constitutes a significant contribution to a sustainable development in the plastics production. This strategy is based on the advantages given by Nature synthesis potential and Green Chemistry principles. In this sense, polymers obtained from renewable raw materials have raised some interest in the last years. The development of polymers synthesized from agricultural products, such as starch, cellulose, sugars or lignin has been considerably increased in the last two decades [1]. Amongst all the possible natural sources for polymers, vegetable oils are considered one of the cheapest and abundant in Nature [2]. They can be used as an advantageous chemical platform to polymer synthesis, by their inherent biodegradable condition and low toxicity to humans and the environment. In this context, many efforts are currently going on to propose a great variety of chemical methods to prepare thermoplastics and thermosets based on vegetable oils. This wide range of chemical methods applicable to these natural materials gives rise to many different monomers and polymers with many applications.

Fatty acids are the major chemical entities present in vegetable oils. They are valuable compounds to design specific monomers in the search of polymers with particular properties without any need of important modifications in their native structure. This is an advantageous issue, not only in sustainability terms, but also in industrial applicability and competitiveness in terms of cost and properties [2–5].

Vegetable oils are mainly formed by triglycerols or triglycerides, mainly composed by three fatty acids bonded to a glycerol molecule. Fatty acids constitute 94–96% of the total triglycerides weight in a vegetable oil and the number of carbon units in their structure is normally between 14 and 22, with zero to three double bonds by fatty acid molecule. The contents in fatty acids in some of the most common vegetable oils are indicated in Table 1.

TABLE 1 Fatty acid distribution in vegetable oils (g fatty acid/100 g oil).

Fatty acid	C:DB	Cotton	Rapeseed	Sunflower	Linseed	Corn	Olive	Palm	Castor	Soybean
Myristic	14:0	0.7	0.1	0.0	0.0	0.1	0.0	1.0	0.0	0.1
Myristoleic	14:1	0.0	0.0	0.0	0.0	0.0	0.0	0.0	0.0	0.0
Palmitic	16:0	21.6	4.1	6.1	5.5	10.9	13.7	44.4	1.5	11.0
Palmitoleic	16:1	0.6	0.3	0.0	0.0	0.2	1.2	0.2	0.0	0.1
Stearic	18:0	2.6	1.8	3.9	3.5	2.0	2.5	4.1	0.5	4.0
Oleic	18:1	18.6	60.9	42.6	19.1	25.4	71.1	39.3	5,0	23.4
Linoleic	18:2	54.4	21.0	46.4	15.3	59.6	10.0	10.0	4.0	53.2
Linolenic	18:3	0.7	8.8	1.0	56.6	1.2	0.6	0.4	0.5	7.8
Ricinoleic	18:1	0.0	0.0	0.0	0.0	0.0	0.0	0.0	87.5	0.0

TABLE 1   *(Continued)*

Fatty acid	C:DB	Cotton	Rapeseed	Sunflower	Linseed	Corn	Olive	Palm	Castor	Soybean
Arachidic	20:0	0.3	0.7	0.0	0.0	0.4	0.9	0.3	0.0	0.3
Gadoleic	20:1	0.0	1.0	0.0	0.0	0.0	0.0	0.0	0.0	0.0
Eicosadienoic	20:2	0.0	0.0	0.0	0.0	0.0	0.0	0.0	0.0	0.0
Behenic	22:1	0.2	0.3	0.0	0.0	0.1	0.0	0.1	0.0	0.1
Erucic	22:1	0.0	0.7	0.0	0.0	0.0	0.0	0.0	0.0	0.0
Lignoceric	24:0	0.0	0.2	0.0	0.0	0.0	0.0	0.0	0.0	0.0
DB/triglyceride		3.9	3.9	4.7	6.6	4.5	2.8	1.8	2.7	4.6
Iodine index (I)		104-117	91-108	110-143	168-204	107-120	84-86	44-58	82-88	117-143

C, number of carbon atoms; DB, number of C = C double bonds.

The use of fatty acids and vegetable oils either in polymer synthesis or as additives comes from some decades, not only by the raising interest in the search for alternatives to fossil fuels but also by the particular chemical characteristics that make them adequate for polymerization processes. Triglycerides are molecules with low reactivity and this fact is a disadvantage in their potential application in polymer synthesis. Nevertheless, the introduction of different functionalities in their reactive sites increases largely the synthetic possibilities of triglycerides [3].

At least three different uses of vegetable oils in polymer formulations can be proposed: (i) as polymer additives (plasticizers, stabilizers, etc.);

(ii) as building blocks to get polymers from them; and (iii) as units for the thermosets synthesis. Much work on the use of vegetable oils as additives [6–11] and as thermosets precursors [12–18] has been reported, but the development of thermoplastic polymer matrices is still in an early stage of research. Thermoplastics can be easily processed and recycled giving them possibilities in many different applications.

The synthesis of thermoplastics from vegetable oils is still in an early stage of the study and development because of the experimental difficulties to be afforded to get reasonable yields in this process. It is known that thermosets obtained from vegetable oils have been largely studied with many reported work [12–22]. This fact is partially due to the own composition of vegetable oils, formed by triglycerides containing different fatty acids with variable number of chain instaurations. Thus, seeds oils are rich in poly-unsaturated fatty acids, giving highly reticulated, rigid and temperature resistant materials [21]. The carbon chains forming the oils can be easily cross-linked by their double bonds. As indicated in Table 1, it can be concluded that most seed oils have fatty acids with 2 or 3 unsaturated bonds as the main component in their lipid profile, except castor, olive and rapeseed oils, which show a mono-unsaturated fatty acid as their main component. Therefore, thermosets can be easily synthesized from oils rich in poly-unsaturated fatty acids, such as those from soya, sunflower or linseed getting polymers with high mechanical and thermal resistance. Castor oil shows high content in ricinoleic acid (87.5% in total oil weight) and the active site is occupied by an alcohol, while olive and rapeseed oils show a main content in mono-unsaturated oleic acid (71.1 and 60.9% in total oil weigh, respectively).

It is known that the potential monomers or polymer building blocks should have at least one (in the case of addition polymerization) or two double bonds in their structure (in the case of condensation polymerization) to get thermoplastic materials. Therefore, triglycerides should be modified of functionalized before polymerization. Nevertheless, there are some examples of thermoplastic biomaterials obtained from naturally functionalized castor oil with homogeneous composition and acceptable polymerization yields. The main thermoplastic materials already synthesized from vegetable oils are thermoplastic polyurethanes (TPUs), polyamides (PA), thermoplastic polyesters, polyesteramides and polyanhydrides.

2.1.1 TPUS: CHEMISTRY, STRUCTURE AND PROPERTIES

Polyurethanes are generally synthesized by addition polymerization between a polyalcohol and a poly-isocyanate. This is an exothermic reaction caused by the release of a proton from the alcohol group followed by a general molecular rearrangement by the formation of the urethane bond [23]. If both reagents are bi-functional linear polyurethanes are obtained, while if functionalities are increased some cross-linked chains are formed, with the formation of reticulated structures. In summary, one of the most common synthesis routes for TPUs consists basically of the reaction of three main components.

1. Polyols with polyester or polyether functionalities with hydroxyl end groups.
2. Di-isocyanate.

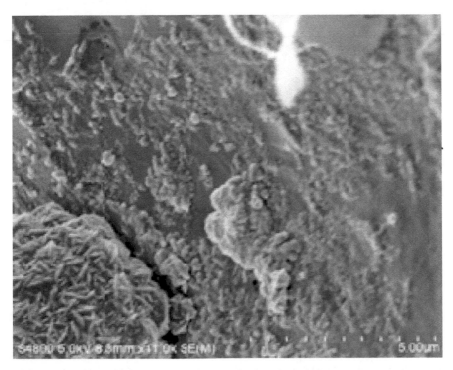

FIGURE 1   Segmented structure of a thermoplastic polyurethane.

However, the phase separation is not complete in TPUs at the molecular scale, and it is possible to find soft segments inside the hard region and vice-versa (Fig. 2), as was reported by Tawa et al. [26]. They indicated that urethane groups from neighbour chains could form hydrogen bonds very easily. This inter-molecular bonding leads to the formation of aggregates acting as physical reticulation nodes with crystalline regions dispersed into the soft area in the polymer structure. This would lead to crosslinking with the final result of the increase in the overall rigidity in the TPU. The phase separation between hard and soft segments depends on, among other factors, their affinity, their relative mobility, the chain extender and the isocyanate structural symmetry [25].

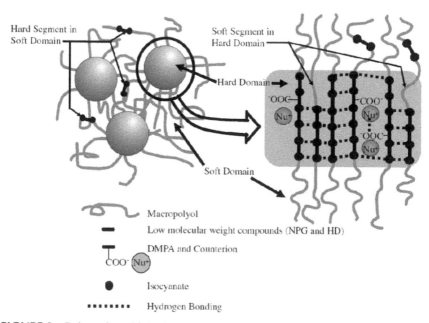

FIGURE 2   Polyurethane biphasic structure at molecular scale [26], with permission.

TPU structures are generally linear, since the relative amount of hard segments is small. Therefore, the most relevant properties of these polymers are conditioned by the secondary and inter-molecular interactions (mainly Van der Waals) between the soft segments [24, 27]. The elastic properties of polyurethanes mainly depend on the polyol chains mobility, which is dependent on their chemical nature and length of the soft segments. The higher the molar mass of the soft segments, the higher tensile strength and elongation to break and consequently the more flexible TPU [24, 25].

On the other hand, when the hard segment content is high, the plastic deformation and the polymer general softening are observed after the application of mechanical stresses at high temperature. The thermal stability of polyurethanes is determined by the temperature range where the rigid segments start to melt and consequently their phase separation and their segmented structure. Polyurethanes will show thermoplastic behaviour at higher temperatures than the melting point of the hard regions [28]. Another important feature of the hard/soft ratio in TPUs is the decrease of molar masses and low phase separation in polymers with high values of that ratio, conditioning their rheological and mechanical properties, giving rise to more rigid materials. In addition, the isocyanate structure also influences the final TPU properties. High-volume di-isocyanates lead to polyurethanes with high elastic modulus and tensile strength [25].

The development of TPUs from vegetable oils is the main goal of this study. In the next sections some work performed by our research group for TPUs synthesized from rapeseed oil dimer fatty acids is presented.

2.2 EXPERIMENTAL

2.2.1 MATERIALS

The bio-based polyester polyol used in this study was kindly supplied by Croda (Yorkshire, UK) and it is based on dimmer fatty acids from rapeseed oil, with purity higher than 98% and weight average molar mass (M_w) around 3000 g mol^{-1}. Hydroxyl and acid values are 40 mg KOH g^{-1} and 0.253 mg KOH g^{-1} respectively, as given by the supplier. 4,4'-diphenylmethane di-isocyanate (MDI) was supplied by Brenntag (Rosheim, France). 1,4-butanediol (BDO), dibutylamine, toluene and hydrochloric acid were purchased from Sigma Aldrich (Lyon, France). All reagents were used without any further purification step.

2.2.2 TPU SYNTHESIS

Four different TPUs were prepared with a NCO/OH ratio equal to 1 and increasing HS content 10–40 wt%, named TPU10–TPU40, respectively), by the two-step prepolymer process for TPU polymerization (Table 2). In a first step, the polyol reacted with an excess of MDI (ratio 2:1) for 2 h in a five-necked round-bottom flask having the provision for nitrogen flushing, mechanical stirring and temperature control at 80°C. During the synthesis, samples were extracted each 30 min in triplicate and diluted in 20 ml of a standard solution of dibutylamine 0.05 M in toluene for the reaction between the residual diisocyanate and the amine, to control the NCO consumption during the reaction. Each solution was then stirred at room temperature for 10 h to ensure the complete reaction of NCO groups with dibutylamine. The excess amine was titrated back with standard aqueous HCl 0.05 M solution using bromophenol green as indicator. For each titration, 25 ml of isopropyl alcohol were added to the solution to ensure compatibility between dibutylamine and the HCl solution. It was calculated from these experiments that 56.1% of NCO groups were consumed at the end of the prepolymer synthesis. This result was used for the addition of the precise amount of diol groups in the processing step. The prepolymer was further melt-blended in a second step by reactive processing with the adequate amount of polyol and chain extender, depending on the HS content targeted in the final TPUs. The prepolymer synthesized in the first step and the calculated amount of polyol for a NCO/OH = 1 ratio were directly introduced in the feeding zone of an internal mixer (counter-rotating mixer Rheocord 9000, Haake, USA) equipped with a pair of high-shear roller-type rotors, at 80°C, with a rotation speed of 50 rpm and 15 min processing time. Then, the adequate amount of BDO chain extender was added and the temperature was immediately increased to 180°C for 8 min, without any catalyst. After polymerization, all systems were cured overnight in an oven at 70°C to ensure the complete reaction of NCO groups, which was further checked by attenuated total reflection Fourier-transform infrared spectroscopy (ATR-FTIR). TPUs were subsequently compression-molded in a hot press at 200°C by applying 200 MPa pressure for 5 min and further quenched between two steel plates for 10 min to obtain sheets with 1.5 mm thickness for each system. The expected HS length for each TPU sample was calculated from data in Table 2 by following a protocol already described by Petrovic et al. [29]

to determine the average polymerization degree in segmented polyurethanes. This method was applied to the TPUs synthesized in this study resulting in an HS average polymerization degree of 5–7, 7–9, 11–13 and 19–21 units for TPU10, TPU20, TPU30 and TPU40, respectively. These results could give an estimation of the length of HS for each TPU.

TABLE 2 Calculated hard segment (HS) percentage and reactants amounts for each TPU blend [30], with permission.

Sample	HS (wt%)	Polyol (g)	MDI (g)	BDO (g)
TPU10	10	49.5	5.1	0.36
TPU20	20	44.0	9.1	1.94
TPU30	30	38.5	13.0	3.52
TPU40	40	33.0	16.9	5.11

2.3 MATERIALS CHARACTERIZATION

2.3.1 ATR-FTIR SPECTROSCOPY

ATR-FTIR was used to screen the complete reaction between NCO and OH groups and, consequently, to evaluate the adequate curing of the TPUs. Infrared spectra were collected on a TA Instruments SDT Q600 (Thermo-Nicolet, New Castle, DE, USA) at a resolution of 4 cm^{-1} and 64 scans per run. The ATR accessory was equipped with a Germanium ($n = 4$) crystal and it was used at a nominal incidence angle of 45°, yielding 12 interval reflections at the polymer surface.

2.3.2 THERMOGRAVIMETRIC ANALYSIS (TGA)

The four TPU systems as well as the bio-based polyol were analyzed in dynamic mode by using a TGA/SDTA 851e Mettler Toledo (Schwarzenbach, Switzerland) equipment. Approximately 7 mg samples were weighed in alumina pans (70 µl) and they were heated from 30°C to 700°C at 10°C min^{-1} under nitrogen atmosphere (flow rate 30 ml min^{-1}). In the case of TPUs, the initial degradation temperature was calculated as the temperature where 5 wt% of the initial mass was lost ($T_{5\%}$).

2.3.3   DIFFERENTIAL SCANNING CALORIMETRY (DSC)

Thermal and structural characterization of TPUs and the polyol was carried out by using a TA Instruments Q2000 (New Castle, DE, USA) equipment. Approximately 5 mg of each sample were weighed in aluminum pans (40 □L) and they were subjected to a first heating stage from 30°C to 240°C, with a further cooling from 240°C to –90°C and a subsequent heating from –90°C to 240°C. All steps were carried out at 10°C min^{-1} under nitrogen (flow rate 50 ml min^{-1}). All tests were performed in duplicate. Glass transition temperatures (T_g) were determined on the second heating scan. T_g of the polyol was determined by using modulated differential scanning calorimetry (MDSC) during a cooling scan from 30°C to –90°C at 2°C min^{-1}, with 60s period and heat-only mode.

2.3.4   UNIAXIAL MECHANICAL TESTS

Tensile properties of TPUs were determined with an Instron tensile testing machine (model 4204, USA), at 25°C and 50% relative humidity at a rate of 20 mm min^{-1}, using dumbbell specimens (dimensions: $30 \times 10 \times 1.5$ mm^3). For each formulation at least five samples were tested.

2.4   RESULTS AND DISCUSSION

2.4.1   ATR-FTIR SPECTROSCOPY

The adequate curing of all TPUs is a key-point prior to the materials characterization, since the presence of residual NCO groups in the final polymer gives an indication of an incomplete synthesis. In this work ATR-FTIR was used to confirm the complete reaction between NCO and OH groups. TPUs spectra are shown in Fig. 3 and they could be used to highlight the main structural differences between them. No peak was found at 2270 cm^{-1} (NCO stretching band) suggesting that the reaction was complete in all cases. As expected, the main variations are related to the increasing content in HS and consequently the higher concentration in urethane groups (-NH-CO-O). In Fig. 3, vibrations at 3335 and 1550 cm^{-1} corresponded to -NH stretching and bending, respectively. Besides, the peak for the C=O stretching from the urethane group could be observed at ≈1700 cm^{-1} [31–33]. All these bands, assigned to the urethane groups, increased in their intensity from TPU10 to TPU40, with confirmation of the higher concentration in carbamate groups at higher HS contents. Nev-

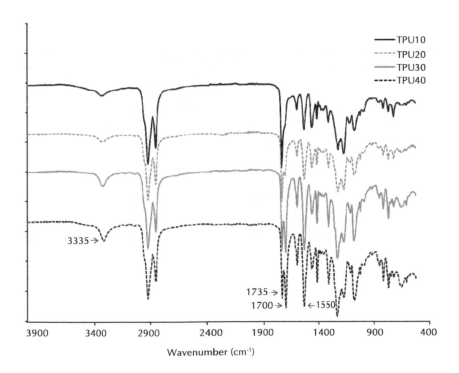

FIGURE 3   ATR-FTIR spectra of TPU 10–40 wt% of HS and main peak assignments (cm^{-1}) [30], with permission.

ertheless, the absorption band at 1735 cm^{-1} was assigned to the C=O group stretching in the polyol, and it was similar for all TPUs, except for TPU10 where this band was broader and almost no discernible from the band at 1700 cm^{-1}, certainly due to the lower content in urethane groups in this material.

2.4.2 THERMOGRAVIMETRIC ANALYSIS (TGA)

The thermal stability of the polyol and all bio-based TPUs were studied by dynamic TGA. Figure 4 shows their mass percentage and derivative curves. The bio-based polyol showed a narrow derivative peak (Fig. 4b, left) due to its purity, while Fig. 4b (right) clearly shows the derivative curves of TPUs with lower intensity peaks. It is known that degradation of polyurethanes is a complex and multistep process, as observed in Fig. 4. An important parameter, the degradation onset, is dependent on the thermal stability of the less thermally stable part on the polyurethane chains [34, 35]. Initial degradation temperatures ($T_{5\%}$) of each step were also studied in TPUs and, together with mass loss percentages, allowed to study the differences between samples depending on their HS content. It was noted that the $T_{5\%}$ value of the pure rapeseed oil-based polyol was higher than in the case of TPU systems, as it was expected, but also higher than the $T_{5\%}$ value reported for castor oil [36] and cashew nut shell liquid-based polyols [37]. TPUs also showed higher thermal stability than polymers with similar structures [38]. It has been reported that their first degradation stage is related to urethane bond decomposition into isocyanate and alcohol with possible formation of primary and secondary amines [39, 40]. Nevertheless, the complexity of this stage is also related with the HS content. In this way, when this content increases $T_{5\%}$ decreases, making those materials more susceptible to degradation and suggesting that the starting point of degradation takes place predominantly within hard segments.

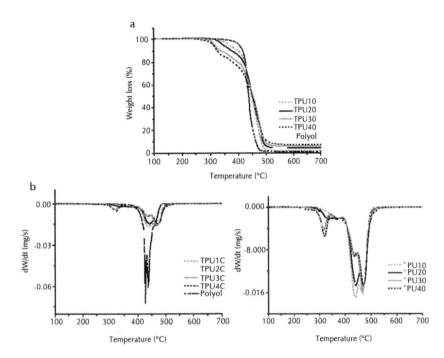

FIGURE 4 TGA curves for mass percentage (a) and derivative (b) vs. temperature for bio-based polyol and TPUs [41], with permission.

2.4.3 DIFFERENTIAL SCANNING CALORIMETRY (DSC)

The structure of TPU samples was investigated by DSC while the bio-based polyol was studied both by DSC and MDSC. The main results of this study are shown in Table 3 and Fig 5. This technique is valuable for a precise determination of T_g and from these values it is possible to estimate the real amounts of HS in the unorganized and organized microphases. It has been indicated that HS in TPUs do not fully belong to the hard domains, since some of them can be found in the soft regions and vice versa. Moreover, when MDI is not bonded to the chain extender but to the polyol, presumably in systems with high polyol amount [42], such as in the case of TPU10, this trend is more clearly observed. This phenomenon was also clearly evidenced in the thermal stability of the TPUs, since the initial degradation temperatures ($T_{5\%}$) of TPU30 and TPU40 fell significantly with respect to those obtained for TPU10 and TPU20. This behavior could be associated to the higher HS content. Moreover, it should be mentioned that the mass loss associated with this first degradation stage could be also correlated with the HS content. In samples with higher HS content, such as TPU30 and TPU40, a peak and a shoulder in their derivative curves were observed, both associated with the first stage of the thermal decomposition of urethane bonds (Fig. 4b).

TABLE 3   Glass transition temperatures (T_g) and hard segment content (%) for TPUs.

Sample	T_{gs} SS (°C)	T_{gh} HS (°C)
TPU10	− 47.0	---
TPU20	− 47.8	122.8
TPU30	− 50.0	120.7
TPU40	− 51.3	118.1

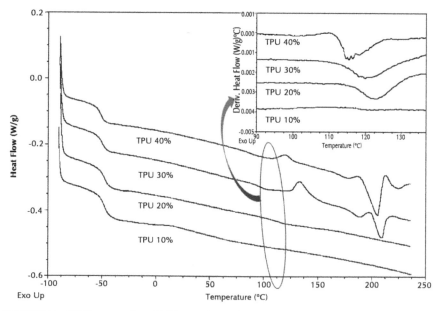

FIGURE 5   DSC curves from TPU samples and zoom in the zone of Tg of HS [30], with permission.

In a preliminary step the bio-based polyol was studied by conventional DSC but the glass transition was not clearly observed since some melting transitions were superposed in this temperature range, probably associated to the crystalline polymorphism of some oils and fats, such as the case of rapeseed oil. Modulation of DSC data (MDSC) is a powerful tool to separate transitions and to get higher resolutions in particular thermal events. In the case of the polyol, MDSC was used during the cooling cycle and the T_g transition was determined in the reversing phase curve at $-61.8°C$. It was also observed that glass transition temperatures of the SS for all TPUs were higher than that of the polyol (Table 3). T_g values of SS were slightly lower with increasing HS concentrations, as it was reported by Xu et al. [43]. Soft domains have higher mobility when larger HS are present, and this fact could be related with a better microphase separation at high HS content. Besides, the TPUs with lower molar masses and the highest concentrations in HS could contribute to the decrease in T_g values. Higher amount of end chains could result in higher free volume, increasing the mobility of the amorphous phase. Moreover, as it is shown in Fig. 5, T_g transition of the MDI-BDO HSs was very difficult to detect due to their stiffness and low mobility [42]. T_g of HS of TPUs slightly increased when their concentration decreased. This result could be attributed to the higher concentration in HS inside the soft domains, as previously explained. Moreover, the intensity of the glass transition of HS is larger at higher HS content but the T_g value is lower, suggesting some interactions between hard and soft phases, that is the low T_g of SS observed for TPU40 which was previously attributed to the microphase separation, could also help to the decrease in T_g of HS in this material due to the higher interaction of soft domains in the hard phase.

2.4.4 UNIAXIAL MECHANICAL TESTS

Tensile properties were determined by uniaxial tensile tests and data are summarized in Table 4 and Fig. 6. Results showed that the increase in HS content lead to a brittle material with higher tensile modulus and lower elongation at break (around 25% for TPU40), as expected, while this material showed lower tensile strength than TPU30 and TPU20. This behavior could be attributed to the fragility provided by the crystalline macro-structures with sizes between 20 and 30 μm that were observed in a previous morphological study [30]. The presence of these macro-structures in TPU40 increased the segregation phase size and may promote points of stress concentration, due to their boundary impingement, which could influence properties such as tensile strength and elongation at break [43]. As expected, TPU40 also exhibited the higher modulus (11.1 MPa), leading to the conclusion that the HS higher crystallinity has a significant effect on the mechanical properties of these TPUs.

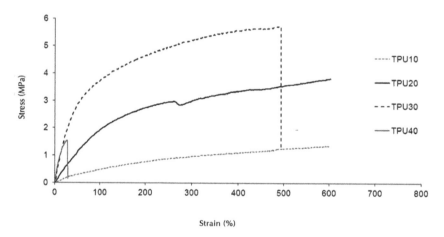

FIGURE 6   Stress-strain curves of TPUs [41], with permission.

TABLE 4   Uniaxial tensile properties of TPUs [41], with permission.

Sample	Tensile strength (MPa)	Elongation at break (%)	Young modulus (MPa)
TPU10	1.3 ± 0.1	> 600	0.7 ± 0.0
TPU20	3.7 ± 0.1	> 600	2.5 ± 0.1
TPU30	5.6 ± 0.3	430 ± 40	8.6 ± 0.5
TPU40	1.5 ± 0.4	25 ± 7	11.1 ± 1.2

Figure 6 shows the elastomeric behavior of those TPUs with lower HS content, as demonstrated by their lower moduli. It was also observed in Fig. 6 that TPU20 was the only material with a yield point in the range of deformation 250–300%. This particular behavior, presented as a rubbery region in Fig. 6, could be related to a good resistance to the permanent deformation.

2.5 CONCLUSIONS

Bio-based TPUs were synthesized from a di-functional dimmer fatty acid-based polyol obtained from rapeseed oil, MDI and BDO at four HS contents, i.e., 10–40 wt%. The polyol characteristics determined the structure and properties of TPUs. FTIR-ATR spectra confirmed that all the isocyanate groups reacted with hydroxyl groups (from polyol or BDO) during the TPUs synthesis. Thermal studies carried out by TGA, DSC and MDSC revealed some interactions between hard and soft domains for all TPUs and a degradation behavior closely linked to their HS concentration. Stress-strain uniaxial tests showed that the increase in HS content in TPUs lead to higher tensile modulus and lower elongation at break. TPU10 and TPU20 showed a strong elastomeric behavior with very high elongation at break (>600%) and very low elastic modulus.

In summary, TPUs partially synthesized from vegetable oils are very promising materials in good agreement with the current tendency for sustainable development, making them very attractive since they are expected to show specific properties which can be easily tailored by selecting the appropriate HS concentration. These materials could also fulfill many industrial requirements for different fields, such as construction, automotive, textile, adhesive and coatings.

ACKOWLEDGEMENTS

Authors thank the Spanish Ministry of Economy and Competitiveness (MAT2011–28648–C02-01) for financial support.

KEYWORDS

- carbamate
- polyol
- stress-strain curves
- thermoplastic polyurethane

REFERENCES

1. Lligadas, G.; Ronda, J. C.; Galià, M.; Cádiz, V. Plant Oils as Platform Chemicals for Polyurethane Synthesis, Current-State-of-the-Art. *Biomacromolecules,* **2010,** *11,* 2825–2835

2. Biermann, U.; Friedt, W.; Lang, S.; Luhs, W.; Machmuller, G.; Metzger, J. O.; Klaas, M. R. G.; Schafer, H. J.; Scheider, M. P. *New Syntheses with Oils and Fats as Renewable Raw Materials for the Chemical Industry.* Angewandte Chemie International Edition, **2000,** *39,* 2206–2224.

3. Montero de Espinosa, L.; Meier, M. A. R. Plant oils, the perfect renewable resource for polymer science! *European Polymer Journal,* **2011,** *47,* 837–852

4. Behr, A.; Gomes, J. P. The refinement of renewable resources, new important derivatives of fatty acids and glycerol. *European Journal of Lipid Science and Technology,* **2010,** *112(1),* 31–50.

5. Kockritz, A.; Khot, S. N. M.; Lascala, J. J.; Can, E.; Morye, S. S.; Williams, G. I.; Palmese, G. R.; Kusefoglu, S. H.; Wool, R. P. Development and application of triglyceride-based polymers and composites. *Journal of Applied Polymer Science,* **2001,** *82(3),* 703–723.

6. Benaniba, M. T.; Belhaneche-Bensemra, N.; Gelbard, G. Stabilizing effect of epoxidized sunflower oil on the thermal degradation of poly(vinyl chloride). *Polymer Degradation and Stability,* **2001,** *74,* 501–505.

7. Boussoum, M. O.; Atek, D.; Belhaneche-Bensemra, N. Interactions between poly(vinyl chloride) stabilised with epoxidised sunflower oil and food simulants. *Polymer Degradation and Stability,* **2006,** *91,* 579–584.

8. Choi, J. S.; Park, W. H. Thermal and mechanical properties of Poly(3-hydroxybutyrate-co-3-hydroxyvalerate) plasticized by biodegradable soybean oils. *Macromolecular Symposium,* **2003,** *197,* 65–76.

9. Semsarzadeh, M. A.; Mehrabzadeh, M.; Arabshahi, S. S. Mechanical and Thermal Properties of the Plasticized PVC-ESBO. *Iranian Polymer Journal,* **2005,** *14(9),* 769–773.

10. Ali, F.; Chang, Y. W.; Kang, S. C.; Yoon, J. Y. Thermal, mechanical and rheological properties of poly(lactic acid)/epoxidized soybean oil blends. *Polymer Bulletin,* **2009,** *62(1),* 91–98.

11. Karmalm, P.; Hjertberg, T.; Jansson, A.; Dahl, R. Thermal stability of poly(vinyl chloride) with epoxidised soybean oil as primary plasticizer. *Polymer Degradation and Stability*, **2009**, *94*, 2275–2281.

12. Meier, M. A. R.; Metzger, J. O.; Schubert, U. S. Plant oil renewable resources as green alternatives in polymer science. *Chemical Society Reviews*, **2007**, *36*, 1788–1802.

13. Güner, F. S.; Yagci, Y.; Erciyes, A. T. Polymers from trygliceride oils. *Progress in Polymer Science*, **2006**, *31*, 633–670.

14. Lu, Y.; Larock, R. C. Novel polymeric materials from vegetable oils and vinyl monomers, Preparation, properties, and applications. *ChemSusChem*, **2009**, *2(2)*, 136–147.

15. Khot, S. N.; Lascala, J. J.; Can, E.; Morye, S. S.; Williams, G. I.; Palmese, G. R.; Kusefoglu, S. H.; Wool, R. P. Development and application of triglyceride-based polymers and composites. *Journal of Applied Polymer Science*, **2001**, *82(3)*, 703–723.

16. Hablot, E.; Zheng, D.; Bouquey, M.; Averous, L. Polyurethanes Based on Castor Oil, Kinetics, Chemical, Mechanical and Thermal Properties. *Macromolecular Materials and Engineering*, **2008**, *293(11)*, 922–929.

17. Liu, Z.; Erhan, S. Z. "Green" composites and nanocomposites from soybean oil. *Materials Science and Engineering A*, **2008**, 483–484, 708–711.

18. Doll, K. M.; Erhan, S. Z. The improved synthesis of carbonated soybean oil using supercritical carbon dioxide at a reduced reaction time. *Green Chemistry*, **2005**, *7*, 849–854.

19. Meiorin, C.; Aranguren, M. I.; Mosiewicki, M. A. Smart and structural thermosets from the cationic copolymerization of a vegetable oil. *Journal of Applied Polymer Science*, **2011**, *124(6)*, 5071–5078.

20. Kong, X.; Omonov, T. S.; Curtis, J. M. The development of canola oil based bio-resins. *Lipid Technology*, **2012**, *24(1)*, 7–10.

21. Kim, J. R.; Sharma, S. The development and comparison of bio-thermoset plastics from epoxidized plant oils. *Industrial Crops and Products*, **2012**, *36(1)*, 485–499.

22. Quirino, R. L.; Woodford, J.; Larock, R. C. Soybean and linseed oil-based composites reinforced with wood flour and wood fibers. *Journal of Applied Polymer Science*, **2012**, *124(2)*, 1520–1528.

23. Seymour, R. B.; Carraher, C. E. *Polymer Chemistry - An introduction*. 2nd Edition. Marcel Dekker, Basel, Suiza, **1988**.

24. Ionescu, M. *Chemistry and Technology of Polyols for Polyurethanes*. Rapra Technology Limited, Shropshire, United Kingdom, **2005**.

25. Meier-Westhues, U. *Polyurethanes. Coatings, adhesives and sealants*. Vincentz Network, Hannover, Alemania, **2007**.

26. Tawa, T.; Ito, S. The role of hard segments of aqueous polyurethane-urea dispersion in determining the colloidal characteristics and physical properties. *Polymer Journal*, **2006**, *38(7)*, 686–693.

27. Wool, R. P.; Sun, X. S. *Bio-Based Polymers and Composites*. Elsevier Academic Press, Burlington, MA, United States, **2005**.

28. Yamasaki, S.; Nishiguchi, D.; Kojio, K.; Furukawa, M. Effects of Polymerization Method on Structure and Properties of Thermoplastic Polyurethanes. *Journal of Polymer Science B, Polymer Physics*, **2007**, *45*, 800–814.

29. Petrovic, Z. S.; Cevallos, M. J.; Javni, I.; Schaefer, D. W.; Justice, R. Soy-oil-based segmented polyurethanes. *Journal of Polymer Science B*, **2005**, *43*, 3178–3190.

30. Bueno-Ferrer, C.; Hablot, E.; Perrin-Sarazin, F.; Garrigós, M. C.; Jiménez, A.; Avérous, L. Structure and morphology of new bio-based thermoplastic polyurethanes obtained from dimer fatty acids. *Macromolecular Materials and Engineering*, **2012**, *297(8)*, 777–784.

31. Irusta, L.; Fernandez-Berridi, M. J. Aromatic poly(ester-urethanes), effect of the polyol molecular weight on the photochemical behaviour. *Polymer*, **2000**, *41*, 3297–3302.

32. Irusta, L.; Iruin, J. J.; Mendikute, G.; Fernández-Berridi, M. J. Infrared spectroscopy studies of the self-association of aromatic urethanes. *Vibrational Spectroscopy*, **2005**, *39*, 144–150.

33. Silva, B. B. R.; Santana, R. M. C.; Forte, M. M. C. A solventless castor oil-based PU adhesive for wood and foam substrates. *International Journal of Adhesion and Adhesives*, **2010**, *30*, 559–565.

34. Javni, I.; Petrovic, Z.; Guo, A.; Fuller, R. Thermal stability of polyurethanes based on vegetable oils. *Journal of Applied Polymer Science*, **2000**, *77*, 1723–1734.

35. Król, P. Synthesis methods, chemical structures and phase structures of linear polyurethanes. Properties and applications of linear polyurethanes in polyurethane elastomers, copolymers and ionomers. *Progress in Materials Science*, **2007**, *52*, 915–1015.

36. Corcuera, M. A.; Rueda, L.; Fernandez d'Arlas, B.; Arbelaiz, A.; Marieta, C.; Mondragon, I.; Eceiza, A. Microstructure and properties of polyurethanes derived from castor oil. *Polymer Degradation and Stability*, **2010**, *95*, 2175–2184.

37. Bhunia, H. P.; Nando, G. B.; Chaki, T. K.; Basak, A.; Lenka, S.; Nayak, P. L. Synthesis and characterization of polymers from cashewnut shell liquid (CNSL), a renewable resource II. Synthesis of polyurethanes. *European Polymer Journal*, **1999**, *35*, 1381–131.

38. Yeganeh, H.; Mehdizadeh, M. R. Synthesis and properties of isocyanate curable millable polyurethane elastomers based on castor oil as a renewable resource polyol. *European Polymer Journal*, **2004**, *40*, 1233–1238.

39. Hablot, E.; Zheng, D.; Bouquey, M.; Avérous, L. Polyurethanes based on castor oil, kinetics, chemical, mechanical and thermal properties. *Macromolecular Materials and Engineering*, **2008**, *293*, 922–929.

40. Hojabri, L.; Kong, X.; Narine, S. S. Fatty acid-derived diisocyanate and biobased polyurethane produced from vegetable oil, synthesis, polymerization and characterization. *Biomacromolecules*, **2009**, *10*, 884–891.

41. Bueno-Ferrer, C.; Hablot, E.; Garrigós, M. C.; Bocchini, S.; Avérous, L.; Jiménez, A. Relationship between morphology, properties and degradation parameters of novative biobased thermoplastic polyurethanes obtained from dimer fatty acids. *Polymer Degradation and Stability*, **2012**, *97(10)*, 1964–1969.

42. Bagdi, K.; Molnar, K.; Pukanszky, B. Jr.; Pukanszky, B. Thermal analysis of the structure of segmented polyurethane elastomers. *Journal of Thermal Analysis and Calorimetry*, **2009**, *98*, 825–832.

43. Xu. Y.; Petrovic, Z.; Das, S.; Wilkes, G. L. Morphology and properties of thermoplastic polyurethanes with dangling chains in ricinoleate-based soft segments. *Polymer*, **2008**, *49*, 4248–4258.

CHAPTER 3

NEW CONCEPT OF USING GENETICALLY ENGINEERED MICROBIAL BIOSENSORS

ANAMIKA SINGH and RAJEEV SINGH

CONTENTS

3.1   INTRODUCTION

A biosensor is an automatic device, which can help us by transforming biological (enzymatic) response to electrical in the form of fluorescent light. The emphasis of this topic concerns enzymes as the biologically responsive material, but it should be recognized that other biological systems may be utilized by biosensors, for example, whole cell metabolism, Ligand binding and the antibody-antigen reaction. Biosensors represent a rapidly expanding field, at the present time, with an estimated 60% annual growth rate; the major impetus coming from the health-care industry (e.g., 6% of the western world are diabetic and would benefit from the availability of a rapid, accurate and simple biosensor for glucose) but with some pressure from other areas, such as food quality appraisal environmental monitoring and defense purposes.

Before designing biosensors we have to make sure about the following important feature must be considered:

1.   The biocatalyst should be specific for particular purpose.
2.   The reaction should be independent of parameters like pH and stirring, etc.
3.   The response should not be affected with concentration or dilution and it should be free from electrical noise.
4.   Biosensor, which is used for clinical situations, the probe must be biocompatible and having no toxic or antigenic effects. The biosensor should not be fouling or proteolysis.
5.   The complete biosensor should be cheap, small, portable and capable of being used as operators.
6.   There should be a market for the biosensor.

In general biosensors must have five components: Biocatalyst, Transducer, Amplifier, Processor and Displaying devices. Biocatalysts are made by using immobilization techniques. The most common immobilization techniques are physical adsorption, cross-liking, entrapment, covalent bond, or some combination of all of these techniques the stability of the immobilization techniques determines the sensitivity and reliability of the biosensor signal. Transducers, required pieces of analytical tools that convert the biological and chemical changes into the useful electronic data, make use of these immobilization techniques to provide different types of transducers, for example, optical, piezoelectric, etc. The analyte selective interface is a biological active substance such as enzyme, antibody, DNA (deoxyribonucleic acid) and microorganism. Such substances are cable of recognizing their specific analyses and regulating the overall performance (response time, reliability, specificity, selectivity and sensitivity) of the biosensors. Biosensors are analytical tools that consist of a substrate and a selective interface in closed proximity or integrate with transducer; therefore, the substrates and transducers are important components of the analytical tools, which contain an immobilized biologically active compound that can interact with specific species (substrates) produces a products in the form of a biological or chemical substance, heat, light, or sound: then a transducer such as a electrode, semiconductor, thermistor, photocounter or sound detector changes the product of the reaction into usable data. The data will be further amplified in the presence of reference data and then data will be processed and displayed in the user-friendly manner (Fig. 1). In this chapter we are giving more emphasis on microbes as biosensors. Microbes utilized in biosensors can also be called as Bioreporters.

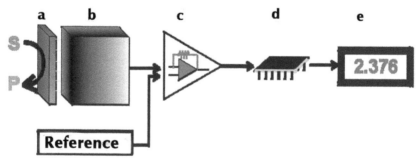

Curtsey: http://vanibala.150m.com

FIGURE 1 Schematic diagram showing the main components of a biosensor. The biocatalyst (a) converts the substrate to product. This reaction is determined by the transducer that (b) converts it to an electrical signal. The output from the transducer is amplified (c), processed (d) and displayed (e).

Bioreporters refer to intact, live microbial cells, which are genetically engineered to produce a detectable signal in response to particular agent in their environment (Fig. 2). Bioreporters are having two essential genetic elements, a promoter gene and a reporter gene. The promoter gene is turned on (transcribed) when the target agent is present in the cells environment. The promoter gene in a normal bacterial cell is linked to other genes that are then likewise transcribed and then translated into proteins that help the cell in either combating or adapting to the agent to which it has been exposed. In the case of a bioreporter, these genes, or portions thereof, have been removed and replaced with a reporter gene. Consequently, turning on the promoter gene now causes the reporter gene to be turned on. Activation of the reporter gene leads to production of reporter proteins that ultimately generate some type of a detectable signal.

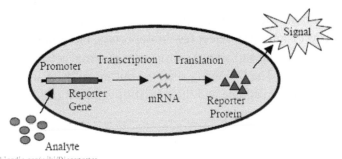

Curtsey: http://en.wikipedia.org/wiki/Bioreporter

FIGURE 2   Steps in Biorepoter. Upon exposure to specific analyte, the promoter/reporter gene complex is transcribed into messenger RNA (mRNA) and then translated into reporter protein that is ultimately responsible for generating a signal.

Therefore, the presence of a signal indicates that the bioreporter has sensed a particular target agent in its environment.

Originally developed for fundamental analysis of factors affecting gene expression, bioreporters were early on applied for the detection of environmental contaminants1 and have since evolved into fields as diverse as medical diagnostics, precision agriculture, food-safety assurance, process monitoring and control, and bio-microelectronic computing. Their versatility stems from the fact that there exist a large number of reporter gene systems that are capable of generating a variety of signals. Additionally, reporter genes can be genetically inserted into bacterial, yeast, plant, and mammalian cells, thereby providing considerable functionality over a wide range of host vectors.

3.2   REPORTER GENE SYSTEMS

Several types of reporter genes are available for use in the construction of bioreporter organisms, and the signals they generate can usually be categorized as either colorimetric, fluorescent, luminescent, chemiluminescent or electrochemical. Although each functions differently, their end product always remains the same- a measurable signal that is proportional to the concentration of the unique chemical or physical agent to which they have been exposed. In some instances, the signal only occurs when a secondary substrate is added to the bioassay (*luxAB*, Luc, and aequorin). For other bioreporters, the signal must be activated by an external light source (GFP and UMT), and for a select few bioreporters, the signal is completely self-induced, with no exogenous substrate or external activation being required (*luxCDABE*). The following sections outline in brief some of the reporter gene systems available and their existing applications.

3.2.1   BACTERIAL LUCIFERASE (LUX)

Luciferase is a generic name for an enzyme that catalyzes a light-emitting reaction. Luciferases enzyme founds in bacteria, algae, fungi, jellyfish, insects, shrimp, and squid. In bacteria, light emitting gene have been isolated and used for light emitting reactions for different organisms. Bioreporters that emit a blue-green light with a maximum intensity at 490 nm (Fig. 3) Three variants of *lux* are available, one that functions at < 30°C, another at < 37°C, and a third at < 45°C. The *lux* genetic system consists of five genes, *luxA*, *luxB*, *luxC*, *luxD*, and *luxE*. Depending on the combination of these genes used, several different types of bioluminescent bioreporters can be constructed.

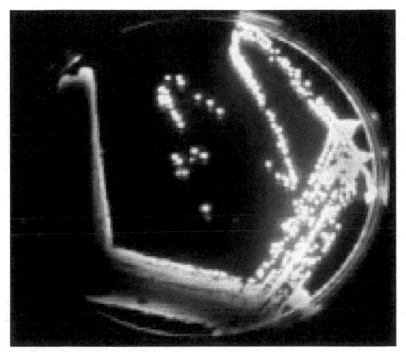

FIGURE. 3.   Genetically engineered bacteria with lux gene glows in culture.

3.2.2   LUXAB BIOREPORTERS

luxAB bioreporters are actually *luxA* and *luxB* genes, which together generat the light signal. However, to fully complete the light-emitting reaction, a substrate must be supplied to the cell. *luxAB* bioreporters have been constructed within bacterial, yeast, insect, nematode, plant, and mammalian cell systems. Few bioreporters having luxAB are bacterial biofilm, bacterial biomass, cell viable counts which perform different functions.

3.2.3   LUXCDABE BIOREPORTERS

Bioreporters can contain all five genes of the *lux* cassette, thereby allowing for a completely independent light generating system that requires no extraneous additions of substrate nor any excitation by an external light source. So in this bioassay, the bioreporter is simply exposed to a target analyte and a quantitative increase in bioluminescence results, often within less than one hour. Due to their rapidity and ease of use, along with the ability to perform the bioassay repetitively in realtime and on-line, makes *luxCDABE* bioreporters extremely attractive. Consequently, they have been incorporated into a diverse array of detection methodologies ranging from the sensing of environmental contaminants to the real-time monitoring of pathogen infections in living mice (e.g., there are few Lux CDABE based bioreporter which detect 2,3 Dichlorophenol,2,4,6, Trichlorophenol, 2,4, D,3-Xylene, etc.).

3.2.4   FIREFLY LUCIFERASE (LUC)

Firefly luciferase catalyzes a reaction that produces visible light in the 550–575 nm range. A click-beetle luciferase is also available that produces light at a peak closer to 595 nm. Both luciferases require the addition of an exogenous substrate (luciferin) for the light reaction to occur. Numerous *luc*-based bioreporters have been constructed for the detection of a wide array of inorganic and organic compounds of environmental concern.

3.2.5   AEQUORIN

Aequorin is a photoprotein isolated from the bioluminescent jellyfish *Aequorea victoria*. Upon addition of calcium ions (Ca2+) and coelenterazine, a reaction occurs whose end result is the generation of blue light in the 460–470 nm range.

3.2.6   GREEN FLUORESCENT PROTEIN (GFP)

Green fluorescent protein (GFP) is also a photoprotein isolated and cloned from the jellyfish *Aequorea victoria*. Variants have also been isolated from the sea pansy *Renilla reniformis*. GFP, like aequorin, produces a blue fluorescent signal, but without the required addition of an exogenous substrate.

All that is required is an ultraviolet light source to activate the fluorescent properties of the photoprotein. This ability to autofluoresce makes GFP highly desirable in biosensing assays since it can be used on-line and in real-time to monitor intact, living cells. Additionally, the ability to alter GFP to produce light emissions besides blue (i.e., cyan, red, and yellow) allows it to be used as a multianalyte detector.

Consequently, GFP has been used extensively in bioreporter constructs within bacterial, yeast, nematode, plant, and mammalian hosts (Fig. 4). Some examples of GFP applications in mammalian cell systems, where its use has revolutionized much of what we understand about the dynamics of cytoplasmic, cytoskeletal, and organellar proteins and their intracellular interactions. Some other applications of GFP are monitoring tumor cells in gene therapy protocols, identification of HIV in infected cells and tissue, monitoring of protein-protein interactions in living cells etc.

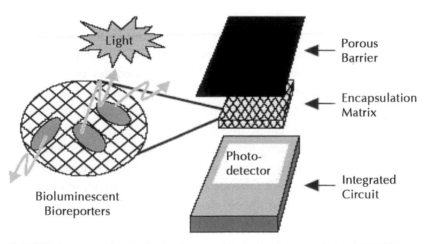

FIGURE 4   Assembly of a bioluminescent bioreporter integrated circuit (BBIC) sensor.

These bioluminescent bioreporter integrated circuits (BBICs) consist of two main components; photo-detectors for capturing the on-chip bioluminescent bioreporter signals and signal processors for managing and storing information derived from bioluminescence (Fig. 4). Remote frequency (RF) transmitters can also be incorporated into the overall integrated circuit design for wireless data relay. Since the bioreporter and biosensing elements are completely self-contained within the BBIC, operational capabilities are realized by simply exposing the BBIC to the desired test sample.

3.2.7   UROPORPHYRINOGEN (UROGEN) III METHYLTRANSFERASE (UMT)

UMT catalyzes a reaction that yields two fluorescent products, which produce a red-orange fluorescence in the 590–770 nm range when illuminated with ultraviolet light. So as with GFP, no addition of exogenous substrates is required.

3.3   DETECTING THE OPTICAL SIGNATURE

Using light as the terminal indicator is advantageous in that it is an easily measured signal. Optical transducers such as photo multiplier tubes, photodiodes, micro channel plates, or charge-coupled devices are readily available and can be easily integrated into high throughput readers. As these usually consist of large, tabletop devices, demand for smaller, portable light readers for remote monitoring has resulted in the development of battery-operated, hand-held photo multiplier units. Recently, The Center for environmental Biotechnology and Oak Ridge National Laboratory (ONRL) have taken steps towards genuine miniaturization of optical transducers and have successfully developed integrated circuits capable of detecting bioluminescence directly from bioreporter organisms.

In addition to incorporation in a BBIC format, the whole-cell bioreporter matrix can also be immobilized on something as simple as an indicator test strip. In this fashion, a home water quality indicator, for example, could be developed to operate in much the same way as a home pregnancy test kit.

Genetically modified microbes are used extensively in various fields like Defense, Medical, Environment, Food and Agriculture, etc.

3.3.1 BIOSENSORS IN THE FIELD OF DEFENSE

3.3.1.1 LAND MINES

Every year in world, due to land mines accidents more than 25,000 people have been killed. Land mines are extremely difficult to find once they are buried in the ground. Plastic mines are almost impossible to locate because they elude metal detectors. Fortunately, most land mines leak slightly and leave traces of explosive chemicals such as TNT shortly after they are installed. ORNL has developed a clever way of using bacteria to detect this faint explosive signature.

Bob Burlage, a microbiologist in ORNL's (Environmental Sciences Division) has genetically engineered microorganisms to emit light in the presence of TNT. As they recognize and consume TNT, the engineered bacteria produce a fluorescent protein that appears as a green light when they are illuminated by ultraviolet (UV) light. Landmines leaks small amount of TNT over time (Fig. 5).

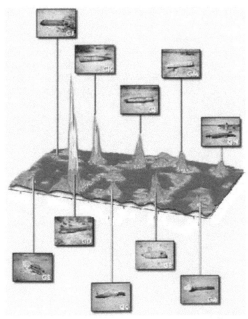

FIGURE 5   Land mines mapping by using illuminated with UV light.

When the bacterial strain of *Pseudomonas putida* encounter the TNT, they will scavenge the compound as a food source, activating the genes that produce proteins needed to digest the TNT, because green fluorescent protein gene obtained from jellyfish to these activated genes and included a regulatory gene that recognizes TNT. As a result, the attached gene will also be turned on. It will produce the green fluorescent protein, which emits extremely bright fluorescence when exposed to UV light (Fig. 6). It is one of the most rapid and advanced methods for landmines detection. The field has been sprayed with bacteria during the day using a agriculture sprayer. At night rolling towers and helicopters by looking for glowing microbes on soil illuminated UV light, ONGL will get the green light to develop a bacteria remote sensing method for land mines detection. This technique offers several advantages. It is inexpensive, it poses no hazard to operators and it is virtually the only mine detector technology which wide areas of fields.

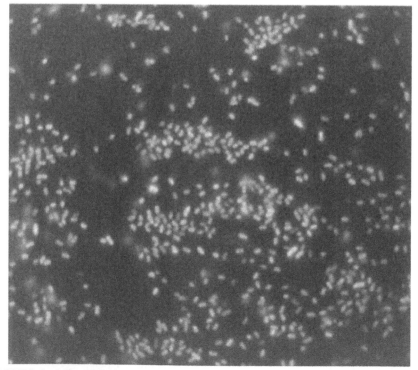

FIGURE 6   Microbial biosensors with GFP after Microbial biosensors.

3.4 BIOSNIFFING

Biosniffing another development in biosensors, its nothing but recognizing the airborne pathogens by using sensors. The major defense concern in USA after September 9/11 is given to control bioterrorism by detecting airborne pathogens like bubonic plague, anthrax, botulism, small pox, tularemia etc. by using wireless Biosensors [1].

3.5 SPACE

At the University of Tennessee's Center for Environmental Biotechnology, researchers are developing wireless biosensors that can be used to monitor microbial contamination and radiation exposure in future manned space missions. The researchers are using "bioluminescent biosensor organisms" built directly onto wireless integrated circuits. They call them Bioluminescent Bioreporter Integrated Circuits (BBIC). Bioluminescent biosensors are bacteria that glow when they come into contact with a specific toxic compound. The more poison the bacteria receive, the more brightly they shine. The BBIC uses this light to determine the level of toxicity present in the spacecraft, sending a wireless message to a central computer. "The ultimate goal," say the researchers, "is to create an array or network of small, unobtrusive, low cost, low power BBICs for intelligent distributed monitoring of the space craft environment, as well as for planetary-based surface habitats." It may sound like science fiction, but this is one technology that's coming down to earth, and not a moment too soon.

3.5.1 BIOSENSORS IN THE FIELD OF MEDICAL

3.5.1.1 CHONDROCYTES

One of the main open questions in chondrocyte transplantation is the fate of the implanted cells in vivo. Intended to establish prerequisites for such studies in animal models and to show the feasibility of this approach in rabbits. Isolated articular chondrocytes were retro virally marked using green fluorescence protein (GFP) as a cell-specific marker in order to allow an in vivo follow-up of these cells [2].

3.6   CANCER

Recently, there has been tremendous interest in developing techniques such as MRI, micro-CT, micro-PET, and SPECT to image function and processes in small animals. These technologies offer deep tissue penetration and high spatial resolution, but compared with noninvasive small animal optical imaging, these techniques are very costly and time consuming to implement. Optical imaging is cost-effective, rapid, easy to use, and can be readily applied to studying disease processes and biology in vivo. In vivo optical imaging is the result of a coalescence of technologies from chemistry, physics, and biology. The development of highly sensitive light detection systems has allowed biologists to use imaging in studying physiological processes. Over the last few decades, biochemists have also worked to isolate and further develop optical reporters such as GFP, luciferase, and cyanine dyes. This article reviews the common types of fluorescent and bioluminescent optical imaging, the typical system platforms and configurations, and the applications in the investigation of cancer biology [3].

In cancer diagnosis the Insertion of the *luc* genes into a human cervical carcinoma cell line (HeLa) illustrated that tumor-cell clearance could be visualized within a living mouse by simply scanning with a charge-coupled device camera, allowing for chemotherapy treatment to rapidly be monitored on-line and in real-time4. In another example, the *luc* genes were inserted into human breast cancer cell lines to develop a bioassay for the detection and measurement of substances with potential estrogenic and antiestrogenic activity.

Optical imaging is a modality that is cost-effective, rapid, easy to use, and can be readily applied to studying disease processes and biology in vivo. For this study, we used a green fluorescent protein (GFP)- and luciferase-expressing mouse tumor model to compare and contrast the quantitative and qualitative capabilities of a fluorescent reporter gene (GFP) and a bioluminescent reporter gene (luciferase) [4]. Biosensors in medical field grow extensively from previous few years ornithine decarboxylase and

S. adenosylmenthionine decarboxylase (AdometDC) are two enzymes of polyamine biosynthesis in cells induces metagenesis in cells which leads cancer development, ODC inhibitor DFMO induces polyamine depletion result cytostatic growth inhibitor [5].

3.7   HIV (AIDS)

Highly active retroviral therapy (HAART) which consists of inhibitor of viral enzyme (reverse transcriptase (RT) and proteases) which is also a part of biosensor, i.e., induction of inhibitor viral enzyme into host and its use as biosensor against HIV [6]. In field of medicine there are nucleic acid biosensor, which helps, in medical diagnosis [7].

3.8   HUMAN B CELL LINE

Aequorin has been incorporated into human B cell lines for the detection of pathogenic bacteria and viruses in what is referred to as the CANARY assay (Cellular Analysis and Notification of Antigen Risks and Yields). The B cells are genetically engineered to produce aequorin. Upon exposure to antigens of different pathogens, the recombinant B cells emit light as a result of activation of an intracellular signaling cascade that releases calcium ions inside the cell [8].

3.8.1   BIOSENSORS IN THE FIELD OF ENVIRONMENT

Heavy metals are an important class of environmental hazards, and as the heavy metals in industry continues to increase, larger segments of biota including human beings, will be exposed to increasing levels of the toxicants, detection of individual metals in the environment may event be possible using biosensor consisting of genetically engineered microorganis [9]. Genetically engineered *Pseudomonas species* (Shk1) with bioluminescence capacity isolated from activated sludge, are able to report or detect heavy metal toxicity [10].

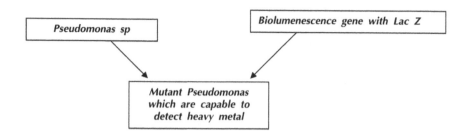

Mainly *Pseudomonas fluorescence* and bakers yeast (*Saccharomyces cerevisiae*) were used for heavy metal detection. Lac-z based reporter gene transposon Tn5B20 was performed random gene fusions in the genome of the common soil bacteria. *Pseudomonas fluorescence* strains ATCC13525 were used to create a bank of 5000 *P. fluorescence* mutants. This mutant bank in heavy metal detection was screened for differential gene expression in the presence of toxic metal Cadmium. One of mutant c-8 shows increase gene expression in the presence of ethanol. Mutant c11 is hypersensitive, to cadmium and zinc oins, it has a sensor/regulator protein pair which is also found in czcRs of *Ralstonia eutropha*, czrRs of *Pseudomonas aeruginosa* and copRs of *Pseudomonas syringae*.

3.8.1.1   NONSPECIFIC LUX BIOREPORTERS

Nonspecific *lux* bioreporters are typically used for the detection of chemical toxins. They are usually designed to continuously bioluminesce. Upon exposure to a chemical toxin, either the cell dies or its metabolic activity is retarded, leading to a decrease in bioluminescent light levels.

Except this, UMT has also been used as a bioreporter for the selection of recombinant plasmids, as a marker for gene transcription in bacterial, yeast, and mammalian cells, and for the detection of toxic salts such as arsenite and antimonite.

Microbial bioreporters play an important role in environmental monitoring and ecotoxicology. Microorganisms that are genetically modified with reporter genes can be used in various formats to determine the bioavailability of chemicals and their effect on living organisms. Cyanobacteria are abundant in the photosynthetic biosphere and have considerable potential with regards to broadening bioreporter applications. Two recent studies described novel cyanobacterial reporters for the detection of environmental toxicants and iron availability [11].

The dose response relationship between seven commonly used herbicides and four luminescence-based bacterial biosensors was characterised. As herbicide concentration increased the light emitted by the test organism declined in a concentration dependent manner. These dose responses were used to compare the predicted vs. observed response of a biosensor in the presence of multiple contaminants. For the majority of herbicide interactions, the relationship was not additive but primarily antagonistic and sometimes synergistic. These biosensors provide a sensitive test and are able to screen a large volume and wide range of samples with relative rapidity and ease of interpretation. In this study biosensor technology has been successfully applied to interpret the interactive effects of herbicides in freshwater environments [12].

3.8.2   UV DETECTION

Microbial biosensors not only gives a significance response in pollutant food or Agriculture but can also report UV variation in ecosystem. E.coli strains containing plasmid borne fusions of the recA promoter/operator region to the *Vibrio fischeri* lux gene were previously shown to increase their luminescence in the presence of DNA damage hazards. Lux genes also found in *Photorhabdus luminescense*, host may be *Salmonella typhi murium* instead of *E. coli*, use of *Salmonella* as host or *P. luminescence* reporter genes gives fastest response however *E. coli* shows highg sensitivity because it has single copy of the *V. fischeri* lux fusion was integrated into the bacterial chromosomes [13].

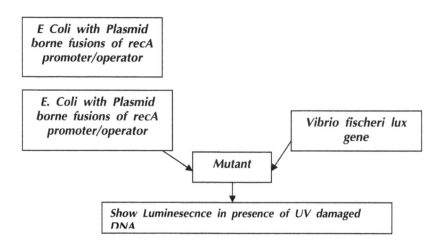

Hydrocarbon degrading *Acinetobacter baumanni* strain S30, isolated from crude oil contaminated soil was inserted with a *lux* gene from the luciferase gene cassette *lux* CDABE. The modified bacteria helped to decide the level of soil contamination with crude oil [14].

3.8.3   FOOD AND AGRICULTURE

Immunodiagnostics and enzyme biosensors are two of the leading technologies that have a greatest impact on the food industry. The use of these two systems has reduced the time for detection of pathogens such as *Salmonella* to 24 hr and has provided detection of biological compounds such as cholesterol or chymotrypsin [15]. Biosensors analyses Beta lactams in milk and presence of urea in milk that lead to production of "synthetic milk," the biocomponent part of the urea biosensor is an immobilized urease yielding bacterial cell biomass isolated from soil and is coupled to the ammonium ions selective electrodes of a potentiometric transducer. The membrane potential of all types of potentiometric cell based probes is related to the activity of electrochemically-detected product [16].

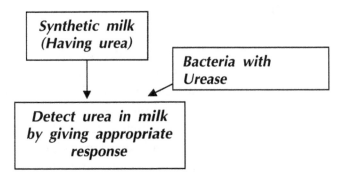

Ralstonia eutropha strain a whole-cell biosensor developed for the detection of bioavailable concentrations of Ni2+ and Co2+ in soil samples, the regulatory genes are fused to the bioluminescent lux CDABE reporter system. This modified bacteria used to study the accumulation of above said metals [17].

A biosensor for detecting the toxicity of polycylic aromatic hydrocarbons (PAHs) contaminated soil has been successfully constructed using an immobilized recombinant bioluminescent bacterium, GC2 (lac$_{luxCDABE}$), which constitutively produces bioluminescence. This biosensor system using a biosurfactant may be applied as an in-situ biosensor to detect the toxicity of hydrophobic contaminants in soils and for performance evaluation of PAH degradation in soils [18].

A bioluminescent whole-cell reporter *Ralstonia eutropha* strain developed by using with a tfdRP(DII)-luxCDABE fusion gene for detection of weedicides like 2,4-dichlorophenoxyacetic acid and 2,4-dichlorophenol in the contaminated soils [19].

A Synechococcus PglnA$_{luxAB}$ fusion for estimation of nitrogen bioavailability to freshwater cyanobacteria was done by fusing the promoter of the glutamine synthetase-encoding gene, P glnA, from *Synechococcus* sp. strain PCC7942 to the luxAB luciferase-encoding genes of the bioluminescent bacterium *Vibrio harveyi* [20].

Tagging potato leaf roll virus with the jellyfish green fluorescent protein gene helped to glow the plant immediately after infection [21].

3.9   CONCLUSION

Biosensors represent a rapidly expanding field, at the present time, with an estimated 60% annual growth rate. Conventional electrochemical biosensors have been widely used in various fields of science. New concept of using genetically engineered microbial biosensors (bioreproters) are one of the most expanding field because it is highly efficient by emitting easily detectable, long duration signals. Recently developed bioreporter technology providing a robust, cost-effective, quantitative method for rapid and selective detection and monitoring of physical and chemical and biological agents in applications as far ranging as medical diagnostics, precision agriculture, environmental monitoring, food safety, process monitoring and control, space and research, defense, etc.

Bioreporters attractiveness lies in the fact that they can often be implemented in real-time, on-line bioassays within intact, living cell systems. Since molecular biology and microbiology are growing with rapid pace in near future we can able to see new and wide rage of genetically modified microbes as biosensors.

KEYWORDS

- aequorin
- bacterial luciferase
- firefly luciferase
- luxAB bioreporters
- luxCDABE bioreporters
- *Pseudomonas fluorescence*
- *Pseudomonas putida*

REFERENCES

1. Mark, Frauenfelder. *Biosniffing Wireless Data*, Jan **2003**.
2. Cancer, Hirschmann F.; Verhoeyen, E.; Wirth, D.; Bauwens, S.; Hauser, H.; Rudert M. Vital marking of articular chondrocytes by retroviral infection using green fluorescence protein, *Osteoarthritis Cartilage.*, Feb. **2002**, *2*, 109–118.
3. Choy, G.; Choyke, P.; Libutti, S. K. Current advances in molecular imaging, noninvasive in vivo bioluminescent and fluorescent optical imaging in cancer research, *Mol. Imaging.*, Oct. **2003**, *2(4)*, 303–12.
4. Choy, G.; O'Connor, S.; Diehn, F. E.; Costouros, N.; Alexander, H. R.; Choyke, P.; Libutti, S. K. Comparison of noninvasive fluorescent and bioluminescent small animal optical imaging, *Biotechniques.*, Nov. **2003**, *35(5)*, 1022–6, 1028–1030.
5. Bachmann, A. S. The role of polyamines in human cancer, prospects for drug combination therapies, *Hawaii Med J.*, Dec. **2004**, *63(12)*, 371–374.
6. Wolkwicz, R.; Nolna, G. P. Gene therapy progress and prospects, Noval gene therapy approches for AIDS, *gene therapy*, Feb. **2005**, (Epud ahead of print).
7. Piunno, P. A.; Krull, U. J. Trends in the development of nucleic acid biosensors for medical diagnostics. *Anal. Bioanal. Chem.*, Feb. **2005**, (Epub ahead of print)
8. David A.; Relman, M. D. Shedding Light on Microbial Detection, *The New Journal of Medicine.*, Nov. **2003**, *349*, 2162–2163
9. Babich, H.; Devanas, M. A.; Stotzky, G. The mediation of mutagenicity and clastogenicity of heavy metals by physicochemical factors, *Environ. Res.*, Aug. **1985**, *37(2)*, 253–286.
10. Ren, S.; Frymier, P. D. Toxicity of metals and organic chemicals evaluated with bioluminescence assays, *Chemosphere.*, Feb. **2005**, *58(5)*, 543–550.
11. Bachmann, T. Transforming cyanobacteria into bioreporters of biological relevance, *Trends Biotechnol.*, Jun. **2003**, *21(6)*, 247–249.
12. Strachan, G.; Preston, S.; Maciel, H.; Porter, A. J.; Paton, G. I. Use of bacterial biosensors to interpret the toxicity and mixture toxicity of herbicides in freshwater, *Water Res.*, Oct. **2001**, *35(14)*, 3490–3495.
13. Rosen, R.; Davidov, L. A.; Rossa, R. A.; Belkin, S. Microbial sensors of ultraviolet radiation based on rec, , lux gene fusion, *Appl. Biochem. Biotechnol.*, Nov-Dec. **2000**, *89(2–3)*, 151–160.
14. Mishra, S.; Sarma, P. M.; Lal B. Crude oil degradation efficiency of a recombinant Acinetobacter baumannii strain and its survival in crude oil-contaminated soil microcosm. *FEMS Microbiol. Lett.*, Jun. **2004**, *235(2)*, 323–331.
15. Richter, E. R. Biosensors, Applications for dairy food industry. *J Dairy Sci.*, Oct. **1993**, *76(10)*, 3114–3117.
16. Verma, N.; Singh, M. A disposable microbial based biosensor for quality control in milk. *Biosens. Bioelectron.*, Sep. **2003**, *18(10)*, 1219–1224.
17. Tibazarwa, C.; Corbisier, P.; Mench, M.; Bossus, A.; Solda, P.; Mergeay, M.; Wyns, L.; van der Lelie, D. A microbial biosensor to predict bioavailable nickel in soil and its transfer to plants, *Environ. Pollut.*, **2001**, *113(1)*, 19–26 [ni].

18. Gu, M. B.; Chang, S. T. Soil biosensor for the detection of PAH toxicity using an immobilized recombinant bacterium and a biosurfactant, *Biosens. Bioelectron.*, Dec. **2001**, *16(9–12)*, 667–674 [pah].
19. Hay, A. G.; Rice, J. F.; Applegate, B. M.; Bright, N. G.; Sayler, G. S. A bioluminescent whole-cell reporter for detection of 2, 4-dichlorophenoxyacetic acid and 2, 4-dichlorophenol in soil. *Appl. Environ. Microbiol.*, Oct. **2000**, *66(10)*, 4589–4594.
20. Gillor, O.; Harush, A.; Hadas, O.; Post, A. F.; Belkin, S. A Synechococcus PglnA, , luxAB fusion for estimation of nitrogen bioavailability to freshwater cyanobacteria, *Appl. Environ. Microbiol.*, Mar. **2003**, *69(3)*, 1465–1474.
21. Nurkiyanova, K. M.; Ryabov, E. V.; Commandeur, U.; Duncan, G. H.; Canto, T.; Gray, S. M.; Mayo, M. A.; Taliansky, M. E. Tagging potato leafroll virus with the jellyfish green fluorescent protein gene, *J. Gen. Virol.*, Mar. **2003**, *81(Pt 3)*, 617–626.

CHAPTER 4

DEGRADATION AND STABILIZATION OF FUR AND LEATHER

ELENA PEKHTASHEVA and GENNADY ZAIKOV

CONTENTS

4.1   INTRODUCTION

This chapter presents the information about biodamages and protection of leather and fur. The contributors put a special emphasis on rawhide microflora, putrefied hide microflora, effect of prior operations on bioresistance of leather, interesting facts, fur skin structure and properties changes by microorganisms and the methods of leather preservation against microorganism impact.

Solving of polymers degradation and stabilization problems is a very important task for pure and applied chemistry [132]. If we can double the life span of materials it will be the same as doubling the production of these materials. It is very important to exactly predict the materials exploitation and storing time. Being the natural protein material, leather assumes its commercial properties via a multistage treatment by various reagents and represents a culture medium where bacteria and mold fungi can develop [15, 16, 21, 27–32].

Animals' hide has a complex structure and various ways of microorganism permeation into it. Hide of an animal while alive carries a lot of microbes appeared from the environment (water, air, soil), because it has direct contacts with it (dipping, rolling on the ground, dust deposition). If the animal care is insufficient its hide may carry a vast number of microbes (up to 1–2 billion/cm²). Microbes found on the hide after flaying where present on the living animal and partly appeared after flaying due to further contamination. After killing and flaying the hide is secondarily contaminated by microbes sourced from dirty floor and baskets, animal dung and dirt. Specific microbial flora of killing chambers and raw leather warehouses consisting of microbes greatly propagating on both molding production waste and stored raw materials and tools is of great importance. Air in these facilities is usually full of specific microflora, mold spores, salt microbiosis, etc. [32].

Rawhide contamination by microorganisms largely depends on the purity of keeping of appropriate compartments and the hide handling. If the rawhide is poorly handled the number of microbes on it significantly increases that may cause subsequent significant damage to the raw material. The external side of just flayed hides contains a lot of microbes, whereas the external side is sterile. However, already after two hours the rawhide completely loses commercial properties due to microorganism effect. It is therefore immediately preserved by solutions with introduced biocides. However, since before treatment by preservative agents, the hide is attacked by microbes, which can occur on it from both the epidermis side and the living hypoderm. While inspected, their external layers demonstrate various kinds of microorganisms, the upper layer of epidermis being most favorable for their habitation and consisting of separated flat keratinized cells losing connection with one another. The animal papillary dermis represents a dense connective tissue consisting of the main substance, intercellular fibrous formations and cellular elements; it is loose, unstable and non-resistant to microorganisms' impact [1].

The reticular layer primarily consists of complexly and densely tangled collagen fiber yarns; therefore, if the hide is just flayed its reticular layer contains no microbes. However, the internal part of the layer adjacent to subcutaneous fat is loose and more permeable for microbes. Four structural levels of collagen, the dermis protein, are distinguished: primary-polypeptide chain; secondary-spiral (the α-form); tertiary-triple spiral (protofilament); and quaternary-supermolecular structure associated with the regulated packing of protofilaments (fibril). Collagen organization level next to fibrils is represented by fibers. In the dermal layer of the skin, collagen fibers are submerged into glycosaminoglycan (GAG) structures, which act as an interfibril "cement" [27–32].

4.2 RESULTS AND DISCUSSION

Hydrated GAG causes a very strong effect on the structure of collagen fibers: they protect protein fibers of fibrils against fusion and increase their mobility, thus providing fiber integrity.

It is known, for instance, that fairly preserved samples of medieval vellum consist of densely packed collagen fibrils submerged into the interfibril amorphous substance. However, some cases have been described when the vellum transformed to bonded protein due to the damages caused by microorganisms [1, 32]. After storing bovine dermis in water during 9 months, the intense differentiation of fibrils and their isolation into fibrils is observed. This testifies to predominant decay of interfiber and interfibril substance, i.e., basically GAG [17, 32]. At the initial stages, microbiological degradation is similar to hydrolysis forming particles, which consist of amino acid groups. These particles, along with separate amino acids, are rapidly subject to further transformation. Microbiological protein degradation products commonly contain ammonium, fatty acids, amino acids, aldehydes, and amines [30–32].

For the microorganism development, subcutaneous fat is a particularly favorable medium, where they propagate and from which penetrate into dermis. When appearing in the reticular layer, in the interfilament space, microbes may induce the putrefactive process deep in the hide. Meanwhile, blood remained in vessels due to poor bleeding and intercellular substance of the dermis and the reticular layer are proper nutrition for microorganisms. Regarding environmental conditions and sponginess of the reticular layer microbes can form numerous colonies on its surface or penetrate into upper layers, propagating and destroying them. They can move easily and rapidly by interfiber and interfilament spaces destroying the dermis substance and various intercellular components. For example, moving by the fiber, rod-shaped microbes penetrate into collagen filaments and then spread in the surrounding tissues, whereas cocci penetrate into hair bags. Mold mycelium spreads either along collagen fibers or, occurring in the interfilament space, in all directions forming dense entanglement [30–32].

By chemical composition, leather tissue represents a medium favorable for fast microorganism propagation. The rawhide contains both inorganic and organic substances. Inorganic substances in the leather tissue are water (50–70%) and mineral substances (0.35–0.5%). Organic substances in it are lipids (fats and fat-like compounds), carbohydrates, non-protein nitrogen-containing components and proteins forming the histological structure base of the leather tissue. The most important components of this structure are fibrous proteins, such as collagen, keratin, elastin and reticulin. In addition, cutaneous covering contains globular (albumins, globulins) and conjugate proteins. Albumins and globulins, whose sufficient amounts are contained in the rawhide, degrade most easily. Fats are affected by specific lipoclastic microbes [32].

High amounts of proteins present in the leather tissue are one of the factors, which cause its extreme sensitivity to destructive impact of putrefactive microbes. This is also promoted by the medium response (rawhides have pH 5.9–6.2). The skin coating contains vitamins, enzymes catalyzing chemical reactions and affecting development of biochemical processes in tissues, and substances, which increase (activators) and suppress (inactivators) enzyme activity in the hide. If one of activators or inactivators is absent, biochemical processes in the hide change. For example, rawhide aging increases protease activity, which induces protein breakdown and promotes development and propagation of microorganisms.

Increasing fat content in the leather tissue entails relative decrease of water content, which increases leather resistance to the action of various microorganisms. Depending on chemical composition of the leather tissue, i.e., whether amounts of proteins, fats, etc. in it are high or low, its resistance to microbial activity is different. It has been found that the kind of microorganisms inhabiting the rawhide depends on animal feeding. In the absence of vitamin B_2 and biotin, for example, dermatitis and loss of hair by the skin coating are observed, that promotes permeating of microbes into the hide [1, 32].

4.2.1 RAWHIDE MICROFLORA

The so-called putrefactive microorganisms, including cocci and rods, sporous and asporous, aerobes and anaerobes, are abundant in the rawhide. The general feature of these microorganisms is their ability to degrade proteins. Without getting into specifics of every species of microbes, several groups of them most commonly observed in the raw material should be highlighted. The majority of them are rod-like, both sporous and asporous. The group which includes *Proteus represents asporous mobile rods; this group has clearly proteolytic ability and degrades proteins to final products.* A group composed by *E. Coli* bacteria, generally occurring from dung and being short rod shaped, represents the intestinal flora, both mobile and immovable. This group's representatives induce peptone decay to amino acids with the indole formation. The sporogenous group, including *Bac. subtilis, Bac. mesentericus, Bac. mycoides,* and *Bac. megatherium,* mostly represent mobile rods generating high stability spores. These microbes also feature pronounced proteolytic ability and breakdown proteins to final products.

To a lesser extent, the group of cocci, including micrococci and pilchard generally producing pigments (yellow, ochreous, brown, red, and white), is observed. Many of them produce enzymes affecting partly degraded protein. A group of actinomycetes has optimum development at pH 7.0–7.5. They are also able to degrade protein. Bacteria of this species are frequently observed in soils, wherefrom they, apparently, appear in the hide. Sometimes bacteria of the fluorescent group occur on the hide. These are asporous gram-negative rods. Many species dissolve gelatin and decompose fats. They are mostly psychrophilic bacteria. These kinds of microbes most often occur in water, among which *Bact. Fluorescens*, etc. All the above groups of microbes are aerobes. Yeasts observed on the raw material are so-called wild, wide-spread in the nature, namely, white, black and red yeasts [21].

Molds representatives often occur on the rawhide. Many of them have pronounced proteolytic ability. Fungi of *Mucor, Rhizopus, Aspergillus, Penicillium,* and *Oidium* genus are observed on the hide. As mentioned above, microbes commonly occur on the surface of the raw, freshly flayed hide, both at the side reticular layer and epidermis. Microtomy on the rawhide shows the absence of microbes in the tissue, both in near-surface and deep-in layers. Single cocci may only be infrequently observed in hair bags.

4.2.2   PUTREFIED HIDE MICROFLORA

Unpreserved hide is easily putrefied. High temperature and humidity, stocking of uncooled hides and their contamination lead to fast propagation of putrefactive microbes in the hide [32].

Aerobic putrefaction starts from the surface and gradually penetrates deep into the layers. Three putrefaction stages are distinguished. The first stage is characterized by fast microbial propagation on the hide surface causing no visible reflex. The second stage has visible hide changes: sliming, color change, odor, and doze. This period coincides with commencing permeation of microflora into the dermis thickness. The third stage is characterized by intensification of visible displays in line with hair and epidermis weakening and deep microbial permeation and spreading by the hide layers (Fig. 1).

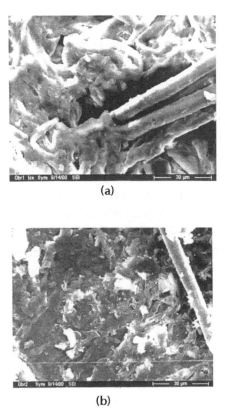

(a)

(b)

FIGURE 1   Microphotographs of the leather tissue of mink's rawhides (1000×): initial sample (a) and 14 days after spontaneous microflora impact (b).

Hide putrefaction demonstrates gradual change of the species composition of the microflora. Coccus species of bacteria, whose significant amount is observed in the rawhide, are gradually substituted by highly propagating rod forms, namely, *Proteus vulgaris, Bac. subtilis, Bac. mesentericus*, clubbing, etc. Microtomy of unpreserved putrefactive rawhides demonstrates particularly rod forms penetrated deep in the hide layers [19].

4.2.3   EFFECT OF PRIOR OPERATIONS ON BIORESISTANCE OF LEATHER

Rawhide preservation. Immediately after flaying, the raw skin is affected by microorganisms, which results in damages and reduces quality of rawhides and the yield of leather. Primary signs of the hide decomposition are surface sliming and flesh side color change. Then characteristic putrefactive odor occurs, hair root bonding with their bags is weakened, hairslip occurs followed by exfoliation. Finally, pigmentation occurs and mechanical strength reduces down to massive fracture [17].

All the above necessitates rawhide preservation within two hours after flaying. Otherwise leather will lose its commercial properties. The preservation goal is to create unfavorable conditions for bacterium and enzyme action. This may be reached by moisture elimination and impacting protein substances of the hide by reagents. At the preservation stage, aerobic bacteria of *Bacillus, Pseudomonas, Proteus, and Achromobacter genus possessing proteolytic enzymes manifest the highest activity. These bacteria are capable of damaging the hair side, its globular proteins, lipids, and carbohydrates. Some of them are capable of causing collagen decomposition* [16].

To prevent putrefactive processes, rawhides are preserved by three methods: flint-dried, dry salting and wet-salting. Flint-dried and dry salting cure is based on suppressing vital activity of bacteria and activity of proteolytic enzymes by reducing raw material humidity to 18–20% due to treatment by dry sodium chloride and sodium silicofluoride. Meanwhile, optimal condition of the flint-dried cure is a definite temperature mode 293–308 K, because the process at lower temperature may cause bacterial damages and hide decay due to slow water removal. Special requirements are also set to relative air humidity in the compartment which is to be 45–60%, and good circulation of air is required [15].

Wet-salting cure is performed by sodium chloride salting from the internal side of the hide (fleshing side) or treatment by saturated sodium chloride aqueous solution—brining, with further hides add salting in stock piles. Wet salting generally removes free water from the hides. Meanwhile, greater part of asporous bacteria dies and development and propagation of other microorganisms and action of enzymes is terminated or suppressed.

At dry salting the preservative effect of sodium chloride is based on hide dehydration and at wet salting—on breaking intracellular processes due to sodium chloride diffusion into cells. However, sodium chloride does not provide full protection against microorganisms and may even be a substrate for halophilic (salt-loving) bacteria and salt-tolerant bacteria (*Bacillus subtilis*), which possess the proteolytic capability. To protect against them at brining, sodium metabisulfide is added as bacteriocide [14].

In addition to the above listed preservation methods, rawhide freezing can be used as a temporary measure. At low temperature activity of bacteria and enzymes is terminated; however, procurement organizations must defrost the frozen material and cure it by wet salting. Irradiation is considered to be an effective method of rawhide protection against microorganism action. After rawhide irradiation in 1 kJ/kg dose, it can be stored during 7 days without noticeable signs of bacterial damage. Irradiation by 3 kJ/kg dose extends the storage period to 12 days. In this case, rawhides need no additional chemical curing. A combination of wet salting and irradiation of the rawhides, in turn, almost completely eliminates microflora, whose activity in the rawhides is terminated for 6 months [32].

4.2.3.1 BASIC OPERATIONS OF RAWHIDE PROCESSING TO LEATHER

The rawhide processing to leather is a multistage process. At various stages of this process quite favorable conditions for microorganism growth and development on the leather may be created [13]. The danger of leather bacterial damage occurs already at the initial stage (soaking) aimed at preservatives removal from the rawhide and making it as close to just flayed state as possible. Soaking is water treatment at 303 K, mostly with electrolyte adding. Salt concentration in the leather thereby abruptly reduces, that promotes occurrence and development of bacteria, which become active in water, especially at higher temperature. In this case, bacterial damage starts from the grain side, and biologically unstable components at this stage are globular proteins. Among 10 species of bacteria detected during soaking more than a half of bacteria have proteolytic enzymes [12].

At this stage, sodium silicofluoride is used, which is active in neutral and weak acid media. At the next stage of *liming, to remove interfiber protein substances and to loosen fibrous structure of the dermis, hides are treated by caustic lime solution. The aim of liming is to weaken hair and epidermis bonding to dermis, which promotes their further free mechanical removal. Moreover, dermis free from interfibrous proteins becomes more permeable, it volume is formed, and tannin diffusion into the dermis is accelerated.* As regards bioresistance, the liming process is characterized by the fact that asporous bacteria die in the lime bath and the sporous ones stop growing and propagating [11].

Then hides without hairs (pelt) are subject to preliminary tanning operations, which are *deliming* and *drenching*. The first operation is performed at 298–303 K, most often using ammonium sulfate or lactic acid. Drenching in turn represents a short-term pelt processing by enzyme compounds in water at increased temperature 310–311 K. As a result, leather becomes soft and elastic, with interfiber proteins and collagen decomposition products eliminated, and breathable. However, at this stage favorable conditions for bacterium propagation are formed. With respect to drenching fluid composition, bacteria of *Sarcina, Staphilococcus, Pseudomonas, Bacillus, etc. genus have been extracted. Therefore, to avoid biodamages duration of operations should be controlled and preset treatment time should not be exceeded.* Thus at all stages of rawhide processing to leather conditions may be formed that promote microorganism growth and development on the leather [7].

Interesting facts. When using fire, the ancient man has discovered preservative properties of smoke. For tanning hides Indians applied a mixture of chopped liver, brain and fat. After currying the leather was smoked. Clothes from such leather were not hardened under the effect of water and microbes and its odor repelled mosquitoes. Another tanning method also existed. Leather was wrapped around the leg, then wrapped by foot wraps and carried until it gained best properties; or leather was soaked in urine or dug in sheep manure.

Tannage process. *Tannage consists in injection of tanning substances into the dermis structure and their interaction with functional groups of protein molecular chains, which results in formation of additional stable cross bonds.* Tannage is one of the most important processes for leather manufacture. This stage in the leather industry radically changes dermis properties, transforming it to tanned leather. Finally, the dermis property change defines leather behavior during dressing, manufacture and operation of articles from it. As compared with the pelt structure, the tanned leather structure features increased fiber separation assumed during preparation and fixed by tanning. The fiber separation defines a number of basic physical and mechanical properties of leather: tensile strength, compression strength, toughness, hardness, elasticity, etc. [5].

Tannage starts with tannic compound penetration into the collagen structure via capillaries and diffusion of these compounds from capillaries to direct response centers. Primarily, adsorptive interaction happens between tanning compounds and collagen, followed by stronger chemical bonding. Tannage causes strengthening of collagen spatial structure due to formation of cross bonds by tanning substances between molecular chains and the protein structure. It is known that collagen spatial cross-linked structure between molecular chains of the protein structure. Finally, tannage leads to formation of new, more stable cross bonds, whose strength depends on the tanning compound origin [4].

Tannage is accompanied by increasing dermis resistance to hydrothermal impact, i.e., increasing seal temperature. Tannage increases dermis resistance to swelling in water that causes significant impact on performance properties of leather. It also increases dermis resistance to proteolytic enzymes, i.e., leather biostability increases. Moreover, the interaction of tanning substances with functional groups of dermis protein leads to increased elastic properties of collagen and, consequently, reduced deformation rate of moist dermis, reduction of leather shrinkage area and thickness when dried. Tanning compounds are intrinsically divided into two main groups: organic and inorganic. Inorganic ones are compounds of various metals: chromium, zirconium, titanium, iron, aluminum, silicon, phosphorus, etc., among which chromium compounds are of the prime importance. Organic compounds are natural tanning substances—tannides, synthetic tanning matters, aldehydes, water-soluble amino resins, and blubber oils [1].

Inorganic tanning matters. A large number of inorganic compounds possessing tanning properties are known. However, mostly chromium compounds are used as tanning matters, because they allow obtaining of high-quality leather good both in manufacture and operation and storage. Primarily, tanning chromium compounds were mostly applied to leather manufacture for shoe upper. At present, chromium compounds are used in manufacture of almost all kinds of leather, both solely or combined with vegetable tanning matters, with syntans and amino resins. Tannage consists in treatment of pelt by tanning chromium compound solution. Drums and worm apparatuses are used for tannage. When rotated in apparatuses, the semi-product is subject to quite intensive mechanical impact, because wooden wobblers or boards rise and drop the semi-product, which, along with continuous fluid mixing and temperature increase, accelerates tannage. Tannage starts with diffusion of compounds into dermis structure. Primarily, tanning matters penetrate into capillaries, wherefrom chromium compounds diffuse to direct response centers and link to functional groups of protein. Meanwhile, tanning agent penetration into the dermis depends on a number of factors:

- firstly, on sufficient separation of structural components, loosening of collagen in preliminary operations—the more loosened fibers, the higher the diffusion rate of chromium compounds into dermis;
- secondly, on tanning particle size—the smaller the tanning particles, the higher the diffusion rate.

Another important factor is pelt pH. It needs to be 4.5–5.5 to provide normal tanning matter diffusion into the dermis. Chromium tanning compound diffusion is controlled by pelt section color change varied from white to blue or green. Resulting from these procedures, chromium tanned leather assumes exclusive properties. It is highly resistant to acids and alkali and has high seal temperature 413–423 K, high tensile strength (11 MPa), high softness and elasticity, and is resistant to higher temperatures (compared with vegetable tannage leather). Moreover, chromium tannage dried leather cannot be soaked again either in hot or in cold water. Although water penetrates easily into the chromium dermis, its structural elements remain waterproof [25].

These negative properties are explained by the nature of chromium complexes, type of bonds between them, functional groups of collagen and the strength of cross bonds between molecular chains forming bridges. The assumed features in turn explain sufficiently high bioresistance of chromium tanned leather. It is proven that tanned leathers are mainly impacted by microscopic fungi, because development of bacteria is hindered. However, it has been found that chromium tanned leather possess the highest resistance to molds because it is deeply impregnated by oils, wax and fats, and its fibers assume hydrophobic properties [21].

Moreover, chromium salts are weak antiseptics, which also play some role. However, despite relatively high resistance to biodamages of chromium-tanned leather, danger of microorganism development is not completely eliminated. Biodamaging agents extracted from tanning solutions and from the semi-product surface may be bacteria *Bacillus mesentericus* and some fungi: *Aspergillus niger, Penicillium chrysogenum, Penicillium cyclopium.* At this stage, sodium pentachlorophenolate and chloramines B may be used as biocides [32].

Beside usage of chromium as the self-tanning agent, it is also used for more intensive bonding of aluminum compounds with collagen. Aluminum tanning is one of the oldest tannage methods. Basic aluminum sulfate, chloride and nitrate are used as tanning agents. However, this method is applied restrictedly, because these aluminum compounds are unstably bound to dermis proteins. This bond breaks under the water effect and leather becomes detanned. Although aluminum tannage leather differs from other types of tannage due to higher softness and white color, it is undurable and due to aluminum compound washing out becomes wet rapidly and warps after drying [16]. These factors are significant for bioresistance of aluminum-tanned leather; washing out of aluminum compounds makes semi-product and final product accessible for microorganism penetration into dermis and their active propagation in it. However, application of aluminum compounds for retanning of vegetable-tanned inner sole leather, for instance, significantly increases its ability to resist mold development [14].

Organic tanning agents. Organic tanning agents are simple and complex by structure. Simple tanning agents are mostly aliphatic compounds: aldehydes and some kinds of blubber oils (tanning oils). Complex tanning agents are aromatic derivatives and some heterochain polymers: vegetable tanning agents (tannides), synthetic tanning agents (syntans), synthetic polymers (mostly amino resins). Tannides are substances contained in various parts of numerous plants, are extracted by water and capable of transforming dermis into leather while interacting. With regard to species of plants tannides are accumulated in their different parts: cortex, wood, leaves, roots and fruits. In this case, tannide content may vary in a very broad range, from parts to tens of percents.

Leather formation during rawhide tanning by complex organic compounds is the result of permeation of these substances into the semi-product and their bonding with broadly developed inner surface of particular structural elements of collagen via both thermodynamic adsorption and chemical interaction with amino groups and peptide groups of protein, with formation of cross bonds of electrovalent, hydrogen and, apparently, covalent type [13].

Semi-product seal temperature increases during tannage by tannides. This is explained by the fact that collagen has a multistage structure, and tanning particles having large size are incapable of penetrating into the smallest structural elements of the protein. These particles can easily be washed off, accompanied by simplest phenols and acids, which reduce the seal temperature. Moreover, similar to collagen, tannides have many reaction groups. As located on the collagen structure surface, tannide particle reacts with several structural elements of collagen forming cross bonds between them, which, in turn, increases the seal temperature (up to 341–363 K).

It is desirable to use tannage by tannides in cases when volume, hardness and rigidity are to be imparted to the leather or to increase its size stability in case of humidity variations, and when leather with high friction coefficient is required. However, such leather has low tensile strength, because when large tannide particles permeate into fibers they expand them and reduce the number of protein substance per specific cross-section. It should be noted that biostability and performance characteristics of such leather and articles from it abruptly decrease with increasing tannide concentration. In this case, tannides represent nutrition for microorganisms and manifest their impact in the form of hydrolysis of tanning agents, pigment spots, grain roughness [12].

However, tannides that represent phenols derivatives possess some bactericide and fungicide action. Fungicide effect of vegetable tanned leather on *Trichophyton genus fungi is shown. Another kind of complex tannides (syntans) is of two types: tannide substitutes and auxiliary ones.* Syntans of the first type are produced from raw phenols, and alongside with tanning properties they provide some leather protection against biodamages. Syntans of the second type are produced from hydrocarbon petroleum and gas refinery products and have no biocide properties. In the leather industry, the instances of severe damages of the semi-product by mold fungi after vegetable tannage using syntans prepared from hydrocarbon material. Thus neither organic, nor inorganic tanning compounds have the absolute biocide which, along with providing biostability, would promote formation of sufficient physical, mechanical and chemical properties of ready leather [21].

4.2.4 FUR SKIN STRUCTURE AND PROPERTIES CHANGES BY MICROORGANISMS

Microbiological stability of fur skins at different stages of manufacture by the standard technology (tannage, greasing) was assessed on the example of mink skins (rawhide and semi-products). Mink skins as rawhides, chrome non-oil and ready (i.e., greasy) semi-product were studied. Mink skin samples matured under conditions favorable for microorganism development showed clear sensory determined signs of degradation—leather tissue samples became fragile and hairslipping of the fur was observed.

Data on the influence of microorganisms, spontaneous microflora, on such physical properties of materials, as thickness, density and porosity are represented in Table 1.

TABLE 1 The change of mink leather thickness, density and porosity under the effect of spontaneous microflora ($T = 303$–305 K; $\varphi = 100\%$; $n = 10$; $P \leq 0.93$).

Sample	Exposure time, days	Thickness, mm	Density, kg/m³ real	seeming	Porosity, %
Rawhide	init.	0.42	1336	701	47.5
	7	0.41	1340	719	46.3
	28	0.40	1345	747	44.5
Tanned semi-product (non-greased)	init.	0.4	1342	647	51.8
	7	0.37	1357	681	49.8
	28	0.33	1363	719	47.2
Prepared semi-product (greased)	init.	0.58	1353	632	53.3
	7	0.55	1355	755	44.3
	28	0.45	1366	825	39.6

As follows from the data obtained, both rawhide and non-greasy and ready semi-product demonstrate density increase resulted from the spontaneous microflora effect.

Such change of the leather tissue density is apparently associated with the fact that molecular weight of collagen decreases and denser packing of structural elements (frequently, more ordered) becomes easier. This is testified by observed increase of real density of the leather tissue and abrupt reduction of its porosity with increasing duration of microbiological impact on both leather tissue of the rawhide and tanned semi-product. The system packing observed leads to leather tissue thickness reduction [10].

Data on the structure density increase resulting from microbiological impact are confirmed by the data obtained by EPR spectroscopy by radical probe correlation time [1]. The radical probe mobility decreases with the increase of microbiological impact duration. This may testify to packing of the leather tissue structure as a result of microflora impact on the material.

Analysis of the chemical composition of mink's leather tissue before and after impact of microorganisms testifies to reduction of the quantity of fatty matter with simultaneous relative increase of collagen proteins and mineral substances quantity (specifically observed in the raw material) (Table 2).

TABLE 2 Chemical composition of mink fur's leather tissue before and after spontaneous microflora impact ($T = 303$–305 K and $\varphi = 100\%$).

Material	Impact duration, days	Humidity, %	Substance content, in % of abs. dry substance			
			fatty	mineral	collagen protein	non-collagen protein
Raw material	Init.	12.3	10.15	2.28	76.39	11.18
	7	12.8	7.25	2.53	83.06	7.16
	14	14.3	7.14	2.69	84.28	5.29

Material	Impact duration, days	Humidity, %	Substance content, in % of abs. dry substance			
			fatty	min-eral	collagen protein	non-collagen protein
Tanned semi-product (ungreased)	Init.	9.7	2.99	5.88	90.08	1.15
	7	10.1	2.78	6.11	90.10	1.01
	14	10.8	2.58	6.49	89.98	0.95
	28	11.3	2.03	6.67	90.37	0.93
Ready semi-product (greased)	Init.	8.2	13.63	6.54	78.83	1.00
	7	9.8	9.99	6.88	82.14	0.99
	14	10.5	7.62	7.06	84.37	0.95
	28	11.1	6.61	7.06	85.42	0.91

One may suggest that affected by the microflora, non-collagen proteins degrade and are "washed off" from the structure. In this connection, relative content of collagen proteins in the leather tissue increases.

Thus, observed reduction of organic matter content (fats, non-collagen proteins) is obviously associated with the change of leather tissue structure. The destructive effect on the structure of studied materials may also cause alkalinity of the medium, which, as known, increases at natural proteins degradation [32]. Table 3 shows data of pH change of water extract of initial and test samples. In all cases, system alkalinity increase was observed.

TABLE 3 Water extract pH of mink's leather tissue at different stages of their manufacture before and after spontaneous microflora impact ($T = 303-305$ K; $\varphi = 100\%$; $n = 10$; $P \leq 0.93$).

Mink's sample	Water extract pH			
	Microflora wimpact duration, days			
	0	7	14	28
Raw material	5.76	6.40	6.30	Not determined
Tanned semi-product (ungreased)	3.84	4.27	5.24	6.40
Ready semi-product (greased)	3.91	4.83	5.92	7.23

Microscopy study results testify to the fact that observed changes in the material properties are caused by the change of supermolecular structure are electron.

Figures 1 and 2 show electron microscopic images of the leather tissue sample surfaces (raw material and ready semi-product) before and after spontaneous microflora impact.

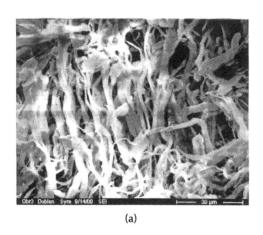

(a)

(b)

FIGURE 2   Microphotographs of mink's fur leather tissue, both tanned and greased (1000□): initial sample (a) and 28 days after spontaneous microflora impact (b).

The images clearly show the fibrous structure of initial samples. Fibrous yarns (within 20–40 μm in diameter), separate filaments forming yarns (up to 20 μm in diameter) and collagen fibrils (0.1–20 μm) are distinguished in collagen. Dermis represents an irregular three-dimensional entanglement of fibers and their yarns, which is clearly seen in the images of initial samples of both mink's raw skins and ready semi-product. Collagen yarns with clear contours are seen, fibers in yarns being higher structured for the semi-product [21].

Samples affected by microorganisms show destruction of fibril formations and their transformation to laminated structures. Hence, it is noted that the leather tissue porosity was slightly reduced as a result of microorganism impact.

Physicomechanical properties, first of all strength and deformation characteristics of the leather tissue, as well as hair connection strength with the leather tissue, are the most important fur properties.

Table 4 shows results of our study of physicomechanical property change of ready mink semi-product (with greasing) impacted by microorganisms—spontaneous microflora, micromycetes *Aspergillus niger* and bacteria *Bac. subilis*.

TABLE 4 The change of physicomechanical properties of ready mink semi-product before and after impact of microorganisms (T = 303–308 K; φ = 100%).

Kind of impacting microflora	Exposure time, days	Breaking strain of leather tissue (σ_p), MPa	Change to init., %	Relative elongation at rupture of leather tissue (ε), %	Change to init., %	Hair bonding strength with leather tissue (σ_p), $1 \cdot 10^{-4}$ N	Change to init., %
	7	11.5	−10.9	50.4	−12.0	64.4	−11.2
Spontaneous microflora	14	11.0	−14.7	45.3	−20.9	53.3	−26.5
	28	9.8	−24.0	37.7	−34.2	34.1	−52.9
	7	9.6	−25.6	48.3	−15.7	57.5	−20.7
Asp. niger	14	4.7	−63.6	43.5	−24.1	45.4	−37.4
	28	2.4	−81.4	35.6	−37.9	29.3	−59.6
	7	7.5	−41.9	47.1	−17.8	53.5	−26.2
Bac. subilis	14	2.6	−79.8	41.5	−27.6	40.3	−44.4
	28	1.9	−85.3	34.3	−40.1	26.7	−63.2

The investigation has determined that the highest reduction of the breaking strain, leather tissue elongation at rupture and hair bonding strength with the leather tissue is observed under the impact of bacteria *Bac. subtilis.*

Thus it is observed that the properties of both rawhide and ready semi-product leather tissue change under the impact of microorganisms. Meanwhile, the true density increases, and porosity decreases due to degradation processes, which simplify packing of structural elements into more ordered formations as a result of steric hindrances reduction. The tensile strength, leather tissue elongation at rupture and hair bonding strength with the leather tissue are also reduced [6, 7].

4.2.5   THE METHODS OF LEATHER PRESERVATION AGAINST MICROORGANISM IMPACT

4.2.5.1   RAWHIDE AND CURED RAW LEATHER PROTECTION

To increase biostability of leather and articles from it, it is recommended to protect leather against microorganisms at all stages of its treatment, starting with the rawhide.

Due to termination of oxygen delivery and metabolism, tissue degradation in just flayed hides is accelerated. The medium response reaches its optimal value for protease action. First breakdown of proteins forming the hide base is initiated and then carbohydrates, fats and other organic compounds degrade. As a result, chemical composition and the structure of tissues change. As impacted by microorganisms and enzymes, rawhides go bad rapidly at the temperature above 291 K, and tissue putrefaction starts [4].

At further development of putrefaction process, epidermis is destroyed and delaminated, and "damaged grain", the absence of grain in some areas, occurs. Putrefactive microbes damage subcutaneous fat. Then occurring in the reticular dermis, they rapidly spread in the interfascicular space, and then degrade collagen and elastin fibers. Due to these processes dermis delaminates which, in turn, leads to complete destruction of leather [2].

To preserve high quality of the raw material at this stage and to make it stable to putrefactive microbe impact the hide has to be thoroughly cured, i.e., all contamination, slices of fat and meat, has to be removed, and proper curing has to be performed. As mentioned above, rawhides are cured by three methods: flint-dried, dry salting and wet salting. The main substance used for curing is sodium chloride. Common salt microflora is represented by microbes developed in salt solutions—brines, and microbes occurring in salt during its production and transportation.

Under natural conditions, this salt possesses microbes of the halophobe group, as well as salt-tolerant species. Moreover, salt contains representatives of sporous microflora, yeasts, mold fungi spores, micrococci, variously dyed bacteria of *Flavobacterium genus*. *When occurring on hides with salt during curing, microorganisms induce various defects; therefore, antiseptic agents are formed alongside with sodium chlorides. They possess bactericide, fungicide, bacteriostatic and fungistatic properties.* Antiseptic agents used for curing must be toxic for microorganisms, well-soluble in water and in sodium chloride solution, cause no negative effect on the hide quality and leather semi-products. In this connection, paradichlorobenzene and sodium silicofluoride are most widespread [1].

When affected by paradichlorobenzene, some microbes developing in the wet-salt rawhide die and development of others is terminated. Hence, gaps between hides are filled with vapors of this substance. These vapors are heavy, slowly removable and hindering microorganism propagation for a long time. Some part of antiseptic agent is dissolved in fat and penetrates into the dermis; therefore, its typical odor is preserved for a long time. Sodium silicofluoride is quite effective. It possesses high bactericide properties and causes no negative effect on dermis. Rawhides brining with simultaneous treatment by sodium silicofluoride ensures long, (over a year) storage of hides without additional salting. However, this antiseptic agent is poisonous and care should be taken when operating it.

It is found that antiseptic agents also give strong effect when combined with each other. Moreover, goods results are reached by application of sodium hypochloride, boric acid, sodium borate, zinc chloride, sodium fluoride, benzene and phenol chlorine derivatives, antibiotics and other antiseptic agents to rawhides curing.

However, beside chemical means of the raw material protection against microorganisms, meeting conditions and technology of curing is of importance. If the raw material was tainted when cured then despite the absence of tissue destruction the microflora of such raw material is richer. The presence of a great layer of reticular dermis, musculature and especially fat inclusions hinders diffusion processes, decelerates curing, has negative impact on the raw material quality and promotes development of microbes [30]. Moreover, the rawhide of flint-dried and dry-salt curing requires ideal conditions during transportation and storage, because high humidity forms favorable conditions for bacterium and mold development [21].

4.2.5.2   PROTECTION OF LEATHER AND LEATHER ARTICLES AGAINST BIODAMAGES

The problem of biological damaging of natural materials, especially upper leather for shoes used in increased humidity conditions, is of great importance. Alongside with direct action associated with leather structure damage, microorganisms also manifest indirect adverse effect on the leather articles. Microscopic fungi promote leather hygroscopic property increase. As a result, relative humidity inside the shoes increases. These causes untimely wear of joints and development of pathogenic microorganisms inside the shoes [17].

In the world practice, rawhide and ready leather is widely protected by the following compounds: phenylmercury, bromo-acetophenone, *n*-chlor-*m*-creosol, alkyl naphthalene-sulphodiacid, sodium borate, zinc oxide, 2-oxydiphenyl, salicylanilide, and some other. However, a wide application of some biocides is restricted by specific requirements to leather protection: biocides must be soluble in fats, thermostable at stuffing temperature and compatible with other components used for leather treatment [15].

It has also been found that most of the above biocides do not manifest a long-term antimicrobial action, because the antiseptic agent injected at the stuffing stage sweats out with fat during the use, and frequently the fungicide simply evaporates. In this connection, the optimal protection may be provided by biocides introduced into finish coating composition and compounds capable of chemically bonding with collagen. In this regard, β-naphthol and β-oxy-naphthaldehyde injected into leather when proofing, have proven their value. Leather treatment after tannage by catamine AB from the class of quaternary ammonium compounds is worthy. Linking to collagen by coordination or salt and adsorptive bonds, catamine AB prevents development of microscopic fungi on the ready leather surface.

Among microorganisms, mold fungi from *Aspergillus* and *Penicillium* genus are the most active and widespread destructors of the real leather. To eliminate mold development much can be done by the temperature regulation and leather humidity decrease below 12% with the help of hygroscopic agents (e.g., silicagel). However, such regulation is not always practical. The most effective means to eliminate mold development on the leather is the use of chemical agents [15, 17, 21, 30].

It is found that natural fungus resistance of the leather is increased by using materials based on organochlorine products; leather with fungicide properties can be obtained by using sulfochlorinated paraffins during stuffing. Both semi-products and the ready leather are effectively protected by benzoguanamine formaldehyde resins (BGAF). These compounds containing 40% of the basic substance are water-soluble, which allows their application at all stages of leather manufacture process.

The ability of BGAF-resins to suppress mold micellium growth was also determined, the resin with higher content of sulphosalicylic acid having higher fungicide activity [32].

4.3 CONCLUSION

The paper presents the information about biodamages and protection of leather and fur. The contributors put a special emphasis on rawhide microflora, putrefied hide microflora, effect of prior operations on bioresistance of leather, interesting facts, fur skin structure and properties changes by microorganisms and the methods of leather preservation against microorganism impact.

KEYWORDS

- **albumins**
- *Aspergillus*
- *cocci*
- *deliming*
- *drenching*
- *globulins*
- *Penicillium*

REFERENCES

1. Emanuel, N. M.; Buchachenko, A. L. "*Chemical physics of degradation and stabilization of polymers*", VSP International Science Publ., Utrecht, **1982**, 354.
2. Bochkov, A. F.; Zaikov, G. E. "*Chemistry of the glycosidic bonds. Formation and cleavage*", Oxford, Pergamon Press, **1979**, 210.
3. Razumovskii, S. D.; Zaikov, G. E. "*Ozone and its reactions with organic compounds*", Amsterdam, Elsevier, **1984**, 404.
4. Emanuel, N. M.; Zaikov, G. E.; Maizus, Z. K. "*Oxidation of organic compounds. Medium effects in radical reactions*", Oxford, Pergamon Press, **1984**, 628.
5. Afanasiev, V. A.; Zaikov, G. E. "*In the realm of catalysis*", Mir Publishers, Moscow, **1979**, 220.
6. Moiseev, Yu. V.; Zaikov, G. E. "*Chemical resistance of polymers in reactive media*", New York, Plenum Press, **1987**, 586.
7. Zaikov, G. E.; Iordanskii, A. L.; Markin, V. S. "*Diffusion of electrolytes in polymers*", Utrecht, VNU Science Press, **1988**, 328.
8. Minsker, K. S.; Kolesov, S. V.; Zaikov, G. E. "*Degradation and stabilization of polymers on the base of vinylchloride*", Oxford, Pergamon Press, **1988**, 526.
9. Aseeva, R. M.; Zaikov, G. E. "*Combustion of polymer materials*", Munchen, Karl Hanser Verlag, **1986**, 389.
10. Popov, A. A.; Rapoport, N. A.; Zaikov, G. E. "*Oxidation of stressed polymers*", New York, Gordon & Breach, **1991**, 336.
11. Bochkov, A. F.; Zaikov, G. E.; Afanasiev, V. A. "*Carbohydrates*", Zeist—Utrecht, VSP Science Press, VB, **1991**, 154.
12. Afanasiev, V. A.; Zaikov, G. E. "*Physical methods in chemistry*", New York, Nova Science Publ., **1992**, 180.
13. Todorov, I. N.; Zaikov, G. E.; Degterev, I. A. "*Bioactive compounds: biotransformation and biological action*", New York, Nova Science Publ., **1993**, 292.

14. Roubajlo, V. L.; Maslov, S. A.; Zaikov, G. E. "*Liquid phase oxidation of unsaturated compounds*", New York, Nova Science Publ., **1993**, 294.

15. Iordanskii, A. L.; Rudakova, T. E.; Zaikov, G. E. "*Interaction of polymers with bioactive and corrosive media*", Utrecht, VSP International Publ., **1994**, 298.

16. "*Degradation and stabilization of polymers. Theory and practice*", Ed. by Zaikov, G. E., New York, Nova Science Publ., **1995**, 238.

17. Polishchuk, A. Ya.; Zaikov, G. E. "*Multicomponent transport in polymer systems*", New York, Gordon & Breach, **1996**, 231.

18. Minsker, K. S.; Berlin, A. A. "*Fast reaction processes*", New York, Gordon & Breach, **1996**, 364.

19. Davydov, E. Ya.; Vorotnikov, A. P.; Pariyskii, G. B.; Zaikov, G. E. "*Kinetic pecularities of solid phase reactions*", Chichester (UK), John Willey & Sons, **1998**, 150.

20. Aneli, J. N.; Khananashvili, L. M.; Zaikov, G. E. "*Structuring and conductivity of polymer composites*", New York, Nova Science Publ., **1998**, 326.

21. Gumargalieva, K. Z.; Zaikov, G. E. "*Biodegradation and biodeterioration of polymers. Kinetical aspects*", New York, Nova Science Publ., **1998**, 210.

22. Rakovsky, S. K.; Zaikov, G. E. "*Kinetics and mechanism of ozone reactions with organic and polymeric compounds in liquid phase*", New York, Nova Science Publ., **1998**, 345.

23. Lomakin, S. M.; Zaikov, G. E. "*Ecological aspects of polymer flame retardancy*", Utrecht, VSP International Publ., **1999**, 158.

24. Minsker, K. S.; Zaikov, G. E. "*Chemistry of chlorine-containing polymers: synthesis, degradation, stabilization*", New York, Nova Science Publ., **2000**, 198.

25. A. Jimenez, ; Zaikov, G. E. "*Polymer analysis and degradation*", New York, Nova Science Publ., **2000**, 287.

26. Kozlov, G. V.; Zaikov, G. E. "*Fractal analysis of polymers*", New York, Nova Science Publ., **2001**, 244.

27. Zaikov, G. E.; Buchachenko, A. L.; Ivanov, V. B. "*Aging of polymers, polymer blends and polymer composites*", New York, Nova Science Publ., **2002**, *1*, 258.

28. Zaikov, G. E.; Buchachenko, A. L.; Ivanov, V. B. "*Aging of polymers, polymer blends and polymer composites*", New York, Nova Science Publ., **2002**, *2*, 253.

29. Zaikov, G. E., , Buchachenko, A. L., , Ivanov, V. B. "*Polymer aging at the cutting adge*", New York, Nova Science Publ., **2002**, 176.

30. Semenov, S. A.; Gumargalieva, K. Z.; Zaikov, G. E. "*Biodegradation and durability of materials under the effect of microorganisms*", Utrecht, VSP International Science Publ., **2003**, 199.

31. Rakovsky, S. K.; Zaikov, G. E. "*Interaction of ozone with chemical compounds. New frontiers*", London, Rapra Technology, **2009**.

32. Pekhtasheva, E. L. "*Biodamages and protections of non-food materials*". Moscow "Masterstvo" Publishing House, **2002**, 224. (in Russian).

CHAPTER 5

RESEARCH NOTES AND DEVELOPMENTS IN MATERIALS CHEMISTRY AND PHYSICS

A. K. HAGHI

CONTENTS

5.1 EFFECTS CARBON NANOFIBERS ON THE STRUCTURE AND PROPERTIES OF VULCANIZED EPDM

5.1.1 INTRODUCTION

This section studies the effect of the structure and composition of carbon nanofibers obtained by co-catalyst washed and not washed from the metal catalyst on the kinetics and properties of vulcanized ethylene propylene diene rubber. It is shown that the fibers obtained on the co-catalyst accelerate the crosslinking of EPDM, improve the physical and mechanical properties, increase the molecular mobility. The purpose of this research—investigation of the carbon nanofibers influence produced by co-catalysts on the physical and mechanical properties and structure of synthetic EPDM.

Currently, one of the promising areas is the co-building of polymer nanocomposites, which are formed by the mixing of polymers and fillers with nanoscale particles (NSP) [1].

Complex physical and mechanical properties of a substance NSP differ from the properties of the same substance in a monolithic sample, and therefore the influence of nanoscale particles (NSP) on the properties of the composite material is different. It should be noted that it is not right to use the term "reinforcing filler in a polymer matrix" to the nanoparticles, because they act at the molecular level rather than the macro level, as in case with regular fillers. Because of this interaction nanocomposites is formed with an ordered structure. Unicity of physical and mechanical properties of nanoparticles is due to their small size, as well as a highly developed and extensive surface area [2].

The properties of nanocomposite polymer materials are mainly due to the properties of the polymer matrix, nature and specific features, as well as the nature, shape and dispersion of nanoparticles. The interaction of the surface of nanoparticles with the elements of the matrix is very essential for these particles. The introduction of more stringent nanoparticles in elastomers and in crystalline polymers, leads to increase the initial modulus of elasticity because of increasing the number of contacts of the polymer matrix and the filler particles [3]. It is known that the mechanical properties of crystalline polymers with the introduction of nanoparticles vary more widely than amorphous ones [4]. In the polymer system to 4–15% of the molecular chains go into the dense phase with a strongly reduced mobility of molecular chains, resulting in increased dimensional stability, thermal conductivity and thermal stability of the polymer composition, and reduced coefficient of thermal expansion [5].

5.1.2 EXPERIMENTAL PART

We used EPDM 505, the ratio of units of ethylene-propylene 60:40 (%), ENB content of 8% and a viscosity of 55 (ML1 +4, 1250S)

For vulcanization was used sulfuric vulcanization system: zinc oxide, sulfur, stearic acid, thiuram D, dibenztiazolildisulfid, curing mode—10 minutes at 160°C. The mixture was prepared in laboratory mill for 10 minutes with a mixture of multiple trimming rolls.

As a filler we used a carbon nanofiber (CNF) prepared in cobalt catalyst (washed and not washed from the metal) demetallization conducted by washing the fibers with concentrated sulfuric acid, and the metal content in the sample was reduced to 38–40% up to 0.2–2% masses, and the specific adsorption surface (Ssp) increased by 2–2.5 times. The concentration of fibers in the rubber is 2% by weight of the rubber. The raw material was used to produce fibers on cobalt catalyst–propane-butane fraction. The main component of synthesis is cobalt oxide. Volumetric flow rate of gas in the synthesis is 1 vol./Hr at a temperature of 7000°C.

Comparison of diameters (Table 1 of Section 5.1) showed that as a result demetallization (Co w/m) the outer diameter of the fibers increases, and internal reduces.

TABLE 1   Specifications nanofibres.

Type of CNF	Specific surface Ssp, nm²		Length, nm	Diameter, nm		Composition
	with calcination	without calcination		Outer (D)	Internal (d)	
Co w/m	85	78	190	30,2	8,7	–
Co	87	77	190	27,9	10,3	Co

To determine the effect of filler on the properties of rubber, the following studies were made: scanning electron microscopy, RPA, DMA, EPR.

Molecular mobility in EPDM vulcanizates before and after the introduction of CNF was determined by electron paramagnetic resonance (EPR) on the device in a temperature of 25°C. Due to the fact that the EPDM are not systems with paramagnetic centers for their research the paramagnetic centers were introduced from outside—the spin probe. As the probe was used radical (TEMPO), because it has a high vapor pressure and easily adsorbed polymers at room temperature.

Curing kinetics was assessed on the instrument RPA2000.

The microstructure of the samples studied by scanning electron microscopy (SEM) with an electron microscope high-resolution at the University of Eastern Finland. Initially, the sample surface of polymeric materials have been processed with the help of oxygen Plasmalab 80 in the following conditions: oxygen flow 20sccm, 100W, and the working pressure 80m Torr, over 30s. Then to increasing the conductivity of the sample surface was covered with a layer of palladium in 2 nm (200Å).

Relaxation transitions in polymer EPDM samples before and after the introduction of CNF in alternating load was studied using dynamic mechanical analysis (DMA) on the instrument Netzsch DMA 242C in a temperature range from −140 to 150°C. We used the special cooler CC 200 L running on liquid nitrogen to obtain low temperatures

Strain-strength characteristics of the materials were determined in uniaxial tension mode on a universal testing machine Instron-1122 at room temperature and constant velocity of the upper beam $v\varepsilon$ = 50 mm/min. The samples present bilateral shoulder with the size of the working part of 4x20 mm. The thickness of the samples was measured with a micrometer with an accuracy of 0.01 mm. The curves load elongation was measured tensile strength σ MPa and elongation at break ε%.

5.1.3 RESULTS AND DISCUSSION

Images show that the fibers are well distributed throughout the volume of rubber and are oriented mainly in the direction of rolling (Fig. 1 of Section 5.1).

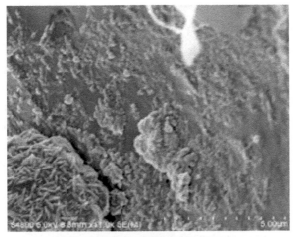

(a) demetallized fiber

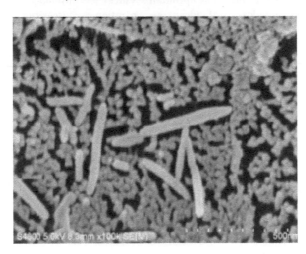

(b) fiber containing cobalt

FIGURE 1 CNF distribution in the matrix EPDM. (a) demetallized fiber; (b) fiber containing cobalt.

Fiber obtained at Co-catalyst has a structure of "Tree", such fibers are smooth and elongated. [6] You can also see the large agglomerates, which are formed mainly with the introduction demetallized fiber, which is the main problem of getting rubbers with a uniform distribution of the fibers throughout the volume. As noted above, the nanoparticles have a high surface energy, and therefore are more likely to form aggregates ranging in size from a few to hundreds of micrometers [7].

From the data of the kinetics of vulcanization (fiber content—2% for all samples) we know that the addition of the fibers, which was obtained on cobalt catalyst regardless of laundered or not it reduces the induction period and the cure time, but fiber, washed from the metal, not much increase the viscosity of the technology (Table 2 of Section 5.1, Fig. 2 of Section 5.1). As we can see from the data on swelling, samples containing fiber swell more than rubber without filler, it's indicative of a lower density of chemical bonds in the vulcanizates containing fiber.

TABLE 2   Influence CNF on the kinetics of binding EPDM.

Sample	Start time of vulca-nization (τ_C, min)	The optimum of vulcaniza-tion (τ_{C90}, min)	Vulca-nization rate (%/min)	Mooney viscosity	Qav, %
Co w/m	2.8	9	16.1	55.6	260
Co	3.9	10	16.3	51.3	210
Without CNF	5.2	12	14.6	52	150

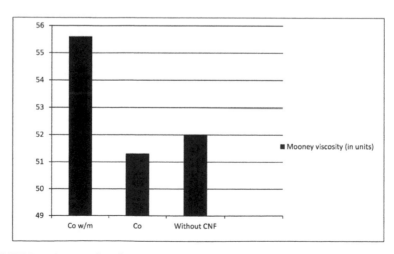

FIGURE 2   Mooney viscosity.

Mechanical testing shows that the addition of nanofibers obtained on co-catalyst, increases the strength of the materials, with the most units at different strains is EPDM, structured fiber, free of metal (Figs. 3, and 4 of Section 5.1), while the EPDM containing crude fiber breaks at a higher aspect ratio.

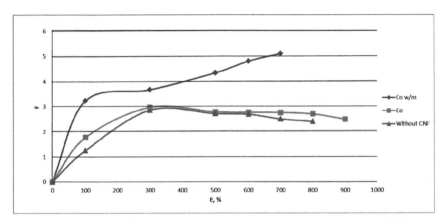

FIGURE 3 Influence of CNF on elastic mechanical properties of vulcanized EPDM.

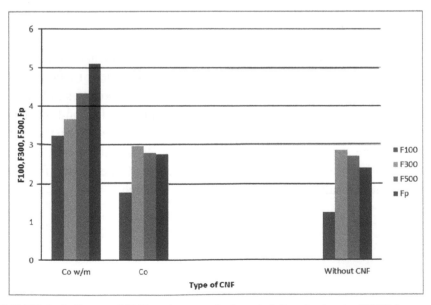

FIGURE 4 Influence of CNF on the strength characteristics of vulcanized EPDM.

The fiber types almost no effects on the residual elongation.

From the results was obtained by EPR, the fibers loosen the structure of rubber, thus increasing the mobility of the macromolecules in the whole volume, compared with the sample containing no filler. Metal content almost no effects on the mobility (Table 3 of Section 5.1).

TABLE 3   The influence of composition on the glass transition temperature and the correlation time of the samples.

Type of CNF	Co w/m	Co	Without CNF
The glass transition temperature, °C	–43.2	–45.3	–40.2
Correlation time × 10^{10}, sec.	3	2.7	4.2

The introduction of these fibers lowers the temperature of the glass-filled rubbers by 5 degrees (Table 3 of Section 5.1), which extends the operating temperature range of the material.

5.1.4   CONCLUSION

From the above it can be concluded that, in general, the nanofibers in the future can be used as a nanofiller affects on the properties of vulcanizates. Fibers obtained by Co-catalyst not strong effect on the kinetics of binding, well-swelled samples in toluene and improve the physical and mechanical properties, molecular mobility. Thus, fiber, adjusted from the metal increases the strength of vulcanized rubber by 40%, but lowers the value of elongation at break, as opposed to a vulcanized fiber, crude from the metal. These results are the basis for further research.

5.2 DETERMINATION OF THE IDENTITY OF THE SIGNS OF NATURAL POMEGRANATE JUICE

5.2.1 INTRODUCTION

In this section, we have carried out comparative studies on the establishment of chromatographic profiles of anthocyanins in fresh pomegranates (freshly squeezed juices) and comparative tests with chromatographic profiles anthocyanins of pomegranate juice from different manufacturers with the purpose of establishing the effectiveness of the method of identification for the detection of forged products.

The main aim of this section is identification of criteria of valuing for the evaluation of pomegranate juice's authenticity on the basis of an assessment of qualitative composition of anthocyanins pomegranate.

Pomegranate juice is widely used for the prevention and treatment of many diseases. This fruit has useful properties because of high content of vitamins, including vitamin K, folic acid, vitamin C, vitamins of B group, vitamin E and unique complex of mineral substances. In this connection it is important to identify the authenticity of drinks from the pomegranate to prevent the distribution of counterfeit pomegranate juice.

Juices from fruits and vegetables are an essential part of a healthy person's diet. The juice has advantages, which, first of all, are necessary to the consumer: natural and fresh taste, convenient packaging. It is important that in addition to good taste and aroma it can effectively quench the thirst and has a high nutritive value due to active substances of fruits and berries, which are necessary for people. A particular importance in the line of juices is given to products with a range of functional properties that determine the important characteristics aimed at the prevention of human diseases. In this field special attention is paid to the pomegranate juice. Useful properties of pomegranate due to a high content of vitamins, including vitamin K that is necessary for blood clotting (coagulation), folic acid (one pomegranate has approximately 25% of the recommended daily norm), vitamin C (about 30% of a daily rate of consumption), vitamins of B group, vitamin E and unique complex of mineral substances. However, according to scientists the main advantage of the pomegranate is the presence of antioxidants, the number of which is several times higher than in any other kind of fruit. The number of these phytonutrients contained in pomegranates is the polyphenols (tanning substances, phytoncides, catechins, and phenol carbonic acid: chlorogenic, neo chlorogenic, p-coumaric, protocatechuic, ellagic acid, anthocyanins, leucoanthocyanins); they determine the color and the taste of juice, possess strong antioxidant, P-vitamin activity. Paramount importance of anthocyanins can be seen in the contents of which vary from 34.0 to 76.5 mg per 100 ml of juice. Pomegranate juice is widely used for the prevention and treatment of many diseases. In this connection it is important to identify the authenticity of drinks from the pomegranate to prevent the distribution of counterfeit pomegranate juice.

We have carried out monitoring of the quality of more than 15 types of pomegranate juice, presented in the retail network of Moscow and represented in different price ranges. It has been established that more than 70% of pomegranate juice in the trading network is forged. Many manufacturers are trying to make profit reducing the cost of raw material, respectively deteriorating the quality of the product, and inventing new methods of falsification.

The aim of this study is the identification of criteria of valuing for the evaluation of pomegranate juice's authenticity on the basis of an assessment of qualitative composition of anthocyanins pomegranate. Several dozen of different anthocyanins were extracted from plants, but for fruit and berries their characteristics are about ten. It is typical for fruits the presence of six most common anthocyanins: delphinidin, cyanidin, malvidin, peonidin, pelargonidin, petunidin, which differ from each other due to radicals R1 and R2. Carbohydrate residues in the molecules of anthocyanins are connected with aglycone (anthocyanidin) through oxygen hydroxyl group in the 3d position of anthocyanidin. Qualitative composition and quantitative ratio of anthocyanins are very specific for each type of plant. This index is stable for every kind of fruit, therefore, is considered to be the hallmark of a particular botanical species. In a pomegranate delphinidin and cyanidin prevail.

We have carried out comparative studies on the establishment of chromatographic profiles of anthocyanins in fresh pomegranates (freshly squeezed juices) and comparative tests with chromatographic profiles anthocyanins of pomegranate juice from different manufacturers with the purpose of establishing the effectiveness of the method of identification for the detection of forged products. The comparison of chromatographic profiles according to the size of the peaks, their mutual arrangement and time for holding signals is carried out.

In the study there were used the following commercial samples: five samples that were marked – 100% pomegranate juice of direct extraction produced without the use of dyes and preservatives. There was a test of the following brands of juice such as: "Packmal", 100% pomegranate juice; "Grunel", natural 100% pomegranate juice of higher grades; "Nar", pomegranate juice, unclarified; "Grante", pomegranate juice of direct extraction; "Noyan", pomegranate juice of direct extraction; 3 samples of recovered juice with the addition of the regulator sugar taste: "Red price", pomegranate juice restored with sugar; "Miri Pak", pomegranate restored with the addition of sugar; "Swell", restored pomegranate. Freshly squeezed juice has been squeezed out exactly on the day of the test in order to preserve the composition of anthocyanins pomegranate. Indicators of the quality of fresh juice were taken as a reference.

It is supposed, that packaged juice, as well as in the restored one and in juices of direct extraction, anthocyanin profile should be the same as the profile of fresh juice.

5.2.2   EXPERIMENTAL PART

Determination of the identity of the signs were carried out on the wave length of 520 nm by high performance liquid chromatography (HPLC). There was used a column of 150 mm x 4.6 mm, sorbent Nucleosil 100 angstroms, C18, precolumn with the same sorbent, at room ambient temperature. The mobile phase A was 0.05% aqueous solution of trifluoroacetic acid, and the mobile phase B acetonitrile for liquid chromatography.

The program elutes the consumption of mobile phase 1 ml/min.

Time, min	A	B
0	90%	10%
3	90%	10%
15	30%	70%

Recorded time of chromatograms – 20 minutes

A spectrophotometric detector, wavelength detection of 520 nm. Regarding standards of dyes there have been used commercially available samples of dyes. Samples were filtered through the chapter filter and then handled in a centrifuge (5000 vol/min 15 minutes). The volume of injection is 20 mcl.

To establish the impact of the botanical varieties of pomegranate and region of its vegetation we have investigated chromatographic profiles of different pomegranate fruit, grown in Turkey, Israel, Uzbekistan, and Azerbaijan (Figs. 1– 3 of Section 5.2).

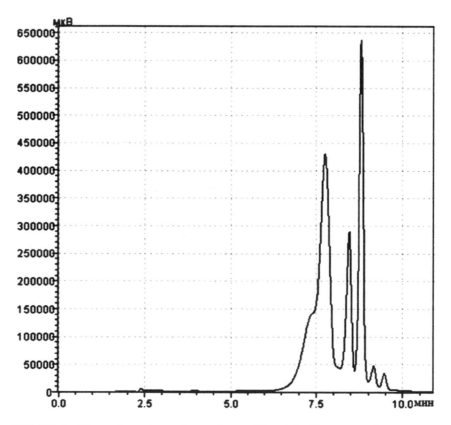

FIGURE 1 Chromatographic anthocyanins profile of freshly squeezed pomegranate juice, production of Turkey.

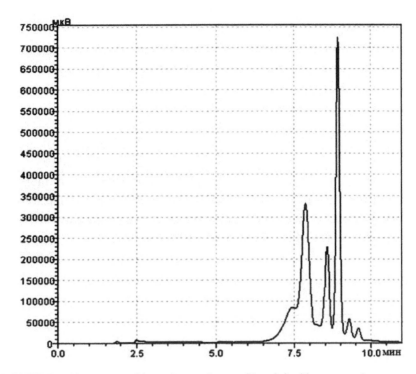

FIGURE 2 Chromatographic anthocyanins profile of freshly squeezed pomegranate juice, production of Israel.

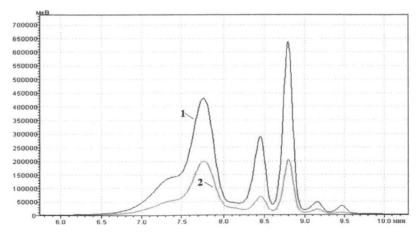

FIGURE 3 Chromatographic anthocyanins profile of freshly squeezed pomegranate, production of Uzbekistan (1) and Azerbaijan (2).

Consider the pomegranate juice (originated from different countries) chromatograms.

According to the above chromatograms it can be concluded that the profiles of the peaks of colored (painted) substances of various pomegranate sorts almost do not differ.

As it can be seen from the profiles of the qualitative composition of anthocyanins in all the investigated samples pomegranate was identical to the installed fluctuations in the number-of individual types of anthocyanins depending on the region of origin, but not in the anthocyanins. Consequently, these chromatographics profiles can be used as a reference in establishing the authenticity of pomegranate juice sold in the trading network. Further, we will draw a comparison of fresh juice and purchased juice chromatograms (Figs. 4 and 5 of Section 5.2).

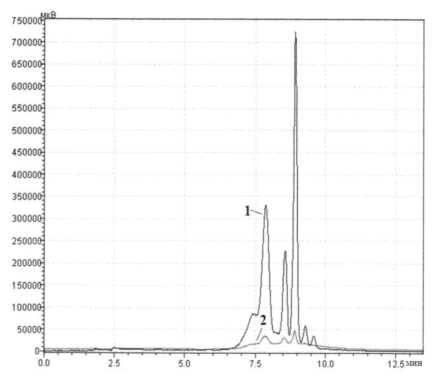

FIGURE 4 Chromatographic anthocyanins profile of freshly squeezed pomegranate juice (1) and pomegranate juice direct extraction "Grante" (2).

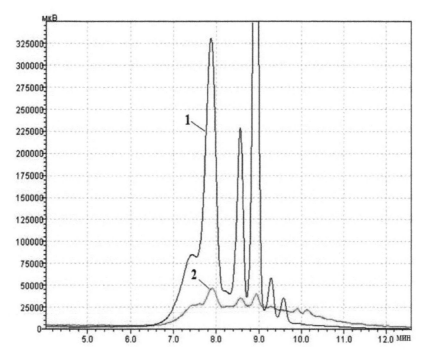

FIGURE 5 Chromatographic anthocyanins profile of freshly squeezed pomegranate (1) and juice "Dumbbell" (2).

It is obvious that anthocyanins are present in these juices; the amount of them is 7 or more times lower, than in the benchmark; this fact allows to assume that juice "Grunel" was diluted in the process of production.

From the chromatograms (Fig. 6 of Section 5.2) we can see that in the juice "Packmal" there are practically no anthocyanins. The only peak by the time that corresponds to the substance, identified on the basis of the standard as the dye E 122 – Carmazin (carmoisine, azorubine). Therefore, it can be assumed that the juice was forged by using food additives: dyes, fragrances and flavors identical to pomegranate juice.

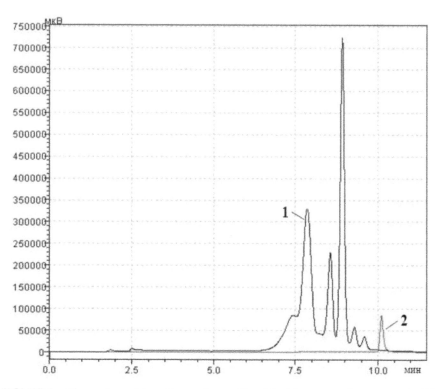

FIGURE 6   Chromatographic anthocyanins profile of freshly squeezed pomegranate (1) and juice "Packman" (2).

In Fig. 7 of Section 5.2 the similar result is presented as in the previous profile—anthocyanins in the commercial juice "Miri Pak", claimed by a manufacturer as pomegranate restored with the addition of sugar. The presence of anthocyanins has not been revealed, but the peak has been set, describing the presence of a large amount of dye E 122 – Carmazin.

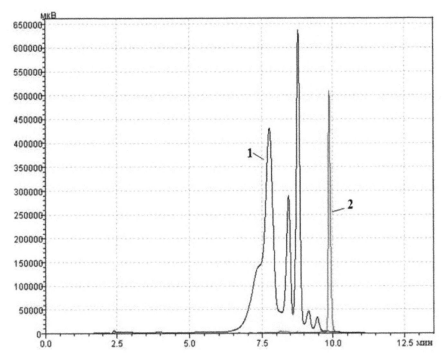

FIGURE 7   Chromatographic anthocyanins profile of freshly squeezed pomegranate (1) and juice "Miri Pak" (2).

5.2.3   RESULTS AND DISCUSSION

According to results obtained by definition of synthetic dyes by high performance liquid chromatography, it can be concluded that in two samples of pomegranate juice ("Packmal" and "Grunel") there has been found synthetic dye E 122 Carmazin with time output as t = 10.0 minutes, which is prohibited in natural juices; and the above samples of juice have also to be attributed to forged ones.

To study the impact of technology of production of reconstituted juice and temperature processing on the content of antioxidants, we have carried out research where the freshly-squeezed pomegranate juice was dried out at the boiling temperature and diluted by adding bidistilled water to the original volume; the juice was heated up during 5 hours. As a result of these experiments, it has been shown that anthocyanins while being heated are partially destroyed, but their contents reduce by no more than 50%. Therefore, in the production of reduced juices chromatographic profile of anthocyanins should remain the same for the qualitative composition of anthocyanins, while their quantitative amount can be reduced to a maximum of 50%.

Other samples of juice have been tested in the same way. It has been found that only two samples—juice of direct extraction "Grante" and recovered juice "Red Price" correspond to the requirements of quality, regulated by the Federal Law No 178 "Technical Regulations for Juice Products from Fruits and Vegetables".

5.2.4 CONCLUSION

According to the research we can conclude that the method of HPLC determination of qualitative composition of anthocyanins profile of pomegranate juice can be effectively used for the identification of their authenticity, identification of juice dilution with water and the introduction of artificial dyes in the process of juice production [1–3].

5.3 DETERMINATION DIFFUSION COEFFICIENT OF BROWNIAN PARTICLES USING VELOCITY OR FORCE AUTOCORRELATION FUNCTION IN MOLECULAR DYNAMIC SIMULATIONS

5.3.1 INTRODUCTION

This section presents method of numerical determination of diffusion D and friction ζ coefficients of Brownian particles from velocity (VACF) and force (FACF) autocorrelation functions in molecular dynamic simulations. Electrostatic parameters of particles in simulation were obtained using modern DFT methods of quantum chemistry. The calculations were carried out with PCS (Patch Clamp Simulation) program package, designed for simulation of neurotransmission processes, using Brownian and molecular dynamic methods, with GPU acceleration support, based on NVIDIA CUDA.

Use of Brownian dynamics for biological modeling suggests involving the averaged action of a solvent on a particle. In classical Langevin model, the averaged action is represented by friction coefficient ζ, which depends on the approximation used. So, Stokes' law is applicable for spherical particles in continuous viscous fluid, theories of Adelman, Onsager [1] take into account the dielectric friction, for systems with small Knudsen number it's proper to use Cunningham correction factor [2], etc.

Even though these approaches works well in the case of simple particles (ions of Na^+, K^+, Cl^-), for polyatomic molecules (NH_4^+, SO_4^{2-}) the situation is much more complicated. This is primarily due to the ambiguity in choice of approximation geometry and charge distribution of particle for these theories. Therefore, for these systems method of MD simulations is one of the few guaranteed ways to obtain the diffusion D and the friction ζ coefficients.

5.3.2 EXPERIMENTAL PART

In this section, the diffusion coefficient D was determined in two ways. The first used the mean square displacement (MSD) of the particle:

$$D = \lim_{t \to \infty} \frac{1}{6Nnt} \left\{ \sum_{t_0=0}^{n} \sum_{j=1}^{N} |\vec{r}_j(t+t_0) - \vec{r}_j(t_0)|^2 \right\}$$

The second used the velocity autocorrelation function (VACF) and Green-Kubo relations [3]:

$$D = \frac{1}{3Nn} \int_0^{\infty} \left\{ \sum_{t_0=0}^{n} \sum_{j=1}^{N} \vec{V}_j(t_0) \cdot \vec{V}_j(t+t_0) \right\} dt$$

The friction coefficient ζ of the Brownian particle was determined by means of the stochastic force autocorrelation function (FACF) [4]:

$$\zeta = \frac{1}{3k_B TNn} \int_0^{\infty} \left\{ \sum_{t_0=0}^{n} \sum_{j=1}^{N} \vec{F}_j(t_0) \cdot \vec{F}_j(t+t_0) \right\} dt$$

The intermolecular potential energy between atomic sites can be calculated by a sum of the Coulomb electrostatic interaction and the Lennard-Jones potential:

$$U_{tot} = \sum_{j \neq i} \left\{ \frac{1}{4\pi\varepsilon\varepsilon_0} \frac{q_i q_j}{r_{ij}} + 4\epsilon_{ij} \left[\left(\frac{\sigma_{ij}}{r_{ij}}\right)^{12} - \left(\frac{\sigma_{ij}}{r_{ij}}\right)^{6} \right] \right\}$$

Water molecules were modeled in terms of SPC approach. Electrostatic parameters of the simulated particles were calculated using the methods of quantum chemistry (DFT). Fig. 1 of Section 5.3 shows the isosurface of electron density of the glutamine molecule. The projection of electron density onto the individual atomic sites was done according to Hirschfeld method.

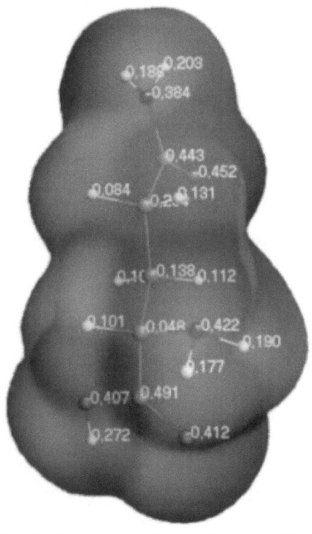

FIGURE 1   Isosurface of electron density of the glutamine molecule (Gln).

5.3.3 RESULTS AND DISCUSSION

First, we calculated the diffusion D and friction ζ coefficients for simple and complex ions. The results of the simulation are shown in Table 1 of Section 5.3. Characteristic velocity autocorrelation function is shown in Fig. 2 of Section 5.3. The results suggest that it can be used for Brownian particles in more complex models: the data are consistent through three different theoretical methods as well as with the experimental values.

TABLE 1 The results of simulation.

Particle	ζ, 10^{-12} kg/s	D, 10^{-9} m²/s			
		MSD	VACF	FACF	Experimental data [5, 6]
Li+	4.33±0.08	0.97±0.04	1.10±0.12	0.97±0.06	1.03
Na+	3.20±0.20	1.33±0.02	1.52±0.06	1.29±0.07	1.33
K+	2.14±0.06	2.16±0.03	1.82±0.07	1.92±0.05	1.96
Cl⁻	2.05±0.13	1.89±0.05	2.03±0.04	1.91±0.12	2.03
NH_4^+	2.15±0.02	1.94±0.04	2.08±0.05	1.89±0.02	1.98
NO_3^-	2.09±0.05	2.05±0.06	2.04±0.05	1.96±0.05	1.90
SO_4^{2-}	3.70±0.08	1.16±0.05	1.17±0.09	1.11±0.02	1.07
Gly	6.01±0.14	0.68±0.04	0.77±0.07	0.81±0.02	–
Gly^z	3.91±0.01	1.01±0.07	0.96±0.05	1.05±0.03	1.06
Gln	7.48±0.04	0.39±0.04	0.51±0.09	0.45±0.08	–
Gln^z	5.91±0.07	0.67±0.06	0.81±0.03	0.70±0.04	0.76

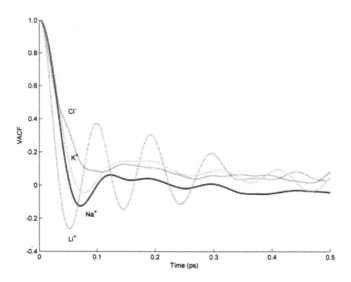

FIGURE 2   The normalized velocity autocorrelation functions of cations and anions in SPC water at 298 K.

Similar calculations were performed for the individual amino acids (Table 1 of Section 5.3). Some amino acids in aqueous solutions at a certain pH are present in the zwitterionic state, which affects the nature of the interaction with the solvent molecules. In this chapter, the molecules of glycine (Gly) and glutamine (Gln) were examined in two states: simple and zwitterion. The geometry of the ground state is determined in the COSMO solvent model. Comparison with experimental data showed that for the non-zwitterionic state the diffusion coefficients are somewhat underestimated, indicating the need for a more accurate accounting of the charge distribution and structural features of the particles.

5.3.4 CONCLUSION

The proposed method extends the applicability of the Brownian dynamics for molecules of arbitrary complexity by eliminating the current shortcomings of crude methods, based on geometric approximation and the complexity of calculating the friction coefficients. To determine the electron density of the complex molecules and the final charge distribution used accurate methods of quantum chemistry, and taking into account the effect of the solvent, which allows to simulate the different states of amino acids in solution.

5.4 IN SILICO SIMULATION OF SILVER AND COPPER IONS INTERACTING WITH FUNGAL CELL WALL

5.4.1 INTRODUCTION

The aim of this section is to find possible mechanism of antifungal activity of silver ions using quantum chemistry simulation for large systems. The complete theory of antifungal activity of these elements is still developing [2, 4].

In computational molecular biology, systems containing metal particles are not well parameterized yet for force field calculations, so precise *ab initio* methods for that class of systems are the rational choice. However the computational complexity of these systems suggests the use of simplified way to choose the most probable geometry of large organic molecules with adsorbed metal atom, to avoid exploring all possible geometries. DFT codes are described in [3, 11].

Interpretation of calculated properties is also a complex problem, so we provide comparative *in vitro* experiment to establish the possible outcome of computed values in real life.

The interaction of silver and copper ions with chitin dimers was modeled using quantum chemistry within the DFT framework. A variation of parameters of hydrogen bonds depending on the presence of different metal atoms was obtained. This variation was compared with the results of *in vitro* study of antifungal activity of corresponding metal ions.

Silver has long been known to serve as antibacterial and antifungal agent. Out of all the metals with similar antimicrobial properties, silver has the least toxicity to human cells. Prior to the introduction of modern antibiotics, silver was commonly used in medical treatments.

The exact mechanism of antifungal activity is still unknown, and is currently the subject of many researches. There is three suggested qualitative concepts: DNA association, cell wall damage and deactivation of enzymes. On the molecular level it is known that silver is capable to form a bond with S, N, and O atoms, and also to catalyze the formation of disulfide bonds.

The new computational quantum chemistry methods make it possible to directly simulate biochemical processes, which involve large molecules. In this work the interaction of silver and copper ions with chitin dimers was modeled using the DFT theory.

5.4.2 EXPERIMENTAL PART

Fungal cell wall simulated as two chitin dimers. To obtain the optimized geometry of ground state, several DFT codes were used, including PC GAMESS, SIESTA and ABINIT (Fig. 1 of Section 5.4).

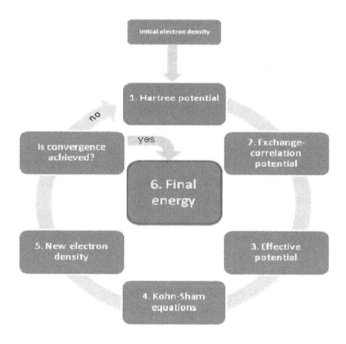

FIGURE 1   Conceptual map of DFT method.

$$\nabla^2 V_{Hartree}(\mathbf{r}) = -4\pi\rho(\mathbf{r})$$

To find the most probable area of metal adsorption on a molecule, Fukui functions scheme was used (Fig. 2 of Section 5.4).

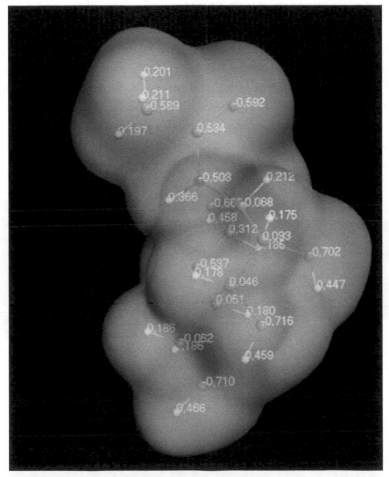

FIGURE 2 Visualisation of the scalar electrophilic Fukui function, mapped onto the isosurface of electron density field for detection of probable adsorption site.

Electron density of ground state chitin system with adsorbed metal atoms was projected on atomic sites, responsible for hydrogen bond formation, using Hirshfeld method.

Calculations were performed on Linux-cluster using parallel MPI (MPICH2) и SMP (OpenMP) parallel paradigms.

In silico calculations were complemented by *in vitro* assay, in which we compared antifungal activity of silver and copper ions in the same concentrations against *Alternaria alternata*.

5.4.3 RESULTS AND DISCUSSION

Chitin dimers form the hydrogen bonds with each other, and there is a variability of parameters of that bond depending on type of adsorbed atom near bond. Adsorbed silver atom increases the symmetry of H-bond, measured by the difference of projected charges on oxygen atoms, involved in H-bond, whereas copper atom decreases that symmetry, comparing to the original system without any metal (Figs. 3–5 of Section 5.4, Table 1 of Section 5.4).

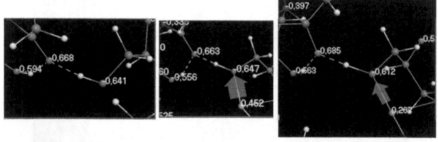

FIGURE 3 Ground state geometry of chitin dimers with silver atoms adsorded. Hydrogen bonds are shown by dash lines. Projected atomic charge is shown near atom.

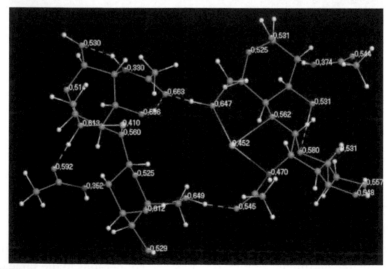

FIGURE 4 Difference between the effects of adsorbed copper and silver on hydrogen bonds between chitin dimers. Arrow depicts the polarity of Me-O bond.

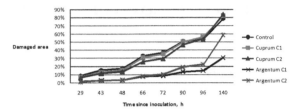

FIGURE 5 Growth dynamic of *Alternaria alternata* under two different silver and copper ions concentrations. C1 = 25 µg/ml (by metal mass) и C2 = 2.5 µg/ml (by metal mass).

TABLE 1 Calculated values of hydrogen bond parameters for original system and system with metal atoms.

System	Total charge, e	Length of H-bond, Å	Charge of farthest oxygen atom, e	Charge of nearest oxygen atom, e	Charge of metal atom, e	Charge of hydrogen atom, e
Chitin	0	2.6	−0.67	−0.64		0.37
Chitin + Ag	+2	2.4	−0.66	−0.65	0.45	0.46
Chitin + Cu	+2	2.4	−0.69	−0.61	0.27	0.46

In vitro study shows increased antifungal activity of silver comparing with copper in the same concentration.

5.4.4 CONCLUSION

The interaction with cell wall and the effect on the mechanical properties of its chitin skeleton could take part in antifungal mechanism of silver ions. It correlates with observed low toxicity of silver for animal cells.

Decrease of the symmetry of hydrogen bonds, induced by copper atom adsorption, correlates with low antifungal activity of copper, comparing to silver atoms, which demonstrate increase of symmetry and high antifungal activity. It could be an important parameter for mechanical properties of chitin skeleton, and changing of that parameter could take part in molecular mechanism of antifungal activities of silver ions [1, 5–10, 12].

5.5   THE ESTIMATION OF ANTIOXIDANTS AS NUTS QUALITY INDEX

5.5.1   INTRODUCTION

A measurement of the content of antioxidants in walnut, hazelnut and pea-nuts by the method of coulometry was made. It has showed that the content of antioxidants is an indicator of oxidative damage of the nut's fat during the process of storage. The main aim of our research is to study the dependence of oxidative damage of the nut's fat on the content of antioxidants.

Antioxidants are natural inhibitors, which slow down the processes, occurring during the storage of food products. When antioxidant's concentration is reduces the quality of products degenerates.

Organic compounds are related to the most common objects of study and analysis. A lot of organic compounds have attracted the attention as toxicants at environmental analytical control recently. There are some problems about achieving rapidity, sensitivity and selectivity of analysis and also we have the needs of automated analysis, which can be solved, with methods of coulometry. For the present coulometry is only method no using the dependence of quality from the concentration of the determined substance, which is characteristic for others analysis. From the point of view of the theoretical foundations and practical aspects the galvanostatic coulometry is very affordable method for the general use in which the main controlled parameters is the time and amperage. The devices, which are used last time, let us to measure these parameters with very high accuracy. All of these factors create the preconditions for development high accuracy methods of coulometrical definition of the different compounds.

The coulometry methods based on the physical law which sets the link between the weights of turned electricity and quantity of spent electricity. In many cases the electro generated coulometrical titrate enters in the oxidation process with organic substratum on the mechanism of reaction with the electron carriers. The most effective carriers are the variable valence metals' ions and its components: oxidants—chrome (VI), manganese (III), cobalt (III), cerium (IV), vanadium (V), copper (II); deoxidants—cobalt (II), chrome (II), vanadium (III), titanium (III), iron (II), copper (I), tin (II). The wide area of practical use of the halide ions (chloride-, bromide-, iodide-ions) highlights them apart from a number of reagents—electron carriers. Halide—ions are used at the generation of free halogens and also for the components in the positive oxidations, for example, iodine and bromide.

In the base of all chemical methods of the organic compounds' quality analysis by functional groups are the definition the substances which are formed and melted by the reaction between the sample and titrate reagent.

Such titrates electrochemical generation give us the obvious advantages: lets us to use the unstable reagents, excludes their storages and standardization, and provides the precision and opportunity of definition the small contents of the substances. In some cases the electro generated titrates can be more reactive than their solutes. For example, this is typical for iodine (I), bromine (I), vanadium (V) and others. The advantages of the coulometry methods are precision, sensitivity, ease, expressivity and wider range of the definable contents compared to the previously known method. The confines of the definable concentrations and between 1 g/ l and 1 mkg/ ml. The relative standard deviation (Sr) are between 0,05–0,002.

The necessary component of living cells is antioxidants—substances, which in the normal physical concentrations supports on the low standard level the free radical auto oxidation process, which are continuously flowing in the cells. Under the rules the processes of the spending and replenishment are balanced. In the violation of such balance the high reactive ability of the free radicals leads to the acceleration oxidation process, which breaks down the molecular basis of the living cells and causes different pathological forms.

In this connection one of the most important problems is the control of the organism's antioxidant system state at the different pathological by way of its further therapeutic correction.

One of the most effective and widespread antioxidants are the tocopherols (vitamin E), polyphenol (catechol, Gallic acid derivatives), flavonoids (routine, quercetin), ascorbic acid. The most perspective objects of the antioxidants are the plant facilities.

5.5.2 EXPERIMENTAL PART

The analysis was made for 12 nuts samples, for three samples it was made an analysis in the system of the accelerated storage. The accelerated storage is carried out at temperature 50°C in the incubator with medium level of air circulation.

It was noted that walnuts have more high concentration of the antioxidant than it was founded in the hazelnuts and peanuts. Also the unroasted hazelnuts have more antioxidants than the roasted one. The reduction of the nuts' antioxidants connected with their oxidative damage (rancidity) that was revealed in the organoleptic analysis.

5.5.3 RESULTS AND DISCUSSION

So, we can state that the antioxidants' content in the nuts reduces very much in the storage process and that coulometry method is able to fix these changes. Also it was noted some taste and smells' letdown in the nuts after the accelerated storage. The index of the antioxidants quantity in the nuts can be used for establishing the degree of oxidative damage.

Also basing on the interim data we make up the following schedule of dependence between the antioxidants' content and during time of accelerated storage Fig. 1 of Section 5.5.

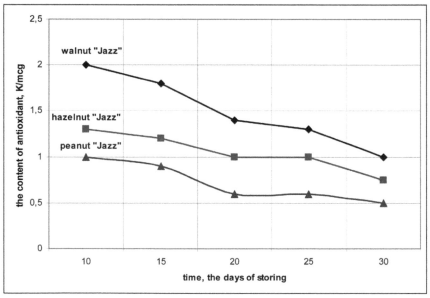

FIGURE 1   The dynamics of antioxidants' contents in the nuts at the accelerated storage.

Also we have made a study of the antioxidants analysis in the seven samples of walnuts from the various producers using the amperometric method. The amperometric method is based on the measurement of electric current, which are come insight at the electrochemical oxidation of the substance or mixture of substances on the surface of working electrode with the certain potential. The detector registers the change of current, which flows through the cells due of the changes in the concentration of antioxidants.

The sensitivity of the amperometric method is determined by the nature of the working electrode's kind and potential applying to it. The amperometric method has a number of advantages like this: the low detection limit, the high selectivity (that means it is determined only those compounds whose molecules can be oxidized), and the small cell size (0.1–5 mkl), ease of maintenance.

The results of the walnuts' antioxidants analysis are represented in the Fig. 2 of Section 5.5. The picture shows that antioxidants reduced in the storage process. The highest antioxidants' content was noted at the first sample, which has got the highest rating according to the organoleptic indicators. The nut's consistency in the sample № 3 hoyrs as become crumbling and dry after 2 weeks of the accelerated storage and because of this we can't pick out the fat.

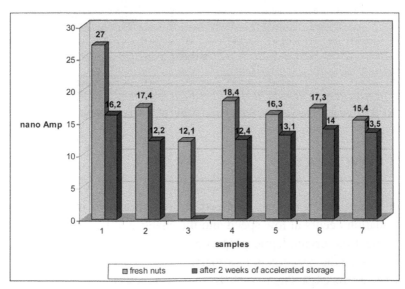

FIGURE 2 The dynamics of antioxidants' contents in the walnuts at the accelerated storage.

The organoleptic test was made by the tasting committee using the five-point scale taking into account the weight coefficients on the four main indicators: the form, color of the kernel on the breaking, taste and smell. During the experiment it was observed the statistical homogeneity of the tasters' rating. The results of the organoleptic test of the walnuts in the storage process are represented in the Fig. 3 of Section 5.5.

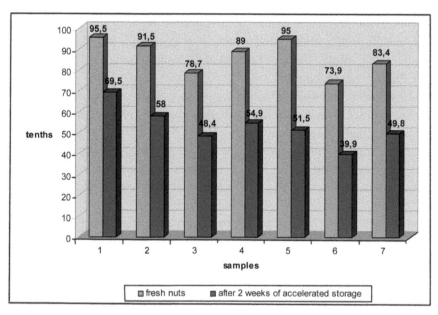

FIGURE 3   Comprehensive indicator of the quality of samples walnut.

5.5.4 CONCLUSION

It can be assumed that there is a certain time threshold in during which the oxidation occurs at low speed and the changes in the fat's compounds don't affect the organoleptic properties of the product and its biological value because the organoleptic properties' changes were fixed only in the last stage of the experiment.

It is recommended to specify the raw materials' place of origin in the marking. The transparent polypropylene packaging is not very good in preserving of nuts quality. Because of this it is better to keep the nuts in the opaque packaging just to avoid the effecting of light on the activation of oxidative processes. It is also very important to specify the date of the raw materials production, the date of goods packing and the expiration date of the goods.

The study found that the oxidation's depends on the type of the packing and the initial condition of the nuts at the packaging. Because of this it's necessary to establish the criteria, which lets us to form an opinion about antioxidantion activity with the goals of forecasting the optimal period of storage depending on the type of the packing [1, 2].

5.6 COMPOSITE MATERIALS BASED ON LDPE WITH THE ADDITION OF WOOD FLOUR AND RUBBERS OF DIFFERENT NATURE

5.6.1 INTRODUCTION

Currently the world's consumption of plastics is growing in this regard there are difficulties in disposing of large amounts of household plastic waste. The solution is to develop biodegradable polymers with a specific expiration date. This work is dedicated to the creation of biodegradable composite material based on LDPE, wood flour and rubbers of different nature.

Research carried out in this chapter, aimed at establishing the possibility of creating composite materials based on polyethylene and natural rubber with biodegradable additive assessment of their compatibility, mechanical and rheological properties as well as their propensity to biodegradation in soil [1–2]. The basis for creating biodegradable composite served as LDPE, natural additive—wood flour and rubbers of different nature. To study the workability (adaptability) of the resulting composition, experiment was carried out on receipt of the melt flow index. In order to determine the physical and mechanical properties of the composite material were obtained indicators such as tensile modulus and elongation at break.

To study the biodegradability of the composite in the environment, field tests were conducted in soil and water absorption determined.

Now there are a lot of work aimed at creating biodegradable composites based on polyolefines and natural supplements [3–7]. Use as a component of a composite material of natural rubber due to the instability of microbiology is new and interesting direction. From a microbiological damage natural rubber faced in the initial stages of its production, as natural rubber latex contains 30–35% of the hydrocarbons in the form of rubber particle size of 0.15–0.5 mm. There are cases of microbial damage vulcanized natural rubber in the operation of underground electrical cables, rubber insulated from NR, sealing materials in contact with the ground, sea and tap water, sewage. Synthetic rubbers are a large group of elastomers that differ in chemical composition and biological stability. The general rule for synthetic rubber, as well as for other polymers, is the growth of biological stability by increasing the length of the macromolecular chain [8]. Among the least biostable synthetic rubber note EPDM.

5.6.2 EXPERIMENTAL PART

The objects of study were selected mixture of low-density polyethylene (LDPE) with wood flour (WF), natural rubber (NR) and ethylene-propylene rubber (EPDM). WF content is 40 wt.%, Rubber injected at 10 and 20 masses %. As composite materials based on LDPE blending performed on a laboratory mixer at a temperature of 140°C for 5 minutes and then get the film samples in a laboratory press. The sample thickness was 100 ± 10 microns.

In order to identify the strength characteristics of the obtained mixtures were conducted physical and mechanical tests in a laboratory tensile machine PM-10 (Table 1 of Section 5.6, Fig. 1 of Section 5.6). Testing, training and payment module samples were assayed according to GOST 9550–81.

TABLE 1   Characteristics of the samples.

Samples	Melt flow rate, g/10min	Modulus, MPa	Elongation at break%
LDPE	2	212	400
LDPE 60–40 WF	1	113	9
LDPE 50–10 NR–40WF	1	22	8
LDPE 40–20 NR–40WF	8	33	8
LDPE 50–10 EPDM–40 WF	–	39	10
LDPE 40–20 EPDM–40 WF	–	48	8

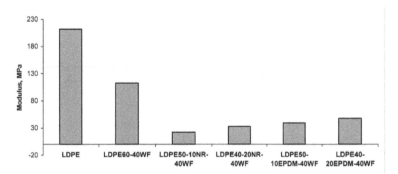

FIGURE 1   Tensile Modulus, MPa.

To study the rheological properties of the compositions used method for determining the melt flow rate (MFR) (GOST 11645–73). The experiment was conducted in a laboratory type IIRT capillary viscometer at 190°C and a load of 2.16 kg, the dwell time in the chamber 5 minutes. The data are presented in Table 1 of Section 5.6. Crushed sample was placed in a heated, covered on one side, the camera device, stand 5 minutes to complete melting of the material. After a time delay, the cap removed, the melt flow was under the load of 2.16 kg. After some time the knife cut off part of the sample (at least three) and weighed, defining and calculating the average weight melt flow for 10 minutes.

Determination of water absorption of composite materials was carried out in accordance with GOST 4650–4680 for 24 hours at 23 ± 2°C. The data presented in Fig. 2 of Section 5.6.

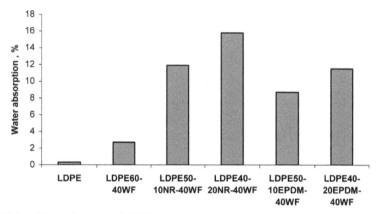

FIGURE 2   Water absorption in 24 hours.

To assess the biodegradation in the soil samples were placed in a combined land, prepared in accordance with GOST 9.060–75. Moisture and soil acidity were kept constant. Data on changes in the mass of film samples for 8 months of exposure are presented in Table 2 of Section 5.6.

TABLE 2   Assessment of the degree of decomposition of film samples in the soil.

Samples	The mass change(%) * For 2 months	The mass change(%) * 3.5 months	The mass change(%) * 8 months
LDPE	0	0	0
LDPE 60–40 WF	+4	–6	–10
LDPE 50–10 NR–40WF	– 6	–17	–20
LDPE 40–20 NR–40WF	– 5	–24	–33
LDPE 50–10 EPDM–40 WF	+2	–15	–17
LDPE 40–20 EPDM–40 WF	+ 10	–13	–20

5.6.3   RESULTS AND DISCUSSION

From these data it is clear that the introduction of the filler significantly reduces the elastic modulus and elongation at break of the material. This is due to the fact that wood flour is fine addition [9]. Flour particles in the matrix of polyethylene are defects, stress concentrators, which results in uneven load distribution in the matrix. Upon loading the sample is detachment of the filler particles from the matrix pores are formed [10], which are then expanded, leading to premature failure of the composite.

Introduction of WF in LDPE results in a slight increase in the viscosity of the melt. The addition of 10% NR is not conducive to change the value of the indicator, while 20% NR plasticizes polymer composite during the heating process, which improves its workability. When adding EPDM observed the opposite result, as in the processing of cross-linked EPDM [11], which leads to a sharp increase in the viscosity of the melt, as a consequence of the material does not flow.

The addition of 40% wood flour increases the rate of 9. With the introduction of rubber (natural and synthetic), water absorption increases due to amorphization of the polymer matrix.

- "+" means an increase in the mass of the sample under study, the symbol
- "–" means a decrease in the mass of the studied sample.

It should be noted that during the action of the soil at the initial stage, the number of competing processes: water absorption, washing additives and degradation. The first leads to an increase in mass of the sample, the second and third—to decrease. These data indicate only that which process dominates in the case of each sample, but cannot serve as a basis for predicting future impacts of soil. Weight loss of a sample of pure PE is zero, but with the introduction of wood dust exposure in 3.5 months is a loss of mass of the sample. The introduction of rubber leads to an acceleration of the process of biodegradation in the soil. It should be noted that the addition of natural rubber has a 10% weight loss increases the filling factor of 2, and at 20%—3 times in 8 months of exposure.

5.6.4 CONCLUSION

As a result of these investigations the possibility of biodegradable material creation based on LDPE and rubbers of different nature was shown. In the course of analyzing the obtained rheological parameters, it can be concluded about how the data processing of composites. For example, for samples with NR should choose extrusion and injection molding, as the main methods of processing. Whereas, for the sample with EVA suitable method pressing.

It should be noted that the introduction of LDPE and rubber wood flour increases the porosity of the polymer matrix. It causes high values of water absorption data of composite materials and facilitates the penetration of microorganisms, accelerating their growth and development in the process of biodegradation in the soil, and thus accelerates the fragmentation pattern.

5.7   FUNGICIDAL INFLUENCE OF IONS OF CU^{2+} ON MICROORGANISMS AS A WAY OF BIOCORROSION PROTECTION OF METAL CONSTRUCTIONS

5.7.1 INTRODUCTION

In the course of metal constructions exploitation by one of determinatives of their durability are protection processes against soil corrosion. The results of microbiological investigations of the effect of increasing concentrations of Cu^{2+} ions on the inhibition of growth and Aspergillus niger, Fusarium moniliforme, Penicillium chrysogenum, Penicillium cyclopium, Trihoderma viride, Actinomycetales sp. were studied.

Corrosion of metals—destruction owing to chemical or electrochemical environment interaction—the main scourge of all metal designs and constructions. Distinguish various kinds of corrosion of metal: by the character of destruction—continuous and local, by the form courses—chemical and electrochemical, by the course conditions—atmospheric, soil, sea, biocorrosion, contact, slot-hole, corrosion by wandering currents, an external current, etc., by the features of conditions of operation—corrosion of the chemical and oil equipment, pipelines, the power equipment, courts, building constructions, materials, etc. [1–4].

The most destructive is soil corrosion. Annual losses of metal owing to soil corrosion reach 4%. Different purpose pipelines, tanks, piles, poles, cables, well casing, all kinds of steel structures operating in the soil are exposed to the soil corrosion [2–6]. Soil corrosion always proceeds on the electrochemical mechanism (the only exception is very dry soils) [7]. Electrochemical corrosion arises during metal contact with surrounding electrolytically conducting medium. An original cause of electrochemical corrosion is thermodynamic instability of metals in environments surrounding them. The oxygen depolarization proceeds at the cathode. On the cathode also can pass and hydrogen depolarization (only in the conditions of acidic soils) [8].

Tremendous influence on soil corrosion of metals is rendered by microorganisms existing in soil. The soil corrosion of metal constructions caused by microorganisms, carries the name of biological (biocorrosion). Waste products of various microorganisms which are present in water, a soil, intensify corrosion process. Microorganisms action on metals can variously occur. First of all, corrosion of metals can cause aggressive microorganisms metabolites—mineral and organic acids and the bases, enzymes and others. They create the corrosion active environment in which in the water presence take place electrochemical corrosion. Colonies of microorganisms can create on the metal surface outgrowths and mycelium films under which take place corrosion as a result of electric potentials difference between metal and microorganisms [9].

The different methods are applied for protection of metal against soil corrosion. The basic methods of protection are: protective coating and insulation products, the creation of the built environment, electrochemical protection, the use of special methods of installation [10, 11].

In the practice of protecting metals from soil corrosion is often applied cathodes protection [12, 13]. This significantly reduces (minimizes) the rate of soil corrosion. To protect drilling platforms, welded metal bases, underground pipes to be connected as a cathode to an external power source. The anode used auxiliary inert electrodes. The negative pole of the direct current is connected to the pipe, and the positive—with the anode. Thus, for cathode protection is a vicious loop current, which flows from the positive pole of the power supply cable to the isolated ground to the anode. The anode grounding current spreads along the ground and into the protected piping, then flows through the pipeline, and the pipeline for insulated cables back to the negative terminal of the power supply. Electrical current coming out of the anode in the form of positive ions of the metal, so the anode is gradually destroyed, and the pipeline is protected from destruction.

In this chapter we conducted comprehensive research the possibility of using copper salt solution to protect against biological corrosion under the action of fungi and actinomycetes. These results will allow us to create an adjustable directional protection structures from biocorrosion.

5.7.2 EXPERIMENTAL PART

For definition of a range of concentration of ions of the copper suppressing growth of microorganisms four solutions of copper (II) sulfate with following normal concentration of ions of copper have been prepared:

Four copper (II) sulfate solution with the following concentrations of copper ions to determine the range of concentrations of copper ions that inhibit the growth of microorganisms were prepared:

1. 10^{-3} H: 5,78 pH;
2. 10^{-2} H: 5,14 pH;
3. 10^{-1} H: 4,33 pH;
4. 1 H: 3,6 pH.

Mycological analysis of the ions concentration effect was performed on the following test cultures in the complex soil microbiota:

Micromycetes: Aspergillus niger, Fusarium moniliforme, Penicillium chrysogenum,

Penicillium cyclopium, Trihoderma viride.

Actinomycetes: Actinomycetales sp.

The considered complex of microorganisms is dominating in agrobiocenosis and causes biological corrosion of metal designs. Inoculation suspensions test cultures were carried out in Petri dishes on medium wort agar at 25°C. Each Petri dish in a culture medium were cut four holes, and places the products with different concentrations of copper ions. Zone suppression agent of growth and development of the fungus was measured as the distance from the center of the hole with drug to fungal spawn Fig. 1 of Section 5.7.

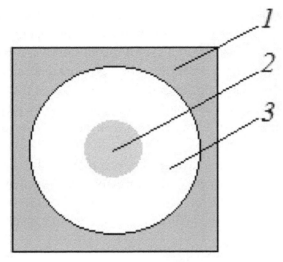

FIGURE 1   Suppression zone of fungi growth, where 1 – fungi spawn, 2 – hole with the drug, 3 – zone suppression.

5.7.3 RESULTS AND DISCUSSIONS

The table shows the values of the suppression the zones radii of microbial growth under the influence of drugs with different concentrations of copper ions Table 1 of Section 5.7.

TABLE 1   Radius of a suppression zone of microorganisms (from the c hole center to the fungi spawn on day 5 exposure, ± 2mm).

Concentration Cu^{2+} Microorganisms	10^{-3} N	10^{-2} N	10^{-1} N	1 N
Aspergillus niger	0	0	5	25
Fusarium moniliforme	0	0	0	30
Penicillium chrysogenum	0	0	0	20
Penicillium cyclopium	0	0	7	20
Trihoderma viride	0	0	5	20
Actinomycetales sp.	10	15	20	35

On the first and second day under the influence of drugs with different concentrations has slowed growth of spores and poor development of mycelium. On day 5 preparations with a concentration of 10^{-3}, 10^{-2} and 10^{-1} H exhibit fungistatic properties when exposed to all strains of micromycetes (i.e. delay and stop the growth of the fungus at the initial stage of exposure, followed by the resumption of growth), and a preparation with the concentration of 1H – fungicides (i.e., exposure leads to the death of the fungus).

The most sensitive microorganisms presented to the copper ions was Actinomycetales sp., growth inhibition is observed even at a concentration of 10^{-3} N (Fig. 2 of Section 5.7). As the concentration of copper ions change the character of the inhibition is a power function ($Y = 33 \cdot X^{0.18}$).

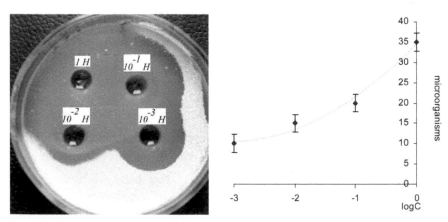

FIGURE 2   Suppression zones of growth Actinomycetales sp. Depending on concentration of ions of copper, (a) photography, (b) graph.

At 15 and 20 days of incubation changes the pigmentation (color) fungi Penicillium chrysogenum and Penicillium cyclopium, while the inhibition observed on Day 5, do not cancel for any fungus.

Its widely known that the development process of mycological corrosion is as follows: the first stage—education mikrocolony, their growth, the emergence of metabolic products and local accumulation of electrolytes with excess hydrogen ions, the second stage—the emergence of electrochemical cells on the surface of the metal, and the third stage—the emergence of the cathode (anode) depolarization the capture of protons (electrons) fungi and their metabolic products.

Inertia fungi to low concentrations of metal ions in the first day due to the occurrence of the metal ions sorption by cell wall of the fungus, which is protective [14]. Getting inside the cells, they inhibit the respiratory chain enzymes, and splits oxidation and oxidative phosphorylation, resulting in the cell dies. With increasing concentration of copper ions in the preparation process is intensification of the death of the fungus. It should be noted that the relationship is nonlinear and is specific to different organisms. During the experiment, such fungi as Penicillium chrysogenum, Penicillium cyclopium, Trihoderma viride demonstrate high adapt capacity to the adverse conditions (drug concentration was $10^{-1}N$) as the formation of stable colonies. At low concentrations of the drug it is also possible for micromicetes to use metal ions as a source of minerals, energy or electron acceptors.

5.7.4 CONCLUSIONS

As a result of studies the use of solutions of copper salts as an effective method of protection against biological corrosion of metal structures of different shapes and metal have found. The adapt capacity of some micromycetes in presence of metal ions in small concentrations was observed. Optimal parameters for the electrochemical method of protection against biological corrosion of metal objects, caused by fungi and actinomycetes are the concentration of copper ions in the solution is not less than 0.1 N.

5.8 INTERFACIAL RHEOLOGY OF HEN EGG WHITE LYSOZYME-5-METHYLRESORCINOL MIXTURES AND THEIR FOAMING PROPERTIES

5.8.1 INTRODUCTION

The capacity of methylresorcinol (MR) to affect on the hen egg white lysozyme (LYS) surface activity and viscoelastisity of its adsorption layers at air/solution interface is compared with the effect of MR on the foam formation ability and foam stability. It was shown that the MR introduction increases multiplicity and stability of foams, the concentration of surfactant had an important role.

Mixtures of proteins and surfactants are often used in many technological applications, including food and pharmaceutical industries, cosmetics, coating processes, and so on. In many of these applications protein-surfactant mixtures are used in the manufacture of the various dispersions. These dispersions contain two or more immiscible phases (aqueous, oil and/or gas phases) in the form of foams and emulsions. Dispersions are inherently unstable systems because of their large interfacial area [5].

The stability of these systems is generally achieved through a protective interfacial layer around the particles (emulsion droplets or foam bubbles). The properties of this interfacial layer are governed by the composition and structure of the adsorbed material and in turn would determine the properties of the dispersion [3].

Wide using of lysozyme in medicine for the treatment of various infections and in food and cosmetic industry to prevent bacterial contamination of the stuffs leads to production of the resistance of microorganisms to lysozyme action. In recent years, a great deal of attention has been devoted to the investigations aimed at studying lysozyme modifications, affecting lysozyme properties while retaining its enzymatic activity. One of the ways to solve this problem is modification of lysozyme structure with using weak nonspecific interaction with MR. It is known numeral effects of MR on the structure and functions of lysozyme. In particularity, in concentration range of 10^{-7}–10^{-3}M MR the specific and nonspecific enzymatic activities raise, its substratum specificity extends and temperature range of lysozyme catalysis expands [9].

Recently we have found out that MR due to diphylic character of its structure is capable to self-organization in a solution and can behave itself similarly to nonionic surfactants [6]. MR changes the adsorption behavior of LYS in their mixed solutions. MR was shown to change thermodynamic affinity of lysozyme to the solvent with the resulting the protein being more surface active and reological properties of its adsorption layers at the air/solution interface are higher in the wide range of MR concentrations [6]. For this reason, we suppose, that using of lysozyme in mixture with MR can be favorably for production of antibacterial pharmacological products of foam type with enhanced physical stability and period of using while increasing its enzymatic activity and wider spectrum of action.

The aim of this chapter is to investigate the effects of MR on the air/water interfacial properties (interfacial dilatational complex modulus and interfacial pressure) of lysozyme comparing with effect on its foam formation ability (volume, multiplicity, stability of foams).

5.8.2 MATERIALS AND METHODS

A sample of hen egg white lysozyme (Sigma-Aldrich, USA) with activity 20,000 U/mg and molecular mass 14,445 Da was used.

Alkyl-substituted hydroxybenzenes, 5-methylresorcinol (5-Methyl-benzene-1,3-diol) (Sigma-Aldrich, USA) with molecular mass 124.14 g/mol (anhydr) was taken.

Tetradecane (Sigma-Aldrich, USA) with molecular mass 198.39 g/mol was used.

All reagents for 0.05M phosphate buffer (pH 6.0) preparation in Milli-Q water were of analytical grade.

5.8.2.1 SOLUTION PREPARATION

Lysozyme protein powder was dissolved in buffer for 2 hour and were centrifuged at 20 000g (Beckman 21, Germany) during 1 hour at room temperature.

The lysozyme concentration was determined by measuring the absorbance at 280 nm and using a specific extinction value 26.4 for a 1 cm$^{1\%}$ [8]. The operating solutions were prepared by mixing equal volumes of protein and methylresorcinol solutions within 40–50 minutes just before using in experiment.

5.8.3 FOAM FORMATION

The initial feed solution volume was 4 ml. The protein solution with constant concentration of 1.0 wt% and MR solutions with varying (0–1.0 wt%) concentrations were used. The comparatively high lysozyme concentration was necessary to obtain fast and sufficient amounts of foam in experiment.

Foam formation was performed by using of homogenizer Heidolph DIAX 900 (Germany). Air was introduced through a porous of homogenizer under constant airflow rate of 23,000 r/min, within 1.5 minutes. Experiments were conducted at the temperature 25°C. The responses selected for evaluating the experimental plan were the foam volume and multiplicity. Evaluation was carried out in the reaction vessel after the completion of foam formation. Foam volume was set by the geometric measurement. Foam multiplicity was calculated using the formula:

$$L = Vn/Vg,$$

where L is the foam multiplicity; Vn is the height of foam column and Vg is the height of the initial solution column.

Foam stability was evaluated using the parameter of half-life. Determined by visual observation is the foam half-life, i.e., the time for the foam height to reduce to 50% of its initial value. Time of the foam destruction was fixed by timer.

The triple preparations were made for each study in all experiments.

5.8.3.1   SURFACE DILATATIONAL PROPERTIES MEASUREMENT

The surface rheological parameters of adsorbed LYS -MR films at the air-water interface were measured with using of an automatic drop Tracker tensiometer (ITC Concept, France). Adsorbed films were formed over 60 000 s to guarantee steady-state. Sample solutions included a lysozyme (0.075 mg/mL) and varying 5-methylresorcinol (0.16–67.2 mM) concentrations in 0.05M phosphate buffer, pH 6.0 at 25°C.

The surface dilatational modulus (E) and the phase angle (θ) were measured as a function of time with constant amplitude ($\Delta A/A$) of 3% and angular frequency (ω). The range of angular frequencies used was 0.007–0.625 rad/s. The sinusoidal oscillation for surface dilatational measurement was made with 3–5 oscillation cycles followed by a time of 3–5 cycles without any oscillation. The average standard accuracy of the surface pressure is roughly 0.1 mN/m. The reproducibility of the results (for at least two measurements) was better than 0.5%.

5.8.4 RESULTS AND DISCUSSION

Proteins are often used as foaming agents because of their ability to unfold at the interface, thus creating films with high surface elasticity and also steric resistance against coalescence of films. Although they decrease the interfacial tension and hence reduce the driving force for disproportionation, often quite high protein concentrations are required for the formation of stable foams. For this reason, mixing with low-molecular weight surfactants can be very efficient and can increase the quality of the foam [1].

Destabilization of foams is a dynamic process that includes dispro-portionation, coalescence in addition to drainage of the thin film between bubbles. In addition to the bulk phase viscosity, all of these processes in-volve interfacial film properties [2, 4, 7, 10, 11]. The greater stabilizing effect may be attributed to a greater enhancement of the local viscosity in a foam lamellae which tends to inhibit film drainage, as well as to increased thickness of the mixed adsorbed layer which tends to enhance steric stabi-lization and inhibit bubble coalescence [3].

Early the dynamic surface tension measurements performed in phos-phate buffer solutions (pH 6.0, I = 0.05 M) at constant lysozyme concen-tration, 5.1×10^{-6} M, and varying concentrations of MR, have shown that the LYS-MR mixture is characterized by more high rate of adsorption and more low quasi-equilibrium surface tension, than pure LYS. MR critical micelle concentration in lysozyme presence is 32.7 mM [6].

The effect of MR on the rheological properties of lysozyme adsorp-tion layers at the air/solution (0.05 M phosphate buffer, pH 6.0) interface at 25°C was studied using a dynamic drop tensiometry and dilatometry methods. The quasi-equilibrium adsorption layers (the formation time of 60,000–70,000 sec) were subjected to compression/dilatation deformation sinusoidally in the field of linear viscoelasticity.

The dependences of complex viscoelastic modulus (**E**) of mixed ad-sorption layers, as well as its elastic (real part, $\mathbf{E_{rp}}$) and viscous (imaginary part, $\mathbf{E_{ip}}$) components as functions of MR concentration at the smallest 0.007 rad/s and the biggest 0.62 rad/s frequency values of the applied de-formation are presented in Figs.1 and 2 of Section 5.8. In Fig.3 of Section 5.8, the dependences of phase angle on MR concentration at the same conditions are shown.

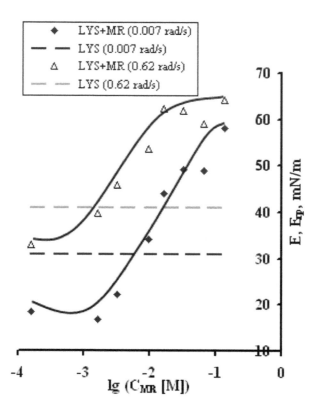

FIGURE 1   Dependence of Lysozyme complex elasticity modulus and it real part on MR concentration at frequency 0.007 and 0.62 rad/s. The drop lines indicate the interval of change corresponding parameters for pure LYS.

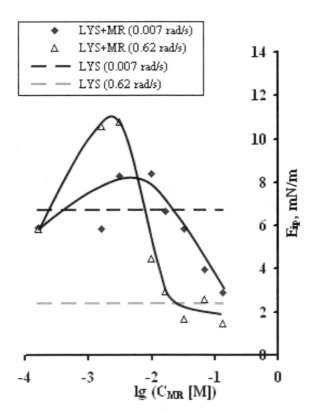

FIGURE 2 Dependence of Lysozyme imagines part of complex elasticity modulus on MR concentration at frequency 0.007 and 0.62 rad/s. The drop lines indicate the interval of change corresponding parameters for pure LYS.

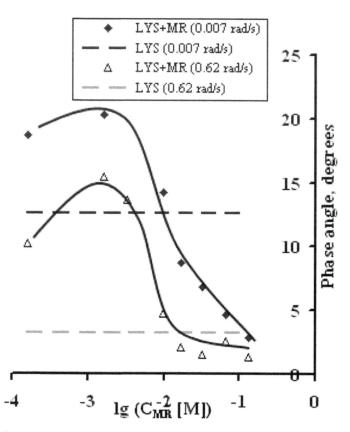

FIGURE 3 Dependence of Lysozyme phase angle on MR concentration at frequency 0.007 and 0.62 rad/s. The drop lines indicate the interval of change corresponding parameters for pure LYS.

The complex modulus of elasticity (**E**) and its elastic component E_{rp} increases with growth of MR concentration from 0.16 to 33 mM (CMC_{MR}). The viscosity component (E_{ip}) and phase angle increase with MR concentration up to 3.21 mM and then decreases up to 33 mM. The rheological behavior of the adsorption layers as well of intact lysozyme as modified ones is solid-like ($E=E_{rp}$, $E_{rp} \gg E_{ip}$, where E_{ip} is elastic component, and low frequency dependence of complex modulus **E**).

As seen, MR can both increase and decrease the viscoelastic parameters of LYS adsorption layers depending on concentration. The effect is more expressed at low frequencies of deformation. So, changing of MR concentration in LYS solution in wide range we can modify interfacial properties and the type of its rheological behavior.

Based on our results and taking into account that there is established positive associations between foam drainage half-life and the interfacial properties (the interfacial pressure and the interfacial elasticity) [12], we have chose the MR concentration for testing of foam formation properties of Lys in mixed solutions: lower and higher CMC_{MR} 32.7 mM (Table 1 of Section 5.8).

TABLE 1   The results of volume and multiplicity foams produced with lysozyme, depending on the concentration of MR.

MR concentration, mM	Foam volume, ml	Multiplicity
0	4.0	0.9
19.5	8.8	2.2
65.0	7.2	1.8

At first the experimental studies the amount of foam and its multiplicity was evaluated.

Thus, it was found that the foaming of lysozyme solution in the 30 and 100-fold molar excess of MR increases the foam total amount in 2.2 and 1.8 times, respectively, and the foam multiplicity increases in 1.8–2.3 times (Table 1 of Section 5.8).

In the second stage of studies the foams stability was measured, which is proportional to the rate of destruction.

Figure 4 of Section 5.8 shows the dynamics of the foam destruction. It implies that the half-life is the same, 45–50 minutes, for control and the sample containing 19.5 mM MR. Half-life of sample containing 65.0 mM MR is more than 4.5 hours (data not shown). The foam trend destruction of lysozyme-MR mixture (19.5 mM) has the character, unusual for the two other samples. This is due to the transition state MR from the molecular to associative form in this concentration. It was shown that MR alone has a critical micelle concentration equal 16.6 mM and in mixture with LYS – 32.7 mM [6].

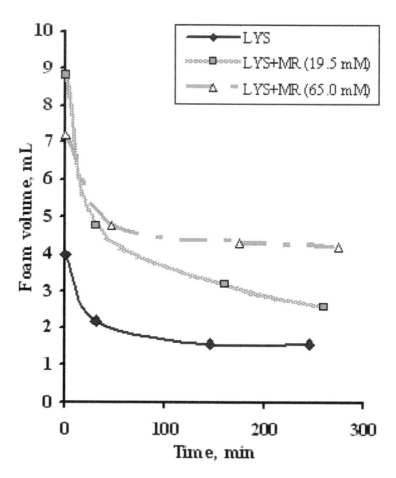

FIGURE 4   Dynamic curves of foams destruction.

So, studies have shown that the MR introduction increases multiplicity and stability of foams. The concentration of surfactant has a crucial importance.

It should to be noted that for all samples the main part of the liquid consumed to the foam formation follows back during the first half-hour of storage. Thus the amount of remaining liquid in the foam was in 1.5–2.0 times higher in MR-containing samples than in the control. Also in these samples, the sizes of foam bubbles are much less, and the density of packing is higher compared with the control (data are not shown)

An important role in foams stability can play liquid viscosity between the films. This is due to the fact that the rate of the liquid drainage is slowed down with viscosity increasing. However, at present time it is generally accepted that the surface viscosity can not be the determining factor for the foams stability, although the correlation between these properties exists.

5.8.5   CONCLUSION

The main factors, which determine the foam formation ability and physical stability of the foams in mixed Lys-MR solutions, are surface activity and complex dilatation modulus. With increase of MR concentration in certain range a surface activity and complex dilatation modulus of interfacial layers increased and phase angle decreased. It means that viscoelasticity of interfacial layers became higher. In these conditions the foam volume and mutiplicity as well as stability of foams were growing. This effect may be used for creation of mixed protein-MR system foam type for pharmaceutical applications with improved physical stability and wide range of antibacterial actions.

5.9   STRUCTURE AND THERMOPHYSICAL PROPERTIES OF BLENDS BASED ON ISOTACTIC POLYPROPYLENE AND LOW-DENSITY POLYETHYLENE

5.9.1   INTRODUCTION

In this section, the structure and thermophysical properties of isotactic polypropylene (PP) and low-density polyethylene (LDPE) blends were studied by differential scanning calorimetry (DSC) and electron paramagnetic resonance (EPR). The experiment shows non-additive changes of the thermal behavior and segmental mobility in the amorphous regions of immiscible PP/LDPE blends in relation to the properties of PP and LDPE initial homopolymers. Based on the results of the study, the probable process of the structure formation for PP/LDPE blends of various compositions described.

Despite the fact that numerous publications are devoted to the subject of polypropylene-polyethylene blends, many questions still remain unclear. Moreover, the results of several studies published in the literature contradict one another. A number of researchers consider that the blends of PP and PE are absolutely immiscible, so PP and PE form independent phases in the blend. Consequently, the structure and properties of the PP/PE blends depend on the composition additively [1–4]. Other authors suppose that the PE component influences the behavior and morphology of PP-PE blends, while the PP component has no impact on the properties of the blend [5–7]. Another point of view is that PP and PE have a mutual influence on the crystallization process and the formation of the phase structure and, consequently, the properties of the blend [8, 9].

For example, Martuscelli et al. [1] revealed three different types of crystallization behavior of PP/HDPE blends at three ranges of crystallization temperatures. Below 125°C, simultaneous crystallization of PP and HDPE was observed. From 125°C to 127°C, two crystallization peaks were observed with PP crystallizing first. Above 127°C, HDPE does not crystallize, and the crystallization of PP was observed only from a melt mixture of PP/HDPE. Wenig and Meyer [2] and Bartczak et al. [3] have determined that the presence of PE did not influence the PP spherulite growth rate or the process of crystallization. In contrast, Ramsteiner et al. [5] found that this growth rate was influenced by the presence of PE. Rybnikar [4] established that the morphology of PP is only temperature-dependent; it remains unaffected by the presence of another polymer. Other authors reported that the shape, size and orientation of LDPE occlusions remain undisturbed during the crystallization of PP spherulites. On the other hand, LDPE occlusions introduce major changes into the internal structure of PP spherulites [6]. Similar results were obtained for the PP/LDPE fibers in the study [7]. According to the authors, the LDPE component has a greater effect on the total crystallinity of the PP/LDPE blends than the PP component. Another standpoint was confirmed by the studies [8, 9]. It was found that LDPE acts as a structure-modifying agent, while PP plays the role of a strengthening (reinforcing) filler.

The aim of this work was to study the effect of the composition of PP/LDPE blends on the phase structure, in particular the quality and quantity of crystallites and amorphous regions of both components in the blends.

The thermophysical and dynamic parameters of the studied systems were obtained using DSC and EPR-spectroscopy.

Creation of new materials by blending polymers implies controlled modification of the component structures. Materials with a set of new features can be created owing to the mechanical mixing of different polymers. In addition, the production of new polymer blends does not require changes in the process of polymers production [10]. Polyolefins is the biggest group of commercial polymers; hence, they are the main component of plastic waste [11]. The blends based on the most popular sorts of polyolefins—low density polyethylene (LDPE) and isotatic polypropylene (PP)—are interesting from the practical and ecological points of view.

It is important to note that certain properties of these blends are better than those of the parent homopolymers. Adding PE to PP allows improvement of the low temperature impact behavior of PP [12]. In turn, the PP content contributes to the increase of toughness, high modulus and heat resistance to PE [13]. The PP/LDPE blends can be used as engineering plastics, consumer items for household use, goods with improved performance, etc.

Moreover, PP/LDPE blends are the products of plastic recycling. Since separation of PP and PE in solid domestic waste (SDW) is very expensive, many recycling companies sell mixtures of reworked PE and PP ("polymer mix") [13]. Thus, the study of the structure of blends of these materials will increase the field of application of recycled plastics. This is the reason for why PP-LDPE blends are the main subject considered in this chapter.

5.9.2 EXPERIMENTAL PART

Isotactic polypropylene (PP) ($M_w = 3.5 \times 10^5$, $M_n = 8 \times 10^4$, $M_w/M_n = 4.4$) and low-density polyethylene (LDPE) ($M_w = 2 \times 10^5$, $M_n = 4 \times 10^4$, $M_w/M_n = 5$) were the objects under investigation. The PP contents in the blends are 0, 5, 10, 20, 30, 40, 50, 60, 70, 80, 90, 95, 100 wt%.

The blend preparation was produced by mechanical mixing of PP and LDPE in a laboratory mixer at a temperature of 190°C for 5 minutes in the nitrogen atmosphere to prevent oxidation of LDPE. Isotropic films were fabricated by pressing with cellophane substrate at a temperature of 190°C and pressure of 7.8 MPa, followed by quenching in water (25°C). The thickness of the PP/LDPE films was 130±10 μm.

The degree of crystallinity χ and the melting temperature T_m of crystallites of each component in the blend were determined by DSC measurements performed on a differential scanning microcalorimeter DSM-10mA operated at a temperature scan rate of 8°C/min [14]. The temperature scale was gauged by Indium (T_m = 156.6°C, ΔH = 28.44 J/g).

The average melting temperatures T_m were measured by the maximums of the melting peaks in DSC thermograms. The melting heats ΔH of samples were calculated from the melting peak areas limited by a baseline. The degree of crystallinity χ_{PP} and χ_{LDPE} was calculated using the following formula Eq. (1):

$$X = \frac{\Delta H_i}{\Delta H_0} \times 100 \tag{1}$$

where ΔH_i—melting heat calculated relatively to the i-polymer content (PP or LDPE) in the blend, $\Delta H_o(PP)$ = 147 J/g—melting heat of a completely crystalline PP [15], $\Delta H_o(LDPE)$ = 295 J/g – melting heat of a completely crystalline LDPE [16].

The amorphous regions were studied using EPR spectroscopy, which allows the estimation of the molecular mobility in disordered regions [17]. A stable nitroxyl radical, 2,2,6,6-tetramethylpiperidine-1-oxyl (TEMPO) was used as a paramagnetic probe. The radical was introduced into the films from vapor at a temperature of 30°C. The rotational mobility of the radical probe was determined by the correlation time τ_c. The following formula is used to calculate Eq. (2) τ_c:

$$\tau_c = 6,65 \times 10^{-10} \left(\sqrt{\frac{I_+}{I_-}} - 1 \right) \times \Delta H \tag{2}$$

where I_+ and I_- are intensities of the first and third peaks in the EPR spectrum; ΔH - half-width of the first peak in the EPR spectrum [17].

The dependence of τ_c on the blend composition is adjusted for isotropic and oriented samples. The orientation stretching was produced by local heating at 70°C (PP/LDPE = 0:100–50:50) and at 90°C (PP/LDPE = 50:50–100:0). The oriented films with a draw ratio λ = 4 were used.

5.9.3   RESULTS AND DISCUSSION

The PP and LDPE are thermodynamically incompatible polymers forming crystalline lattices of different types. The structures of the two polymers in a blend are formed separately because of their different melting temperatures. PE crystallizes into a more stable orthorhombic crystalline lattice, while PP—into a monoclinic crystalline lattice [18]. Our task was to study the crystallization process of both polymers taking into account the influence of either component on the structure formation of the blend.

Figure 1 of Section 5.9 shows the dependence of the degree of crystallinity χ of each component of the blend on the PP/LDPE blend composition. The degree of crystallinity of LDPE changes slightly ($\chi_{LDPE} = 19.4\pm1.5\%$) across the full range of the compositions. An increase in the PP content causes a small decrease of χ_{LDPE}. Thus, the presence of PP does not have a significant effect on the crystallization process of LDPE.

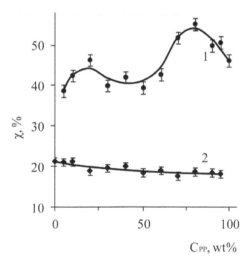

FIGURE 1   Effects of the PP-LDPE blend composition on the degree of crystallinity χ of PP (1) and LDPE (2).

The degree of crystallinity of PP is changed more significantly, depending on the composition of the blend. Higher values of χ_{PP} are observed for the blends with the LDPE content from 10 to 30%, with the maximum of χ_{PP} equal to 55.3% (PP/LDPE = 80:20). Apparently, since LDPE has a lower solidification temperature as compared to PP, the process of crystallization of the PP phase takes place in the presence of molten LDPE. A small amount of molten LDPE in the blend acts as a plasticizer, so the process of crystallization of PP proceeds more perfectly. The increase of the LDPE content by more than 30 wt% prevents the process of crystallization of PP (the volume fraction of the crystalline phase of PP decreases). Blends with the LDPE contents from 30 to 60 wt% and below 5 wt% have a lower value of χ_{PP} ($\chi_{PP} = 41 \pm 1.5\%$). It seems that the phase inversion occurs in the range of compositions containing PP from 30 to 60 wt%.

As the result, both components of the blend have a mutual influence on the crystallization process and the formation of the phase structure and, consequently, the properties of the blend. However, the LDPE component has a greater effect on the process of the structure formation of PP/LDPE blends than the PP component.

The melting temperatures T_m(PP) and T_m(LDPE) for the blends of different compositions vary within the narrow limits of $160.8 \pm 3.1°C$ and $104.1 \pm 0.8°C$, respectively (Table 1 of Section 5.9). It means that the lattice perfection of PP and LDPE is practically constant for the entire range of compositions. The minimum values of T_m(PP) and T_m(LDPE) are observed at the lower content of the corresponding polymer in the blend. Thus, the addition of one polymer to the other causes an insignificant reduction in the value of T_m as compared with the corresponding value for the homopolymer.

TABLE 1   The values of melting temperatures of PP and LDPE for PP/LDPE blends as a function of weight per cent of PP. The measurement error is ± 5%.

C_{pp} wt%	0	5	10	20	30	40	50	60	70	80	90	95	100
T_m (PP), °C		159	163	162	162	162	162	162	162	163	162	162	164
T_m (LDPE), °C	106	106	106	106	106	106	106	106	106	106	106	104	

Figure 2 of Section 5.9 presents the dependence of the correlation time τ_c for isotropic and oriented ($\lambda=4$) samples on the blend compositions. The value of τ_c characterizes the molecular mobility of macrochains in an amorphous phase.

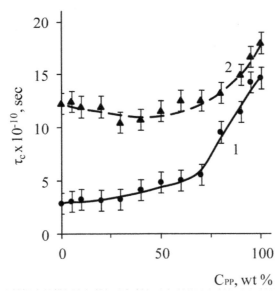

FIGURE 2 Effect of the PP-LDPE blend composition on the correlation time τc for isotropic (1) and oriented (2) samples.

The value of τ_c increases with the increase in the PP content, i.e. the mobility of macrochains in the system decreases. For isotropic samples, there is a small increase of the τ_c value (from 2.9×10^{-10} s to 5.6×10^{-10} s) up to 60–70 wt% of PP in the blend and then the value of τ_c increases sharply with the maximum value (14.7×10^{-10} s) for PP. For oriented samples, there is a far less dramatic increase in the value of τ_c in passing from LDPE to PP (τ_c increases from 12.3×10^{-10} s to 18×10^{-10} s).

Comparison of the values of τ_c for oriented and isotropic samples of the blends of the same composition indicates that a higher content of LDPE in the blend is associated with a greater difference between the values of τ_c for the isotropic and oriented samples. Thus, the τ_c value for an isotropic sample in LDPE is four times greater than that for an oriented sample, while in PP, the difference between the values of τ_c is approximately 3×10^{-10} sec.

Using the area of the EPR spectrum allows the estimation of the amount of stable nitroxyl radicals sorbed by samples of the blends of different composition under identical conditions. The dependence of the areas of the EPR spectra for the blends of various compositions on the content of PP in the blend is presented in Fig. 3 of Section 5.9. The area of the EPR spectrum decreases with the increase of the PP content in the blend. Therefore, the TEMPO radical has a better solubility in the LDPE amorphous regions than in the PP ones. A sharp decrease in this value is over a range of compositions containing from 40 to 60 wt% of PP. This indirectly suggests the occurrence of the phase inversion and the formation of the PP continuous phase in this range of compositions.

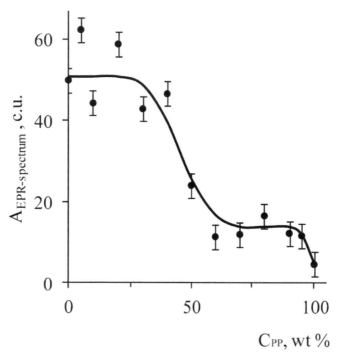

FIGURE 3 Effect of the PP-LDPE blend composition on the area A of the EPR-spectrum calculated as an amorphous region.

These data confirm the fact that LDPE has looser amorphous regions as compared to PP. LDPE tends to sorb up to 60–70 wt% of the PP content from the probe of the blend. Given that the correlation time of a radical rotation is an integral characteristic of the PP/LDPE blend, the value of τ_c grows rapidly only when a certain amount of the radical penetrates into PP. It explains the type of the dependence between the area of the EPR spectrum and the blend composition presented in Fig. 2 of Section 5.9.

Figure 4 of Section 5.9 demonstrates the relationship between the activation energy E_A of the radical probe rotation and blend compositions. A slight increase in E_A is observed at the initial part of this graph (PP content up to 60 wt%). Further, the value of E_A grows more rapidly, increasing from 41 kJ/mol for the composition with PP/LDPE = 60:40 to the maximum value of 56 kJ/mol for PP. The increase in the E_A value indicates that the amorphous regions gain stiffness. The dependence of the E_A value on the radical probe rotation presented here suggests that the LDPE continuous phase transforms to the discrete phase in the range of 40–60 wt% of PP.

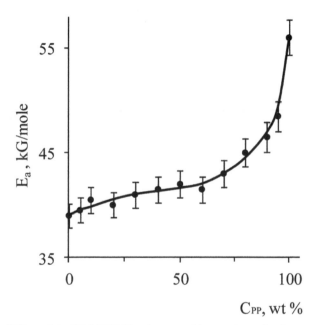

FIGURE 4   Effect of the PP-LDPE blend composition on the activation energy E_A of the radical rotation.

5.9.4 CONCLUSION

Based on the data obtained in the study, it appears possible to describe the probable process of the structure formation for different PP/LDPE blends. After a gradual addition of PP to LDPE, the former begins to form a dispersed phase in the continuous phase of LDPE up to the PP content in the blend of about 40 wt%. Phase inversion with the formation of the PP continuous phase is observed in the range of compositions with the PP content from 40 to 60 wt%. This results in an increase in the volume fraction of the crystalline phase of PP and the overall stiffness of the system. Molten LDPE in the blend acts as a structure-modifying agent, so the crystallization process of PP proceeds more perfectly.

5.10   INVESTIGATING THE EFFECT OF FUNGIES ON THIN FILMS OF POLYETHYLENE, POLYETHYLENE – POLY-3-HYDROXYBUTYRATE AND POLYETHYLENE – D$_2$W

5.10.1   INTRODUCTION

In this section influence of fungi on films of PE—PHB, PE—d$_2$w and pure PE without the additional power supply was investigated. It was established that films of PE—PHB were attacked by fungi more intensively, than other samples. Tests in the climatic chamber were conducted and it was revealed that with growth of PHB content the sizes of cracks increase. This fact is coordinated with the data on degradation of films of PE—PHB in soil. The main idea of current section is focused on investigation the effect of fungies on films of pure polyethylene and polyethylene composites and study of biodegradation of the samples in soil.

PHB (Polyhydroxybutyrate) is an example of PHA polyhydroxyalkanoate), first isolated in 1925 by the French Microbiologist Maurice Lemoigne. The most common form of PHB is poly-3-hydroxybutyrate, denoted P3HB. Unlike its petrochemical alternatives, PHB is based on renewable resources and can be completely metabolized by a wide variety of bacteria found in a range of environments, making it truly biodegradable. A problem posed by non-biodegradable polymers is efficient waste disposal—in 1992 the total volume fraction of plastics in landfill sites had reached 20%, and the proportion continues to rise. PHB holds several advantages over other biodegradable polymers, for example its water insolubility, good oxygen permeability, UV resistance and biocompatibility.

PHB is produced by various microorganisms as an energy storage molecule under conditions of stress. Much current research is dedicated to finding an economically efficient method to produce a high yield in a short amount of time. Studies show successful PHB accumulation from a variety of substrates, including *Bacillus sp, Pseudomonas cepacia, Halomonas boliviensis* [1–3]. The type of microorganisms and feedstock used the method of extraction and the down-processing techniques determine the molecular weight and physical properties of PHB.

The biodegradation of PHB has been extensively studied recently. Previous research investigates the degradation of PHB and its copolymers in several environments, such as soils, compost, fresh water and sewage sludge. The microorganisms which act to digest the polymer have been isolated, and include various strains of *Penicillium* and *Aspergillus*. General consensus is that PHB shows increased degradation in soil environments, although good results in sewage sludge and compost are also reported [4, 5].

Of increased interest are PHB blends. PHB can be blended with many commercially available plastics such as polyethylene (PE). This produces a material with all the beneficial properties of the PE (e.g., cheap, readily available) with the biodegradable properties of the PHB. Investigation is required to determine the necessary blend parameters, as well as to determine microorganisms capable of biodegradation. This is the motivation for this chapter. Similar products on the market include d_2w, produced by Symphony Environmental. d_2w is an additive that changes the molecular structure of a plastic at the end of its useful life, producing a compound which can be bio-assimilated into the environment.

This chapter presents a study on the suitability of various polymer films (PE, PE/PHB blend and PE/ d_2w blend) to the growth of fungi strains, including *Penicillium, Aspergillus* and *Trichoderma*. Subsequent comparison of fungal growth allows conclusions to drawn about the biodegradability of the samples.

5.10.2   EXPERIMENTAL PART

5.10.2.1   SOURCE OF POLYMERS

Three polymers were tested: polyethylene (PE), PE with added d_2w, and the PE/PHB blends. The PHB (of specific structure H[-O-CH(CH$_3$) CH$_2$COO-]$_n$OH and average molecular weight 2.5×10^5 g mol^{-1}) was sourced from Biomer biopolyesters, who use a selected soil bacterium to grow the PHB in an aqueous medium. The PE and d_2w blend was sourced from a readily available plastic carrier bag, where the d_2w was produced by Symphony Environmental.

5.10.2.2   PREPARATION OF POLYMER FILM

PHB (2, 4, 8, 16, 32% mass) and PE were blended and formed into a thin film. All polymer films were cut into strips of approximate dimensions 10 mm by 100 mm. The mass of each sample was recorded.

5.10.2.3   MICROBIAL STRAINS AND INCUBATION

Water suspensions containing spores of *Aspergillus Niger*, *Penicillin Chrysagenum* and *Trichoderma Viride* were prepared. 5–6 drops of each suspension were placed onto the polymer strips. The strips were placed in an incubator held at room temperature (20°C +/– 2°C).

5.10.2.4   OBSERVATION OF SAMPLES

Samples were removed from the incubator after 14 days and 28 days. An Olympus CX41 optical microscope was used to observe the samples, under 400× magnifications. Photographs of the sample were taken using a Canon digital camera.

The climatic measurement was occurred using 'Xentest 150-S" chamber at T=35°C with the day-night program.

5.10.2.5 THERMAL ANALYSIS

Thermal analysis was performed using a DSC differential scanning calorimeter at a heating rate of 8°C min^{-1}, for the PE + d$_2$w sample. The DSC calibration was done with Indium, which has a known melting point of 156.61°C. Both heating and cooling curves were obtained, and the endotherms produced allowed calculation of the melting point, the degree of crystallinity and the temperature of crystallization. The degree of crystallinity α is calculated using the equation

$$\alpha = \frac{\Delta H}{\Delta H^\circ m} \times 100\%,$$

where ΔH is the melting enthalpy of the sample and ΔH°_m is the hypothetical melting enthalpy of a purely crystalline sample.

5.10.3 RESULTS AND DISCUSSION

First of all we hade to investigate the supermarket's sample. The IR-spectroscopy and DSC methods were used. The IR-spectra with some characteristic lines was obtained: 720, 1470, 2850 × 2920 sm^{-1}, so this polymer was polyethylene (Fig. 1 of Section 5.10) [6]. The melting temperature (T$_m$) 134°C and the crystallinity degree (α) 75% obtained by DSC method confirmed it.

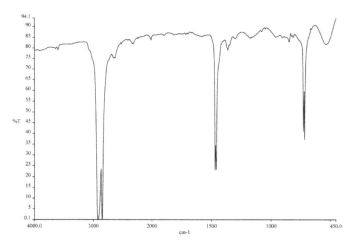

FIGURE 1   IR-spectra of the film sample from a supermarket.

For our investigation the sample 4 wt.% of PHB was taken because the d_2w content approximately was about 5% [7] and we could compare the samples.

Next step was a fungies contamination:

After 14 days:

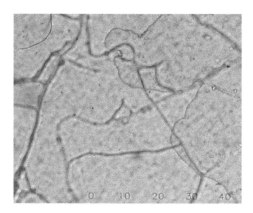

<u>PE</u>

Light microscope image of PE strip with *Penicillium Chrysogenum*. We observe the presence of a developed network of mycelia, indicating that the fungi have grown. The PE has started to biodegrade under the influence of the *Penicillium Chrysogenum*.

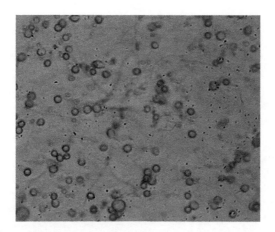

<u>PE + PHB</u>

Image of PHB (4%) and PE (96%) polymer strip, with *Trichoderma Viride*. We observe the early stages of mycelia growth from the spores, signaling the beginning of biodegradation.

After 28 days:

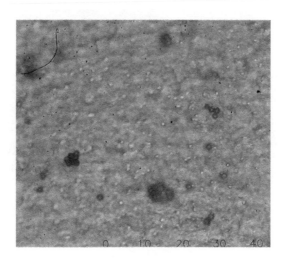

<u>PE + d$_2$w</u>

Light microscope image of PE + d$_2$w with the presence of *Aspergillus Niger*. No mycelia are observed, indicating that the *Aspergillus Niger* fungi has not grown on the PE + d$_2$w sample.

PE + d$_2$w

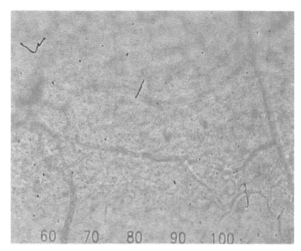

The best growth on the PE + d₂w occurred for the *Trichoderma Viride*. We observe less mycelium growth on this sample for *Penicillium Chrysogenum*, and hardly any growth for the *Aspergillus Niger*.

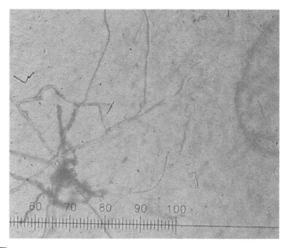

PE + PHB

We observe similar growth to that shown after 14 days. The most advanced mycelium growth is found on the sample of PE + 4% PHB, with the *Penicillium Chrysogenum*, and also the *Trichoderma Viride*. We observe a more complex mycelium structure, with many branches.

The growth of three fungi strains *Aspergillus Niger, Penicillium Chrysagenum* and *Trichoderma Viride* on thin film samples of PE, PE+d$_2$w and PE+4% PHB was investigated. The samples were subjected to water suspensions of the three kinds of spores, and incubated for 14 days and 28 days. Comparison of light microscope images allowed us to determine the extent of mycelia growth. It was found the best mycelium growth occurs on PE + PHB for both *Penicillium Chrysagenum* and *Trichoderma Viride*.

After fungi contamination the degradation in soil of pH= 6.0–6.2 at room temperature (20°C +/– 2°C) was studied. Biodegradation was monitored every 3 months for approximately 12 months by measuring residual mass. For this, the buried samples were recovered, washed with distilled water and dried at room temperature before being weighed and then buried again in their respective trays. The residual mass was calculated as the ratio between the final and initial weights. The samples of pure PE and PE-PHB (2 wt.%) had no mass changes even after 12 months. The significant weight loss was observed for samples with 16 and 32 wt.% of PHB: 6 and 10% respectively. These data are confirmed the climatic chamber measurement.

As a result of climate processes of ageing in polymer composites can precede processes of structure formation, which result in various flaws. It may be gaps of macromolecules, the emergence of cracks in the volume and on the surface of the material. From the Table 1 of Section 5.10 one can see that the craze size increased with growing PHB content.

TABLE 1   The craze sizes of PE—PHB films.

PE—PHB wt.%	Length of crazes, μm	Width of crazes, μm
100:0	No	No
98:2	No	No
96:4	60–1000	0.5–10
92:8	180–1000	3–10
84:16	200–1000	15–30
68:32	200–1000	30–45

The presence of defects in the samples, as a rule, leads to the growth of colonies of microorganisms, when the material loses its physic-mechanical characteristics and the integrity of the matrix. Based on the above, it can be assumed that in natural conditions to the extent of the impact of environmental factors on the surface of the film will be formed cracks, the presence of which will contribute to the penetration of the dispute mold fungi and other microorganisms. Microorganisms, eating PHB, will be developed, to allocate metabolites and destroy the film, fragmenting them up to a size commensurate with the size of soil particles.

KEYWORDS

- *Alternaria alternate*
- **Anthocyanins**
- *Aspergillus*
- **Fukui functions**
- **Grante**
- **Grunel**
- **Hirschfeld method**
- **Packmal**
- *Penicillium*
- *Penicillium Chrysagenum*
- **Red Price**
- *Trichoderma*
- *Trichoderma Viride*

REFERENCES

SECTION 5.1

1. Gusev, A. I. Nanomaterials, nanostructures, and nanotechnology. *Physmathlit.*, **2005**, 416.
2. Suzdalev, I. P. *Nanotechnology: Physics and chemistry of nanoclusters, nanostructures and nanomaterials.* M.: KomKniga., **2006**, 596 s.
3. Lipatov, Yu. S. *Physical chemistry of filled polymers.* M.: Chemistry, **1977**, 304
4. Pomogailo, A. D.; Rosenberg, A. S.; Uflyand, A. S. *Metal nanoparticles in polymers.* Chemistry, **2000**, 672.
5. Pomogailo, A. D.; Rosenberg, A. S.; Uflyand, A. S. *Metal nanoparticles in polymers.* Chemistry, **2000**, 672.
6. Baker, R. T. K. *Catalytic growth of carbon filaments// Carbon.*, **1989**, *27*, 315–323.
7. Chmutin, I. A.; Rvykina, N. G.; Dubnikova, I. L. etc. *Control of homogeneity of distribution of nanoparticles in the polymer matrix; Abstracts of the conference "Nanotechnologies—Production—2007."* Fryazino, **2007**, 229–234.

SECTION 5.2

1. The Federal Law of the Russian Federation of October 27, **2008**, *178* "*Technical Regulations for Juice Products from Fruits and Vegetables*".
2. Methodical instructions 4.1/4.2.2486–09 "Methodical instructions on Eden identifying, including for the purpose of revealing of falsification, juice production of fruit and/or vegetables". *Methodical instructions.* M.: Federal Centre of hygiene and epidemiology of Rospotrebnadzor, **2009**.
3. National standard GOST R 53137–2008 "*Juices and juice products. Identification. General provisions*" M. : Publishers standards, **2008**.

SECTION 5.3

1. Chen, J. H.; Adelman, S. A. Macroscopic model for solvated ion dynamics. *The Journal of Chemical Chem. Physics Phys.*, **1980**, *72*, 2819.
2. Cunningham, E. On the Velocity of Steady Fall of Spherical Particles through Fluid Medium. Proceedings of the Royal Society A: Mathematical, *Physical and Engineering Sciences*, **1910**, *83*, 357.
3. Uchida, H.; Matsuoka, M. Molecular dynamics simulation of solution structure and dynamics of aqueous sodium chloride solutions from dilute to supersaturated concentration. *Fluid Phase Equilibria*, **2004**, *219*, 49.

4. Espanol, Pep.; Zuniga, Ignacio. Force autocorrelation function in brownian motion theory. *The Journal of Chemical Chem. Physics Phys.*, **1993**, *98(1)*, 574–580.
5. Yuan-Hui, Li.; Gregory, Sandra. Diffusion of ions in sea water and in deep-sea sediments. *Geochimica et Cosmochimica Acta*, **1974**, *38(5)*, 703–714.
6. Longsworth, L. G. Diffusion measurements, at 25'C, of aqueous solutions of amino acids, peptides and sugars. *Journal of the American Chemical Society*, **1953**, *75(22)*, 5705–5709.

SECTION 5.4

1. Damm, C.; Münstedt, H.; Rösch, A. 'The antimicrobial efficacy of polyamide 6/silver-nano- and microcomposites', *Materials Chemistry and Physics*, **2008**, *108(1)*, 61–66.
2. Fabrega, J.; Luoma, S. N.; Tyler, C. R.; Galloway, T. S.; Lead, J. R. 'Silver nanoparticles: Behaviour and effects in the aquatic environment', *Environment International*, **2011**, *37(2)*, 517–531.
3. Gonze, X. et al.; ABINIT : First-principles approach of materials and nanosystem properties." *Computer Phys. Commun.*, **2009**, *180*, 2582–2615.
4. Jo, Y.; Kim, B. H.; Jung, G. 'Antifungal Activity of Silver Ions and Nanoparticles on Phytopathogenic Fungi', *Plant Disease*, **2009**, *93(10)*, 1037–1043.
5. Kim, K. J.; Sung, W.; Suh, B.; Moon, S. K.; Choi, J. S.; Kim, J.; Lee, D. 'Antifungal activity and mode of action of silver nano-particles on Candida albicans', *BioMetals*, **2009**, *22*, 235–242.
6. Luo, P. G.; Stutzenberger, F. J. *Nanotechnology in the Detection and Control of Microorganisms*, in Sima Sariaslani Laskin, Allen I.; Gadd, Geoffrey M. ed., Academic Press, **2008**, 145–181.
7. Madhumathi, K.; Sudheesh Kumar, P.; Abhilash, S.; Sreeja, V.; Tamura, H.; Manzoor, K.; Nair, S.; Jayakumar, R. 'Development of novel chitin/nanosilver composite scaffolds for wound dressing applications', *Journal of Materials Science: Materials in Medicine*, **2010**, *21*, 807–813.
8. Monteiro, D.; Silva, S.; Negri, M.; Gorup, L.; de Camargo, E.; Oliveira, R.; Barbosa, D.; Henriques, M. 'Silver nanoparticles: influence of stabilizing agent and diameter on antifungal activity against Candida albicans and Candida glabrata biofilms', *Letters in Applied Microbiology*, **2012**, *54(5)*, 383–391.
9. Monteiro, D. R.; Gorup, L. F.; Takamiya, A. S.; Ruvollo-Filho, A. C.; De Camargo, E. R.; Barbosa, D. B. 'The growing importance of materials that prevent microbial adhesion: antimicrobial effect of medical devices containing silver.', *International Journal of Antimicrobial Agents*, **2009**, *34(2)*, 103–110.
10. Morones, J. R.; Elechiguerra, J. L.; Camacho, A.; Holt, K.; Kouri, J. B.; Ramírez, J. T.; Yacaman, M. J. 'The bactericidal effect of silver nanoparticles', *Nanotechnology*, **2005**, *16(10)*, 2346.
11. Soler, J. M., Artacho, E., Gale, J. D., García, A., Junquera, J., Ordejón, P., & Sánchez-Portal, D. The SIESTA method for ab initio order-N materials simulation. *Journal of Physics: Condensed Matter*, **2002**, *14*, 2745.

12. Wales, D. S.; Sagar, B. F. Recovery of Metal Ions by Microfungal Filters. *Journal of Chemical Technology & Biotechnology*, **1990**, *49(4)*, 345–355.

SECTION 5.5

1. Voskoboinikov V. A.; Ilyina N. E. Features store kernels. *Your food*, **2001**, *1*.
2. Komissarenkov A. A.; Dmitrievich I. N.; Fedorova O. V. *Coulometric methods of analysis: Teaching aid*, **2009**.

SECTION 5.6

1. Popov, A. A.; Koroleva, A. V. "Biodegradable polymer composites based on polyolefin and natural polymers*", Modern Problems in Biochemical Physics*: New Horizons, **2012**, 123–132.
2. Bazunov, M. V.; Prochukhan, Y. "Disposal of waste polymers", *Journal of Bashkir University*, **2008**, *13(4)*, 880.
3. Lukanina, J. K.; Khvatov, A. V.; Kolesnikova, N. N.; Popov, A. A. Structure and properties of biodegradable polymer composite materials. *Progress in chemical and biochemical physics, kinetics and thermodynamics*, **2008**, 209–218.
4. Lukanina, Yu. K.; Kolesnikova, N. N.; Khvatov, A. V.; Likhachev, A. N.; Popov, A. A. Influence of the structure of a composite material's polymeric template on the development of micromycetes. *Handbook of Chemistry, Biochemistry and Biology*: Nova Science Publishers **2010**, 357–363.
5. Khvatov, A. V.; Shatalova, O. V.; Krivandin, A. V.; Lukanina, Yu. K.; Popov, A. A. Structure and mechanical properties of biodegradable materials based on LDPE and natural supplements. *Deformacia I Razrushenie materialov*. **2012**, *8*, 45–48.
6. Lukanina, Yu. K.; Kolesnikova, N. N.; Khvatov, A. V.; Likchachev, A. N.; Popov, A. A. Influence of polypropylene structure during micromycete growth on their composition. *Journal of the Balkan tribological association*. **2012**, *18(1)*, 142–148.
7. Lukanina, Yu. K.; Kolesnikova, N. N.; Likhachev, A. N.; Khvatov, A. V.; Popov, A. A. Effect of the polymer matrix structure the on the micromycetes development on mixed compositions of polyolefins and cellulose. *Plasticheskie massi*, **2010**, *11*, 56–59.
8. *"The burning, destruction and stabilization of polymers"*, Ed. Zaikova G. E., Publisher "Fundamentals and Technologies," Moscow, **2008**, 73–79.
9. Lipatov, Yu. S. *"Physical chemistry of filled polymers"*, publishing house "Chemistry"*, **1977**, 159.
10. Bazhenov, S. L.; Berlin, A. A.; Kulkov, A. A.; Oshmyan, V. G. *"Polymer composites. Strength and technology"*, Publishing House "Intelligence", Dolgoprudny, **2010**, 147.
11. *"Encyclopedia of Polymer"*, publishing house "Soviet encyclopedia", Moscow, **1977**, *3*. str.1011.

SECTION 5.7

1. Semenova, I. V.; Florianovich, G. M.; Khoroshilov, A. V. *Corrosion and Corrosion Protection,* M. Fismatlit, **2002**, 335.
2. Vigdorovich, V. I.; Shel, N. V.; Krylova, A. G. Features of the atmospheric metals corrosion. Bulletin of the Tambov University. *Series: Natural and Technical Sciences,* **2001**, *6(3),* 279–289.
3. Semenov, S. A.; Gumargalieva, K. Z.; Zaikov, G. E. Characteristics of the processes and characteristics of damage of materials engineering by microorganisms in operation. *Bulletin of MITHT,* **2008**, *3(2),* 1–21.
4. Alehova, T. A.; Alexandrova, A. V.; Zagustina, N. A.; Novozhilova, T. Y.; Romanov, S. Y. Mikrosokopicheskie fungi on the Russian segment of the International Space Station. *Mycology and phytopathology,* **2009**, *43(5),* 377.
5. Mikhailov, A. A.; Strekalov, P. V. Modeling of atmospheric corrosion of metals and types of dose-response functions. *Corrosion: Materials, protection,* **2006**, *3,* 2–13.
6. Mikhailov, A. A.; Strekalov, P. V. Modeling of atmospheric corrosion of metals and types of dose-response functions (continued). *Corrosion: Materials, protection,* **2006**, *4,* 2–10.
7. Karpov, V. A.; Kovalchuk, Y. L.; Kharchenko, U. V.; Beleneva, I. A. Microaccretions impact on marine corrosion of metals and destruction of protective coatings. *Corrosion: Materials, protection,* **2011**, *3,* 11–18.
8. Andreuk, E. I.; Bilai, V. I.; Koval, E. Z.; Kozlova, I. A. *Microbial corrosion and its agents.* Kiev: Nauk. Dumka, **1980**, 286.
9. Zaikina, N. A.; Duganova, N. V. The formation of organic acids, fungi isolated from samples of affected biological corrosion. *Mycology and phytopathology,* **1975**, *9(4),* 303.
10. Glazkov, V. I.; Zinevich, A. M.; Kotik, V. G.; et al. *Corrosion long metal structures.* M.: Nedra, **1969**.
11. *Protection against corrosion, aging, biological damage of machinery, equipment and structures: a guide: in 2 volumes*/ed. A. A. Gerasimenko. Moscow, **1987**.
12. Glazov, N. P. Particular corrosion protection of steel underground piping. *Physical chemistry of surface protection materials.* **2004**, *40(5),* 522–528.
13. Sulimina, E. Y. Investigation of pulsed current cathode protection on the technology of small-diameter pipeline. *Oil and Gas business.* **2011**, *12,* 52–58.
14. Lebedeva, E. V.; Nazarenko, A. V.; Kozlova, I. V.; Tomilin, B. A. Effect of increasing concentrations of copper on soil fungi are. *Mycology and phytopathology.* **1999**, *33,* 257–263.

SECTION 5.8

1. Alahverdjieva, V. S.; Khristov Khr.; Exerowab, D.; Miller R. Correlation between adsorption isotherms, thin liquid films and foam properties of protein/surfactant mix-

tures: Lysozyme/C10DMPO and lysozyme/SDS Colloids and Surfaces A: *Physicochem. Eng. Aspects.* **2008**, *323*, 132–138.

2. Bos, M. A.; van Vliet, T. Interfacial rheological properties of adsorbed protein layers and surfactants: a review. *Advances in Colloid and Interface Science*, **2001**, *91*, 437–471.

3. Dickinson E.; Izgi E. Foam stabilization by protein-polysaccharide complexes. *Colloid and Surfaces.* **1996**, *113*, 191–201.

4. Foegeding, E. A.; Luck, P. J.; Davis, J. P. Factors determining the physical properties of protein foams. *Food Hydrocolloids*, **2006**, *20*, 284–292.

5. Maldonado-Valderrama J.; Patino, J. M. R. Interfacial rheology of protein-surfactant mixtures. *Current Opinion in Colloid & Interface Science*, **2010**, *15*, 271–282.

6. Martirosova, E. I.; Plashchina, I. G. Adsorption behavior of methylresorcinol and its mixtures with lysozyme at air-water interface. *In Biochemistry and Biotechnology.* Nova Science Publishers. N.-Y. **2012**, 95–104.

7. Murray, B. S. Stabilization of bubbles and foams. *Current Opinion in Colloid and Interface Science*, **2007**, *12*, 232–241.

8. Page, C. N.; Vajdos F.; Fee L.; Grimsley G.; Gray Th. How to measure and predict the molar absorption coefficient of a protein. *Prot. Sci.*, **1995**, *4*, 2411–2423.

9. Petrovskii, A. S.; Deryabin, D. G.; Loiko, N. G.; Mikhailenko, N. A.; Kobzeva, T. G.; Kanaev, P. A.; Nikolaev Yu. A.; Krupyanskii Yu. F.; Kozlova, A. N.; El'-Registan, G. I. Regulation of the functional activity of lysozyme by alkylhydroxybenzenes. *Microbiology.* **2009**, *78(2)*, 146–155.

10. Rodriguez Patino, J. M. R.; Sanchez, C. C.; Rodriguez Nino, M. R. Implications of interfacial characteristics of food foaming agents in foam formulations. *Advances in Colloid and Interface Science*, **2008**, *140*, 95–113.

11. Wilde, P. J. Interfaces: their role in foam and emulsion behavior. *Current Opinion in Colloid and Interfaces Science*, **2000**, *5*, 176–181.

12. Xin Yang, E. Allen Foegeding. *Effects of sucrose on egg white protein and whey protein isolate foams: Factors determining properties of wet and dry foams (cakes) Food Hydrocolloids*, **2010**, *24*, 227–238.

SECTION 5.9

1. Martuscelli E.; Pracella M.; Avella M.; Greco R.; Ragosta G. Properties of polyethylene-polypropylene blends—crystallization behavior. *Makromolekulare Chemie-macromolecular chemistry and physics.* **1980**, *181*, 957–967.

2. Wenig W.; Meyer K. Investigation of the crystallization behaviour of polypropylene—polyethylene blends by optical microscopy. *Kolloid-Zeitschrift & Zeitschrift für Polymere.* **1980**, *258(9)*, 1009–1014.

3. Bartczak Z.; Galeski A.; Pracella M. Spherulite nucleation in blends of isotactic polypropylene with high density polyethylene. *Polymer.* **1986**, *27*, 537–543.

4. Rybnikar F. Crystallization and morphology in blends of isotactic polypropylene and linear polyethylene. *Journal of Macromolecular Science. Part B. Physics.* **1988**, *27*, 125–129.

5. Ramsteiner F.; Kanig G.; Heckmann W.; Gruber W. Improved low temperature impact strength of polypropylene by modification with polyethylene. *Polymer.* **1983**, *24,* 365–370.

6. Galeski A.; Pracella M.; Martuscelli E. Polypropylene spherulite morphology and growth rate changes in blends with low density polyethylene. *Journal of Polymer Science. Part B. Polymer Physics.* **1984**, *22,* 739–747.

7. Ujhelyiova A.; Marcincin A.; Legen J. DSC analysis of polypropylene-low density polyethylene blend fibres. *Fibres and Textiles in Eaten Europe.* **2005**, *13(5)(53),* 129–132.

8. Livanova, N. M.; Popova E. S, Ledneva, O. A.; Popov, A. A. Properties and phase structure of the poly(propylene)-low-density poly(ethylene) blends. *Polymer Science. Ser. A.,* **1998**, *40(1),* 41–46.

9. Prydilova O.; Petronyuk U.; Tertyshnaya U.; Ledneva O.; Popov A. The correlation of structure and properties of PE-PP blends. *Chemistry and computer modeling.* **2002**, *11,* 103–106.

10. Paul, D. R.; Newman S. *Polymer Blends.* Academic Press, New York. **1978**, 189–204.

11. White, James L.; David Choi. *Polyolefins: Processing, Structure Development and Properties.* Hanser-Gardner Publications. Cincinnati., **2005**, 1–2.

12. Karger-Kocsis, J. *Polypropylene Structure, blends and Composites: Vol. 2 Copolymers and Blends.* **1994**, 57–59.

13. Nwabunma D.; Kyu T. *Polyolefin blends.* John Wiley & Sons, Inc. Canada. **2007**, 3–4, 59–66.

14. Bershtein, V. A., Egorov, V. M. *Differencialnaya scaniruyushya kalorimetriya v fizikohimii polimerov (Differencial scanning calorimetry in physycochemistry of polymers).* L.: Himiya. **1990**, 10–25.

15. Vunderlih, B. *Fizika makromolekul (Physics of macromolecules).* M.: Mir., **1984**, 124.

16. Godovskiy, U. K. *Teplofizicheskie metody issledovaniya polimerov (Thermophysical methods of polymers investigation).* M.: Himiya, **1976**, 67.

17. Vasserman, A. M.; Kovarskiy, A. L. *Spinovye metki i zondy v fizikohimii polimerov (Spin labels and probes in physicochemistry of polymers).* M.: Nauka, **1986**, 5–40, 124–152.

18. Kuleznev, V. N. *Smesi polimerov. Structura i svoystva (Polymer blends. Structure and properties).* M.: Himiya, **1980**, 78–90.

SECTION 5.10

1. Vishnuvardhan Reddy, S.; Thirumala, M.; Mahmood, S. K. A novel Bacillus sp. accumulating poly (3-hydroxybutyrate-co-3-hydroxyvalerate) from a single carbon substrate, *Journal of Industrial Microbiology and Biotechnology,* **2009**, *36,* 837.

2. Ramsay, B. A.; Ramsay, J. A.; Copper, D. G. Production of poly-β-hydroxyalkanoic acid by Pseudomonas cepacia. *Appl Environ Microbiol.,* **1989**, *55,* 584.

3. Quillaguaman, J.; Delgado, O.; Mattiasson, B.; Kaul, R. H. Poly (β -hydroxybutyrate) production by a moderate halophile, Halomonas boliviensis LC1. *Enzyme Microb Technol.,* **2006**, *38,* 148.

4. Manna, A.; Paul, A. K. Degradation of microbial polyester poly(3-hydroxybutyrate) in environmental samples and in culture. *Biodegradation,* **2000,** *11,* 323.
5. Tertyshnaya Yu. V.; Feofanova, E. S.; Popov, A. A. Biodegradation of the films poly-3-hydroxybutyrate and its blends in soil and water. *Modern problems in biochemical physics.* Nova Science Publ., Inc. **2012,** 117.
6. Dechant, J.; Danz, R.; Kimmer, W.; Schmolke, R. *Ultrarotspektroskopische untersuchungen an polymeren.* Akademie-Verlag. Berlin. **1972,** 472.
7. Esthappan, S. K.; Kuttappah, S. K.; Joseph, R. Effect of titanium dioxide on the thermal ageing of polypropylene. *Polymer Degrad. and Stab.,* **2012,** *97(3),* 615.

CHAPTER 6

QUANTUM-CHEMICAL CALCULATION IN CHEMICAL REACTION

V. A. BABKIN, A. K. HAGHI, and G. E. ZAIKOV

CONTENTS

6.1 QUANTUM-CHEMICAL CALCULATION OF SOME MOLECULES OF STEROLS BY METHOD MNDO

6.1.1 INTRODUCTION

Quantum-chemical calculation of molecules of o-allyloksistyrene, p-allyloksistyrene, trans-izosafrol was done by method MNDO. Optimized by all parameters geometric and electronic structures of these compound was received. The universal factor of acidity was calculated (pKa=32). Molecules of o-allyloksistyrene, p-allyloksistyrene, trans-izosafrol pertain to class of very weak H-acids (pKa>14).

The aim of this work is a study of electronic structure of molecules o-allyloksistyrene, p-allyloksistyrene, trans-izosafrol and theoretical estimation its acid power by quantum-chemical method MNDO. The calculation was done with optimization of all parameters by standard gradient method built-in in PC GAMESS [1]. The calculation was executed in approach the insulated molecule in gas phase. Program MacMolPlt was used for visual presentation of the model of the molecule [3].

6.1.2 METHODICAL PART

Geometric and electronic structures, general and electronic energies of molecules o-allyloksistyrene, p-allyloksistyrene, trans-izosafrol was received by method MNDO and are shown on Figs. 1–3 and in Tables 1–3, respectively. The universal factor of acidity was calculated by formula: pKa = 49, 4–134, 61*$q_{max}H^+$ [2] (where, $q_{max}H^+$ – a maximum positive charge on atom of the hydrogen (by Milliken [1]) R=0. 97, R- a coefficient of correlations, q_{max}^{H+}=+0, 07, +0, 07, +0, 08, respectively pKa=32.

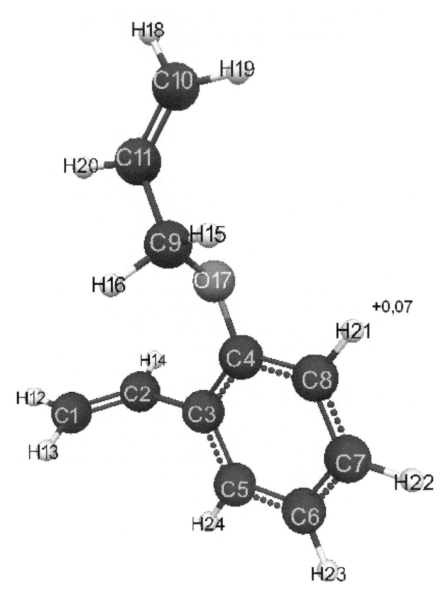

FIGURE 1 Geometric and electronic molecular structure of o-allyloksistyrene ($E_0 =$ −182959 kDg/mol, $E_{el} =$ −958061 kDg/mol).

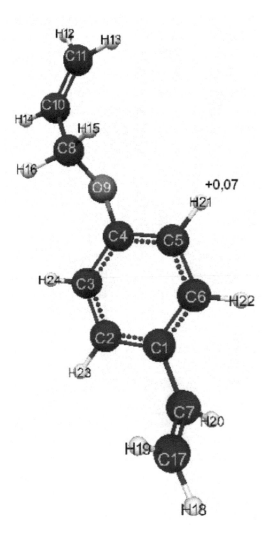

FIGURE 2 Geometric and electronic molecular structure of p-allyloksistyrene. (E_0= −182964 kDg/mol, E_{el}= −928761 kDg/mol).

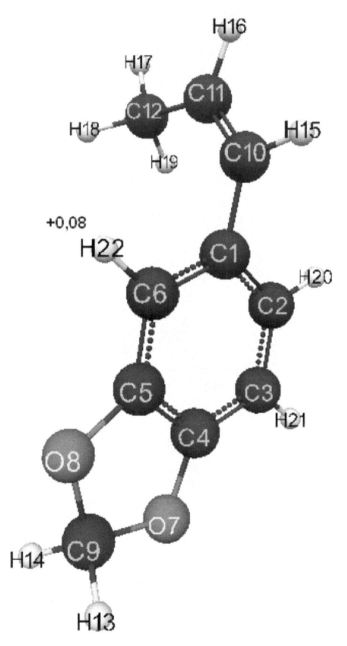

FIGURE 3 Geometric and electronic molecular structure of trans-izosafrol. (E_0= −199066 kDg/mol, E_{el}= −990337 kDg/mol).

TABLE 1 Geometric and electronic structures.

Bond lengths	R, A	Valence corners	Grad	Atom	Charge (by Milliken)
C(1)-C(2)	1.34	C(3)-C(2)-C(1)	126	C(1)	−0.06
C(2)-C(3)	1.48	C(4)-C(3)-C(2)	122	C(2)	−0.05
C(3)-C(4)	1.43	C(8)-C(4)-C(3)	120	C(3)	−0.06
C(3)-C(5)	1.41	C(7)-C(8)-C(4)	120	C(4)	0.11
C(4)-O(17)	1.37	C(4)-C(3)-C(5)	118	C(5)	−0.03
C(5)-C(6)	1.40	C(3)-C(5)-C(6)	121	C(6)	−0.07
C(6)-C(7)	1.40	C(5)-C(6)-C(7)	120	C(7)	−0.04
C(7)-C(8)	1.40	C(6)-C(7)-C(8)	120	C(8)	−0.08
C(8)-C(4)	1.42	C(4)-O(17)-C(9)	120	C(9)	0.22
C(9)-O(17)	1.41	C(9)-C(11)-C(10)	126	C(10)	−0.03
C(10)-C(11)	1.34	O(17)-C(9)-C(11)	108	C(11)	−0.12
C(11)-C(9)	1.51	C(2)-C(1)-H(12)	122	H(12)	0.04
H(12)-C(1)	1.09	C(2)-C(1)-H(13)	124	H(13)	0.04
H(13)-C(1)	1.09	C(1)-C(2)-H(14)	120	H(14)	0.06
H(14)-C(2)	1.10	C(11)-C(9)-H(15)	111	H(15)	−0.01
H(15)-C(9)	1.12	C(11)-C(9)-H(16)	109	H(16)	−0.01
H(16)-C(9)	1.12	C(3)-C(4)-O(17)	120	O(17)	−0.29
H(18)-C(10)	1.10	C(11)-C(10)-H(18)	122	H(18)	0.04
H(19)-C(10)	1.10		124	H(19)	0.04
H(20)-C(11)	1.09	C(11)-C(10)-H(19)	114	H(20)	0.06
H(21)-C(8)	1.09		121	**H(21)**	**0.07**
H(22)-C(7)	1.09	C(9)-C(11)-H(20)	120	H(22)	0.06
H(23)-C(6)	1.09	C(4)-C(8)-H(21)	120	H(23)	0.06
H(24)-C(5)	1.09	C(6)-C(7)-H(22)	120	H(24)	0.06
		C(5)-C(6)-H(23)			
		C(3)-C(5)-H(24)			

TABLE 2 Geometric and electronic structures

Bond lengths	R, A	Valence corners	Grad	Atom	Charge (by Milliken)
C(1)-C(2)	1.41	C(3)-C(2)-C(1)	121	C(1)	–0.06
C(2)-C(3)	1.40	C(4)-C(3)-C(2)	120	C(2)	–0.03
C(3)-C(4)	1.42	C(5)-C(4)-C(3)	120	C(3)	–0.08
C(4)-C(5)	1.42	C(6)-C(5)-C(4)	120	C(4)	0.10
	1.40	C(1)-C(6)-C(5)	121	C(5)	–0.07
C(5)-C(6)	1.41	C(2)-C(1)-C(6)	118	C(6)	–0.03
C(6)-C(1)	1.48	C(2)-C(1)-C(7)	121	C(7)	–0.07
C(7)-C(1)	1.41	C(4)-O(9)-C(8)	119	C(8)	0.22
C(8)-O(9)	1.37	C(3)-C(4)-O(9)	120	O(9)	–0.29
O(9)-C(4)	1.51	O(9)-C(8)-C(10)	108	C(10)	–0.12
C(10)-O(8)	1.34	C(8)-C(10)-C(11)	126	C(11)	–0.03
C(11)-C(10)	1.09	C(10)-C(11)-H(12)	122	H(12)	0.04
C(11)-C(10)	1.09	C(10)-C(11)-H(13)	124	H(13)	0.04
H(12)-C(11)	1.09	C(8)-C(10)-H(14)	114	H(14)	0.06
H(13)-C(11)	1.12	O(9)-C(8)-H(15)	111	H(15)	–0.01
H(14)-C(10)	1.12	O(9)-C(8)-H(16)	111	H(16)	–0.01
H(15)-C(8)	1.34	C(1)-C(7)-C(17)	126	C(17)	–0.05
H(16)-C(8)	1.09	C(7)-C(17)-H(18)	122	H(18)	0.04
H(16)-C(8)	1.09	C(7)-C(17)-H(19)	124	H(19)	0.04
C(17)-C(7)	1.09	C(1)-C(7)-H(20)	114	H(20)	0.05
H(18)-C(17)	1.09	C(4)-C(5)-H(21)	121	**H(21)**	**0.07**
H(19)-C(17)	1.09	C(1)-C(6)-H(22)	120	H(22)	0.06
H(20)-C(7)	1.09	C(1)-C(2)-H(23)	120	H(23)	0.06
H(21)-C(5)	1.09	C(2)-C(3)-H(24)	119	**H(24)**	**0.07**
H(22)-C(6)					
H(23)-C(2)					
H(24)-C(3)					

TABLE 3   Geometric and electronic structures.

Bond lengths	R, A	Valence corners	Grad	Atom	Charge (by Milliken)
C(1)-C(2)	1.41	C(3)-C(2)-C(1)	123	C(1)	−0.05
C(2)-C(3)	1.42	C(4)-C(3)-C(2)	116	C(2)	−0.04
C(3)-C(4)	1.39	C(5)-C(4)-C(3)	121	C(3)	−0.04
C(4)-C(5)	1.44	C(6)-C(5)-C(4)	122	C(4)	0.03
C(4)-O(7)	1.37	O(8)-C(5)-C(4)	108	C(5)	0.02
C(5)-C(6)	1.39	C(1)-C(6)-C(5)	117	C(6)	−0.01
C(6)-C(1)	1.43	C(9)-O(8)-C(5)	108	O(7)	−0.25
O(7)-C(9)	1.42	C(2)-C(1)-C(6)	120	O(8)	−0.25
O(8)-C(5)	1.37	C(5)-C(4)-O(7)	108	C(9)	0.29
C(9)-O(8)	1.42	O(7)-C(9)-O(8)	107	C(10)	−0.04
C(10)-C(1)	1.48	C(4)-O(7)-C(9)	108	C(11)	−0.09
C(11)-C(10)	1.34	C(2)-C(1)-C(10)	120	C(12)	0.06
C(12)-C(11)	1.49	C(1)-C(10)-C(11)	128	H(13)	0.03
H(13)-C(9)	1.12	C(10)-C(11)-C(12)	128	H(14)	0.03
H(14)-C(9)	1.12	O(7)-C(9)-H(13)	110	H(15)	0.05
H(15)-C(10)	1.10	O(7)-C(9)-H(14)	110	H(16)	0.05
H(16)-C(11)	1.09	C(1)-C(10)-H(15)	113	H(17)	0.00
H(17)-C(12)	1.11	C(10)-C(11)-H(16)	118	H(18)	0.00
H(18)-C(12)	1.11	C(11)-C(12)-H(17)	111	H(19)	0.00
H(19)-C(12)	1.11	C(11)-C(12)-H(18)	111	H(20)	0.06
H(20)-C(2)	1.09	C(11)-C(12)-H(19)	112	H(21)	0.08
H(21)-C(3)	1.09	C(1)-C(2)-H(20)	119	H(22)	0.08
H(22)-C(6)	1.09	C(2)-C(3)-H(21)	121		
		C(1)-C(6)-H(22)	121		

Quantum-chemical calculation of molecules o-allyloksistyrene, p-allyloksistyrene, trans-izosafrol by method MNDO was executed for the first time. Optimized geometric and electronic structures of these compound was received. Acid power of molecules o-allyloksistyrene, p-allyloksistyrene, trans-izosafrol was theoretically evaluated (pKa=32). These compound pertain to class of very weak H-acids (pKa>14).

Total Energy (E_0), the maximum charge on the hydrogen atom ($q_{max}H^+$), the universal factor of acidity (pKa)

Molecules of sterols	E_0 (kDg/mol)	$q_{max}H^+$	pKa
o-allyloksistyrene	−182964	+0.07	32
p-allyloksistyrene	−182959	+0.07	32
trans-izosafrol	−199066	+0.08	32

6.2 QUANTUM-CHEMICAL CALCULATION STUDIES THE MECHANISM OF PROTONATION OF ISOBUTYLENE BY METHOD MNDO

6.2.1 INTRODUCTION

For the first time it is researched of classical quantum chemical method MNDO of modeling mechanism protonizataion of isobutylene—monomer of cationic polymerization. Showing, that he considerate some self-usual mechanism connection proton to olefin corresponding Morkovnikov's rule. Reaction exothermic and carry without a barrier character. Prize energy in result of reaction –499 kDg/mol. Theoretical estimation of affinity proton to isobutylene –577 kDg/mol

According to modern conceptions about the mechanism of initialization of cationic polymerization of isobutylene the true catalyst of this reaction is Lewis' aqua acids of the type $AlCl_3 \cdot H_2O$, $AlCl_2C_2H_5 \cdot H_2O$, $BF_3 \cdot H_2O$ and others (i.e., there are always admixtures of water in the system) out of which due to complex coordinated interactions initiating particle $H^{+\delta}$ forms and which is in turn according to Morkovnikov's rule attacks the most hydrogenated atom of carbon C_α [1–3]. Studying the mechanism of isobutylene protonation is the first step in studying of the mechanism of elementary act of initialization of cationic polymerization of this monomer.

In this connection, the aim of the present work is quantum-chemical study of the mechanism of isobutylene protonation by classical semi-empirical method MNDO.

6.2.2 METHODOLOGICAL PART

For studying the mechanism of protonation the classical quantum-chemical method MNDO was chosen with geometry optimization of all the parameters by gradient method built in PC GAMESS [4], as this method specifically parameterized for the best reproduction of the energy characteristics of molecular systems and it is an important factor in analysis of the mechanisms of cation processes. Calculations were done in the approximation of an isolated molecular in a gas phase. In the system H^+ ... C_4H_8 (isobutylene) 13 atoms, $M=2S+1=1$ (where S—the total spin of all electrons of study system is zero (all electrons are paired), M—multiplicity), the total charge of a molecular system Σ qc $= 1$. It was done the calculation of potential energy of proton interaction with isobutylene for studying the mechanism of isobutylene protonation by the following way. The distances from proton H_1 up to C_2 ($R_{H_1C_2}$) and from H_1 up to C_3 ($R_{H_1C_3}$) were chosen as reaction coordinates. The original meanings of $R_{H_1C_2}$ and $R_{H_1C_3}$ were taken as 0.31 nm. Further, changing meanings of $R_{H_1C_2}$ with 0.02 nm step quantum-chemical calculation of molecular system was done changing $R_{H_1C_3}$ meanings with the same 0.02 nm step. According to the received data of energy meanings equipotential surface of proton interaction with isobutylene was built along the reaction coordinates (Fig. 4). The initial model of the proton attack of isobutylene molecule is shown in Fig. 1 of Section 6.2. The affinity of the proton to isobutylene was calculated by the formula:

$$E_{cp} = E_0 (H^+ ... C_4H_9) - (E_0 (H^+) + E_0 (C_4H_9)) \tag{1}$$

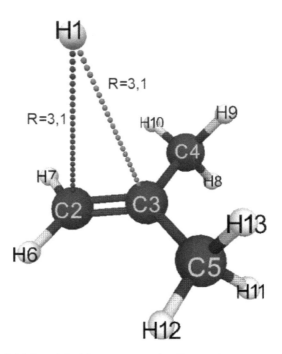

FIGURE 1   The initial model of the proton attack of isobutylene.

The famous program MacMolPlt [5] was used for the visual representation of molecules' models.

6.2.3. THE RESULTS OF CALCULATIONS

The energies of the molecular system H^+ ... C_4H_8 along the reaction coordinates $R_{H_1 C_2}$ and $R_{H_1 C_3}$ were shown in Table 1 of Section 6.2. The final structure of the formed carbcation after proton H_1 attack of α-carbon atom of isobutylene (C_2) and a break of the double bond of isobutylene is introduced in Fig. 2 of Section 6.2. The final structure of the formed carbcation after proton H_1 attack of β-carbon atom of isobutylene (C_3) and a break of the double bond $C_2 = C_3$ is shown in Fig. 3 of Section 6.2. The charges on atoms of the final structures of formed carbcations are introduced in Table 2 of Section 6.2. The changing of the total energy under protonization of isobutylene is shown in Table 1 of Section 6.2, it is seen that the initiating particle $H^{+\delta}$ along the reaction coordinates $R_{H_1 C_2}$ and $R_{H_1 C_3}$ negative value of the total energy of the system $H^+ \ldots C_4 H_8 (E_0)$ is steadily increasing up to the complete formation of carbcation (*see* Table 1 of Section 6.2) on the whole way of proton movement having barrier-free nature as well as under the attack of proton on α- and β-carbon atoms of isobutylene. However, the final structure of the attack of proton on α carbon atom per 90 kJ/mol is energetically more favorable than the final structure of attack of proton β-carbon atom, which is in full accordance with the classical rule of Morkovnikov. As a result of this reaction the energy gain under attack on α-carbon atom is 499 kJ/mol and under attack on β-carbon atom is 409 kJ/mol. The value of affinity of proton to isobutylene was calculated according to the Eq. (1) Ecp = 577 kJ/mol. Moreover, the analysis of the results of quantum-chemical calculations and changing of the bond lengths and valence angles along the reaction coordinate in both cases under the attack of proton on α- as well as on β-carbon atoms of isobutylene testify that the mechanism of protonation of cationic polymerization of isobutylene proceeds according to the classical scheme of joining proton to the double bond of monomer.

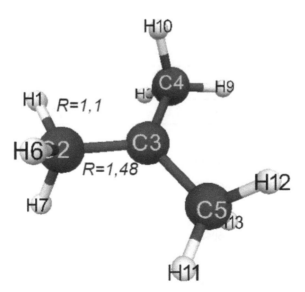

FIGURE 2   The final structure of the formed carbcation after proton H_1 attack of α-carbon atom of isobutylene (C_2).

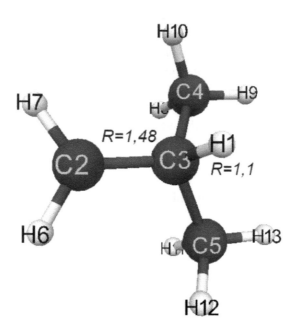

FIGURE 3   The final structure of the formed carbcation after proton H_1 attack of β-carbon atom of isobutylene (C_3).

TABLE 1   Values of energy of the molecular system $H^+ \ldots C_4H_9$. E_o (in kJ/mol) along the reaction coordinates $R_{H_1 C_2}$ and $R_{H_1 C_3}$ (in A)

$R_{H_1 C_3}$	$R_{H_1 C_2}$				
	3, 1	**2, 9**	**2, 7**	**2, 5**	**2, 3**
3, 1	−60369	−60378	−60392	−60411	−60438
2, 9	−60374	−60384	−60397	−60416	−60445
2, 7	−60382	−60392	−60404	−60424	−60452
2, 5	−60393	−60403	−60416	−60434	−60461
2, 3	−60408	−60421	−60435	−60451	−60475
2, 1	−60429	−60448	−60463	−60478	−60499
1, 9	−60650	−60483	−60505	−60521	−60538

TABLE 1 *(Continued)*

$R_{H_1 C_3}$	$R_{H_1 C_2}$				
	3, 1	**2, 9**	**2, 7**	**2, 5**	**2, 3**
1, 7	−60550	−60523	−60560	−60581	−60596
1, 5	−60380	−60548	−60617	−60653	−60672
1, 3	−60174	−60546	−60644	−60711	−60743
1, 1	−	−60502	−60602	−60701	−60763

$R_{H_1 C_3}$	$R_{H_1 C_2}$					
	2, 1	**1, 9**	**1, 7**	**1, 5**	**1, 3**	**1, 1**
3, 1	−60472	−60512	−60542	−60553	−60562	−60543
2, 9	−60484	−60536	−60596	−60646	−60666	−60627
2, 7	−60493	−60550	−60623	−60702	−60756	−60729
2, 5	−60500	−60558	−60636	−60728	−60807	−60817
2, 3	−60511	−60566	−60641	−60736	−60827	−60864
2, 1	−60530	−60577	−60646	−60736	−60828	−60868
1, 9	−60562	−60599	−60655	−60733	−60815	−60856
1, 7	−60613	−60637	−60676	−60733	−60796	−60823
1, 5	−60682	−60693	−60711	−60741	−60775	−60775
1, 3	−60754	−60755	−60752	−60751	−60746	−60706
1, 1	−60778	−60777	−60759	−60729	−60683	−60593

TABLE 2   Charges of atoms of the final models of formed carbcations.

Atom	Charges on atoms of formed carbcation	
	After the attack of H_1 proton a-carbon atom of isobutylene (C_2)	After the attack of H_1 proton β-carbon atom of isobutylene (C_3)
H(1)	+0.06	+0.15
C(2)	–0.06	+0.53
C(3)	+0.36	–0.21
C(4)	–0.06	+0.03
C(5)	–0.07	+0.03
H(6)	+0.11	+0.11
H(7)	+0.11	+0.11
H(8)	+0.10	+0.03
H(9)	+0.07	+0.06
H(10)	+0.10	+0.03
H(11)	+0.07	+0.03
H(12)	+0.10	+0.03
H(13)	+0.10	+0.06

Thus we have studied the mechanism of protonation of isobutylene by quantum-chemical method MNDO for the first time. It is shown that this mechanism is a usual reaction of joining proton to double bond of olefin. The reaction is exothermic and has a barrier-free nature. It is energetically favorable for the reaction to follow the classical scheme in accordance with Markovnikov's rule.

6.3   GEOMETRICAL AND ELECTRONIC STRUCTURE OF MOLECULE 2-(2, 2-DIMETHOXIETHYLSULFANIL)-5-IZOPROPYL-6-METHYL PYRIMIDINE-4(3H)-OH METHOD MNDO

6.3.1   INTRODUCTION

For the first time it is executed quantum chemical calculation of a molecule of 2-(2, 2-dimethoxiethylsulfanil)-5-izopropyl-6-methyl pyrimidine-4(3H)-OH method MNDO with optimization of geometry on all parameters. The optimized geometrical and electronic structure of this connection is received. Acid force of 2-(2, 2-dimethoxiethylsulfanil)-5-izopropyl-6-methyl pyrimidine-4(3H)-OH is theoretically appreciated. It is established, than it to relate to a class of weak H-acids (pKa=+11, where pKa-universal index of acidity).

The Aim of this work is a study of electronic structure of molecule 2-(2, 2-dimethoxiethylsulfanil)-5-izopropyl-6-methyl pyrimidine-4(3H)-OH and theoretical estimation its acid power by quantum-chemical method MNDO. The calculation was done with optimization of all parameters by standard gradient method built-in in PC GAMESS [1]. The calculation was executed in approach the insulated molecule in gas phase. Program MacMolPlt was used for visual presentation of the model of the molecule [2].

6.3.2 METHODICAL PART

Geometric and electronic structures, general and electronic energies of molecule 2-(2, 2-dimethoxiethylsulfanil)-5-izopropyl-6-methyl pyrimidine-4(3H)-OH was received by method MNDO and are shown on Fig. 1 of Section 6.3 and in Table 1 of Section 6.3. The applying factor of acidity was calculated by formula: $pKa = 49, 4-134, 61*q_{max}^{H+}$ [3] (where, q_{max}^{H+} – a maximum positive charge on atom of the hydrogen $q_{max}^{H+} =$ +0. 21 (for 2-(2, 2-dimethoxiethylsulfanil)-5-izopropyl-6-methyl pyrimidine-4(3H)-OH q_{max}^{H+} alike table) successfully use, for example in the works [4–34]) pKa=11.

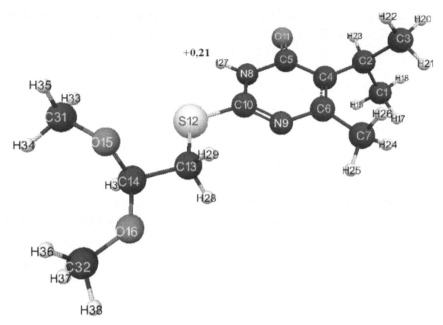

FIGURE 1 Geometric and electronic molecule structure of 2-(2, 2-dimethoxiethylsulfanil)-5-izopropyl-6-methyl pyrimidine-4(3H)-OH.

$$(E_0 = -330721 \text{ kDg/mol}, E_{el} = -2042408 \text{ kDg/mol}).$$

Optimized bond lengths, valence corners and charges on atoms of the molecule of 2-(2, 2-dimethoxiethylsulfanil)-5-izopropyl-6-methyl pyrimidine-4(3H)-OH.

TABLE 1 Geometric and electronic molecule structure.

Bond lengths	R, A	Valence corners	Grad	Atom	Charge (by Milliken)
C(1)-C(2)	1.54	C(1)-C(2)-C(3)	113	C(1)	0.04
C(2)-C(4)	1.53	C(1)-C(2)-C(4)	113	C(2)	−0.01
C(3)-C(2)	1.54	C(2)-C(4)-C(5)	116	C(3)	0.04
C(4)-C(5)	1.49	C(2)-C(4)-C(6)	127	C(4)	−0.20
C(5)-N(8)	1.43	C(4)-C(6)-C(7)	128	C(5)	0.39

TABLE 1 *(Continued)*

C(6)-C(4)	1.39	C(4)-C(5)-N(8)	115	C(6)	0.11
C(7)-C(6)	1.51	C(4)-C(6)-N(9)	122	C(7)	0.08
N(8)-C(10)	1.38	C(5)-N(8)-C(10)	122	N(8)	−0.33
N(9)-C(6)	1.40	C(4)-C(5)-O(11)	129	N(9)	−0.34
C(10)-N(9)	1.32	N(8)-C(10)-S(12)	117	C(10)	0.13
O(11)-C(5	1.23	C(10)-S(12)-C(13)	111	O(11)	−0.35
S(12)-C(10)	1.69	S(12)-C(13)-C(14)	110	S(12)	0.14
C(13)-S(12)	1.74	C(13)-C(14)-O(15)	11	C(13)	−0.07
C(14)-C(13)	1.57	C(13)-C(14)-O(16)	107	C(14)	0.33
O(15)-C(14)	1.41	C(2)-C(1)-H(17)	113	O(15)	−0.37
O(16)-C(14)	1.40	C(2)-C(1)-H(18)	110	O(16)	−0.36
H(17)-C(1)	1.11	C(2)-C(1)-H(19)	111	H(17)	−0.01
H(18)-C(1)	1.11	C(2)-C(3)-H(20)	110	H(18)	−0.00
H(19)-C(1)	1.11	C(2)-C(3)-H(21)	113	H(19)	−0.00
H(20)-C(3)	1.11	C(2)-C(3)-H(22)	111	H(20)	−0.00
H(21)-C(3)	1.11	C(1)-C(2)-H(23)	104	H(21)	−0.01
H(22)-C(3)	1.11	C(6)-C(7)-H(24)	112	H(22)	−0.00
H(23)-C(2)	1.12	C(6)-C(7)-H(25)	112	H(23)	0.03
H(24)-C(7)	1.11	C(6)-C(7)-H(26)	110	H(24)	0.00
H(25)-C(7)	1.11	C(5)-N(8)-H(27)	119	H(25)	0.02
H(26)-C(7)	1.11	S(12)-C(13)-H(28)	109	H(26)	0.01
H(27)-N(8)	1.00	S(12)-C(13)-H(29)	111	H(27)	+0.21
H(28)-C(13)	1.11	C(13)-C(14)-H(30)	109	H(28)	0.05
H(29)-C(13)	1.11	C(14)-O(15)-C(31)	122	H(29)	0.06
H(30)-C(14)	1.13	C(14)-O(16)-C(32)	123	H(30)	0.01
C(31)-O(15)	1.40	O(15)-C(31)-H(33)	113	C(31)	0.22
C(32)-O(16)	1.40	O(15)-C(31)-H(34)	113	C(32)	0.22
H(33)-C(31)	1.12	O(15)-C(31)-H(35)	107	H(33)	−0.02
H(34)-C(31)	1.12	O(16)-C(32)-H(36)	113	H(34)	−0.03
H(35)-C(31)	1.12	O(16)-C(32)-H(37)	113	H(35)	0.02
H(36)-C(32)	1.12	O(16)-C(32)-H(38)	107	H(36)	−0.01
H(37)-C(32)	1.12			H(37)	−0.03
H(38)-C(32)	1.12			H(38)	0.02

Quantum-chemical   calculation   of   molecule   2-(2, 2-dimethoxiethylsulfanil)-5-izopropyl-6-methyl   pyrimidine-4(3H)-OH by method MNDO was executed for the first time. Optimized geometric and electronic structures of these compound was received. Acid power of molecule 2-(2, 2-dimethoxiethylsulfanil)-5-izopropyl-6-methyl pyrimidine-4(3H)-OH was theoretically evaluated (pKa=11). These compound pertain to class of weak H-acids (pKa>9).

6.4   GEOMETRICAL AND ELECTRONIC STRUCRURE OF SOME MOLECULES AROMATIC OLEFINS

6.4.1   INTRODUCTION

Quantum-chemical   calculation   of   molecules   of   α-cyclopropyl-p-izopropylstyrene,   α-cyclopropyl-2, 4-dimethylstyrene,   α-cyclopropyl-p-ftorstyrene was done by method MNDO. Optimized by all parameters geometric and electronic structures of these compound was received. The universal factor of acidity was calculated (pKa=32). Molecules of α-cyclopropyl-p-izopropylstyrene, α-cyclopropyl-2, 4-dimethylstyrene, α-cyclopropyl-p-ftorstyrene pertain to class of very weak H-acids (pKa>14). The aim of this work is a study of electronic structure of molecules a—cyclopropyl-p-izopropylstyrene, α-cyclopropyl-2, 4-dimethylstyrene, α-cyclopropyl-p-ftorstyrene [1] and theoretical estimation its acid power by quantum-chemical method MNDO. The calculation was done with optimization of all parameters by standard gradient method built-in in PC GAMESS [2]. The calculation was executed in approach the insulated molecule in gas phase. Program MacMol-Plt was used for visual presentation of the model of the molecule [3].

6.4.2   METHODICAL PART

Geometric and electronic structures, general and electronic energies of molecules α-cyclopropyl-p-izopropylstyrene, α-cyclopropyl-2, 4-dimethylstyrene, α-cyclopropyl-p-ftorstyrene was received by method MNDO and are shown on Figs. 1–3 of Section 6.4, and in Tables 1–3 of Section

6.4, respectively. The universal factor of acidity was calculated by formula: pKa = 49, 4–134, 61*$q_{max}H^+$ [4], which used with success, for example, in [5–35] (where, $q_{max}H^+$ – a maximum positive charge on atom of the hydrogen (by Milliken [1]) R=0. 97, R– a coefficient of correlations, q_{max}^{H+}=+0.06, +0.06, +0.08, respectively. pKa=30–33.

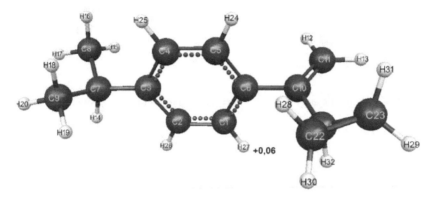

FIGURE 1 Geometric and electronic molecular structure of α-cyclopropyl-p-izopropylstyrene.

(E0= –196,875 kDg/mol, Eel= –1,215,375 kDg/mol).

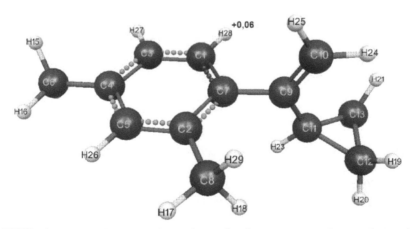

FIGURE 2 Geometric and electronic molecular structure of α-cyclopropyl-2, 4-dimethylstyrene.

(E0= –181,125 kDg/mol, Eel= –1,084,125 kDg/mol).

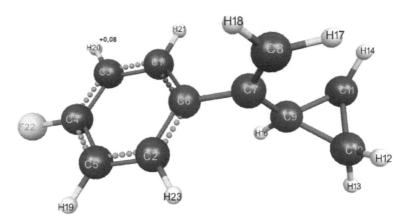

FIGURE 3   Geometric and electronic molecular structure of α-cyclopropyl-p-ftorstyrene.

(E0= –196,875 kDg/mol, Eel= –1,155,000 kDg/mol).

Optimized bond lengths, valence corners and charges on atoms of the molecule of α-cyclopropyl-p-izopropylstyrene.

TABLE 1   Geometric and electronic molecular structure.

Bond lengths	R, A	Valence corners	Grad	Atom	Charge (by Milliken)
C(1)-C(6)	1.40	C(6)-C(1)-C(2)	121	C(1)	–0.04
C(2)-C(1)	1.39	C(1)-C(2)-C(3)	121	C(2)	–0.05
C(3)-C(2)	1.40	C(2)-C(3)-C(4)	119	C(3)	–0.07
C(3)-C(7)	1.50	C(3)-C(4)-C(5)	121	C(4)	–0.05
C(4)-C(3)	1.40	C(4)-C(5)-C(6)	121	C(5)	–0.05
C(5)-C(4)	1.39	C(2)-C(3)-C(7)	120	C(6)	–0.03
C(6)-C(5)	1.40	C(3)-C(7)-C(8)	111	C(7)	–0.02
C(6)-C(10)	1.47	C(3)-C(7)-C(9)	111	C(8)	+0.04
C(7)-C(8)	1.52	C(8)-C(7)-C(9)	110	C(9)	+0.04
C(7)-C(9)	1.52	C(1)-C(6)-C(10)	121	C(10)	–0.06
C(10)-C(11)	1.34	C(6)-C(10)-C(11)	122	C(11)	–0.03
C(10)-C(21)	1.47	C(10)-C(11)-H(12)	122	H(12)	+0.04
H(12)-C(11)	1.10	C(10)-C(11)-H(13)	122	H(13)	+0.04
H(13)-C(11)	1.10	C(3)-C(7)-H(14)	108	H(14)	+0.01

TABLE 1   *(Continued)*

H(14)-C(7)	1.13	C(7)-C(8)-H(15)	110	H(15)	0.00
H(15)-C(8)	1.12	C(7)-C(8)-H(16)	111	H(16)	0.00
H(16)-C(8)	1.12	C(7)-C(8)-H(17)	110	H(17)	–0.01
H(17)-C(8)	1.12	C(7)-C(9)-H(18)	111	H(18)	0.00
H(18)-C(9)	1.12	C(7)-C(9)-H(19)	110	H(19)	0.00
H(19)-C(9)	1.12	C(7)-C(9)-H(20)	110	H(20)	–0.01
H(20)-C(9)	1.12	C(6)-C(10)-C(21)	116	C(21)	–0.07
C(21)-C(22)	1.51	C(11)-C(10)-C(21)	123	C(22)	–0.05
C(22)-C(23)	1.50	C(22)-C(23)-C(21)	60	C(23)	–0.06
C(23)-C(21)	1.51	C(10)-C(21)-C(22)	121	H(24)	+0.06
H(24)-C(5)	1.10	C(21)-C(23)-C(22)	60	H(25)	+0.06
H(25)-C(4)	1.10	C(21)-C(22)-C(23)	60	H(26)	+0.06
H(26)-C(2)	1.10	C(22)-C(21)-C(23)	60	**H(27)**	**+0.06**
H(27)-C(1)	1.10	C(4)-C(5)-H(24)	120	H(28)	+0.04
H(28)-C(22)	1.10	C(3)-C(4)-H(25)	120	H(29)	+0.04
H(29)-C(23)	1.10	C(1)-C(2)-H(26)	120	H(30)	+0.04
H(30)-C(22)	1.10	C(2)-C(1)-H(27)	120	H(31)	+0.04
H(31)-C(23)	1.10	C(21)-C(22)-H(28)	119	H(32)	+0.05
H(32)-C(21)	1.11	C(21)-C(23)-H(29)	118		
		C(21)-C(22)-H(30)	119		
		C(21)-C(23)-H(31)	120		
		C(10)-C(21)-H(32)	111		

Optimized bond lengths, valence corners and charges on atoms of the molecule of α-cyclopropyl-2, 4-dimethylstyrene.

TABLE 2 Geometric and electronic molecular structure.

Bond lengths	R, A	Valence corners	Grad	Atom	Charge (by Milliken)
C(1)-C(7)	1.42	C(1)-C(7)-C(2)	119	C(1)	−0.0508
C(2)-C(5)	1.42	C(7)-C(1)-C(3)	121	C(2)	−0.0835
C(3)-C(1)	1.40	C(1)-C(3)-C(4)	121	C(3)	−0.0412
C(4)-C(3)	1.41	C(2)-C(5)-C(4)	123	C(4)	−0.1009
C(5)-C(4)	1.41	C(3)-C(4)-C(5)	118	C(5)	−0.0280
C(6)-C(4)	1.51	C(3)-C(4)-C(6)	121	C(6)	0.0814
C(7)-C(2)	1.42	C(5)-C(2)-C(7)	119	C(7)	−0.0187
C(7)-C(9)	1.50	C(5)-C(2)-C(8)	119	C(8)	0.0807
C(10)-C(9)	1.35	C(1)-C(7)-C(9)	118	C(9)	−0.0545
C(11)-C(9)	1.50	C(7)-C(9)-C(10)	120	C(10)	−0.0416
C(12)-C(11)	1.54	C(7)-C(9)-C(11)	115	C(11)	−0.0617
C(13)-C(12)	1.52	C(9)-C(11)-C(12)	125	C(12)	−0.0561
C(13)-C(11)	1.54	C(9)-C(11)-C(13)	125	C(13)	−0.0568
H(14)-C(6)	1.11	C(4)-C(6)-H(14)	111	H(14)	−0.0028
H(15)-C(6)	1.11	C(4)-C(6)-H(15)	111	H(15)	−0.0027
H(16)-C(6)	1.11	C(4)-C(6)-H(16)	113	H(16)	−0.0050
H(17)-C(8)	1.11	C(2)-C(8)-H(17)	112	H(17)	−0.0072
H(18)-C(8)	1.11	C(2)-C(8)-H(18)	111	H(18)	−0.0002
H(19)-C(12)	1.10	C(11)-C(12)-H(19)	121	H(19)	0.0389
H(20)-C(12)	1.10	C(11)-C(12)-H(20)	118	H(20)	0.0368
H(21)-C(13)	1.10	C(11)-C(13)-H(21)	121	H(21)	0.0387
H(22)-C(13)	1.10	C(11)-C(13)-H(22)	118	H(22)	0.0370
H(23)-C(11)	1.10	C(9)-C(11)-H(23)	111	H(23)	0.0451
H(24)-C(10)	1.09	C(9)-C(10)-H(24)	124	H(24)	0.0394
H(25)-C(10)	1.09	C(9)-C(10)-H(25)	123	H(25)	0.0425
H(26)-C(5)	1.09	C(2)-C(5)-H(26) C(1)-	119	H(26)	0.0550
H(27)-C(3)	1.09	C(3)-H(27)	119	H(27)	0.0581
H(28)-C(1)	1.09	C(3)-C(1)-H(28)	119	**H(28)**	**0.0600**
H(29)-C(8)	1.11	C(2)-C(8)-H(29)	111	H(29)	−0.0019

Optimized bond lengths, valence corners and charges on atoms of the molecule of α-cyclopropyl-p-ftorstyrene.

TABLE 3   Geometric and electronic molecular structure.

Bond lengths	R, A	Valence corners	Grad	Atom	Charge (by Milliken)
C(1)-C(3)	1.40	C(1)-C(6)-C(2)	119	C(1)	−0.02
C(2)-C(6)	1.41	C(4)-C(5)-C(2)	120	C(2)	−0.02
C(3)-C(4)	1.42	C(5)-C(4)-C(3)	120	C(3)	−0.09
C(4)-C(5)	1.42	C(1)-C(3)-C(4)	120	C(4)	+0.15
C(5)-C(2)	1.40	C(6)-C(2)-C(5)	121	C(5)	−0.09
C(6)-C(7)	1.49	C(3)-C(1)-C(6)	121	C(6)	−0.06
C(7)-C(9)	1.50	C(1)-C(6)-C(7)	121	C(7)	−0.06
C(8)-C(7)	1.35	C(6)-C(7)-C(8)	120	C(8)	−0.04
C(9)-C(10)	1.54	C(6)-C(7)-C(9)	115	C(9)	−0.07
C(10)-C(11)	1.52	C(7)-C(9)-C(10)	124	C(10)	−0.06
C(11)-C(9)	1.54	C(7)-C(9)-C(11)	125	C(11)	−0.06
H(12)-C(10)	1.10	C(9)-C(10)-H(12)	121	H(12)	+0.04
H(13)-C(10)	1.10	C(9)-C(10)-H(13)	118	H(13)	+0.04
H(14)-C(11)	1.10	C(9)-C(11)-H(14)	121	H(14)	+0.04
H(15)-C(11)	1.10	C(9)-C(11)-H(15)	118	H(15)	+0.04
H(16)-C(9)	1.10	C(7)-C(9)-H(16)	111	H(16)	+0.04
H(17)-C(8)	1.09	C(7)-C(8)-H(17)	124	H(17)	+0.04
H(18)-C(8)	1.09	C(7)-C(8)-H(18)	123	H(18)	+0.04
H(19)-C(5)	1.09	C(2)-C(5)-H(19)	120	H(19)	+0.08
H(20)-C(3)	1.09	C(1)-C(3)-H(20)	120	**H(20)**	**+0.08**
H(21)-C(1)	1.09	C(3)-C(1)-H(21)	119	H(21)	+0.07
F(22)-C(4)	1.33	C(3)-C(4)-F(22)	120	F(22)	−0.18
H(23)-C(2)	1.09	C(5)-C(2)-H(23)	119	H(23)	+0.07

Quantum-chemical calculation of molecules α-cyclopropyl-p-izopropylstyrene, α-cyclopropyl-2, 4-dimethylstyrene, α-cyclopropyl-p-ftorstyrene by method MNDO was executed for the first time. Optimized geometric and electronic structures of these compound was received. Acid power of molecules o-allyloksistyrene, p-allyloksistyrene, trans-izosafrol was theoretically evaluated (pKa=30–33). These compound pertain to class of very weak H-acids (pKa>14) (Table 4 of Section 6.4).

TABLE 4   General and energies (E_0, maximum positive charge on atom of the hydrogen ($q_{max}H^+$), universal factor of acidity (pKa).

Molecules of aromatic olefins	E_0	$q_{max}H^+$	pKa
α-cyclopropyl-p-izopropylstyrene.	−196875	+0.06	33
α-cyclopropyl-2, 4-dimethylstyrene	−181125	+0.06	33
α-cyclopropyl-p-ftorstyrene.	−196875	+0.08	30

KEYWORDS

- **MacMolPlt**
- **MNDO**
- **Morkovnikov's rule**
- **Quantum-chemical calculation**

REFERENCES

SECTION 6.1

1. Shmid, M. W.; Baldrosge, K. K.; Elbert, J. A.; Gordon, M. S.; Enseh, J. H.; Koseki, S.; Matsvnaga, N. K.; Nguyen, A.; Su S. J. et al. *J. Comput. CHEM.* **1993,** *14*, 1347–1363.
2. Babkin, V. A.; Fedunov, R. G.; Minsker, K. S. et al. *Oxidation communication,* **2002,** *25(1),* 21–47.
3. Bode, B. M. and Gordon, M. S. *J. Mol. Graphics Mod.,* **1998,** *16*, 133–138.

SECTION 6.2

1. Kennedy, J. Cationic polymerization of olefins. *J. Kennedy. M.,* **1978,** 431.
2. Sangalov, Y. A. Polymers and copolymers of isobutylene. Sangalov, Y. A.; Minsker Ufa, K. S., **2001,** 381.
3. Babkin, V. A.; Zaikov, G. E., Minsker, K. S. *The quantum-chemical aspect of cationic polymerization of olefins,* **1996,** Ufa, 182.

4. Shmidt, M. W. *J. Comput. Chem.* Shmidt, M. W.; Gordon, M. S. [and another]. – **1993**, *14*, 1347–1363.
5. Bode, B. M. *J. Mol. Graphics Mod.* Bode, B. M.; Gordon, M. S. – **1998**, *6*, 133–138.

SECTION 6.3

1. Shmidt, M. W.; Baldrosge, K. K.; Elbert, J. A.; Gordon, M. S.; Enseh, J. H.; Koseki, S.; Matsvnaga, N.; Nguyen, K. A.; Su, S. J. et al. *J. Comput. Chem.* **1993**, *14*, 1347–1363.
2. Babkin, V. A.,; Fedunov, R. G.; Minsker, K. S. et al. *Oxidation communication,* **2002**, *25(1)*, 21–47.
3. Bode, B. M.; Gordon, M. S. *J. Mol. Graphics Mod.,* **1998**, *16*, 133–138.
4. Babkin, V. A.; Dmitriev, V. Yu.; Zaikov, G. E. Quantum chemical calculation of molecule hexene-1 by method MNDO. In: *Quantum chemical calculation of unique molecular system.* **2010**, *I.* Publisher VolSU, c. Volgograd, 93–95.
5. Babkin, V. A.; Dmitriev, V. Yu.; Zaikov, G. E. Quantum chemical calculation of molecule heptene-1 by method MNDO. In: *Quantum chemical calculation of unique molecular system.* **2010**, *I.* Publisher VolSU, c. Volgograd, 95–97.
6. Babkin, V. A.; Dmitriev, V. Yu.; Zaikov, G. E. Quantum chemical calculation of molecule decene-1 by method MNDO. In: *Quantum chemical calculation of unique molecular system.* **2010**, *I.* Publisher VolSU, c. Volgograd, 97–99.
7. Babkin, V. A.; Dmitriev, V. Yu.; Zaikov, G. E. Quantum chemical calculation of molecule nonene-1 by method MNDO. In: *Quantum chemical calculation of unique molecular system.* **2010**, *I.* Publisher VolSU, c. Volgograd, 99–102.
8. Babkin, V. A.; Andreev, D. S. Quantum chemical calculation of molecule isobutylene by method MNDO. In: *Quantum chemical calculation of unique molecular system.* **2010**, *I.* Publisher VolSU, c. Volgograd, 176–177.
9. Babkin, V. A.; Andreev, D. S. Quantum chemical calculation of molecule 2-methyl-butene-1 by method MNDO. In: *Quantum chemical calculation of unique molecular system.* **2010**, *I.* Publisher VolSU, c. Volgograd, 177–179.
10. Babkin, V. A.; Andreev, D. S. Quantum chemical calculation of molecule 2-methyl-butene-2 by method MNDO. In: *Quantum chemical calculation of unique molecular system.* **2010**, *I.* Publisher VolSU, c. Volgograd, 179–180.
11. Babkin, V. A.; Andreev, D. S. Quantum chemical calculation of molecule 2-methyl-pentene-1 by method MNDO. In: *Quantum chemical calculation of unique molecular system.* **2010**, *I.* Publisher VolSU, c. Volgograd, 181–182.
12. Babkin, V. A.; Dmitriev, V. Yu.; Zaikov, G. E. Quantum chemical calculation of molecule butene-1 by method MNDO. In: *Quantum chemical calculation of unique molecular system.* **2010**, *I.* Publisher VolSU, c. Volgograd, 89–90.
13. Babkin, V. A.; Dmitriev, V. Yu.; Zaikov, G. E. Quantum chemical calculation of molecule hexene-1 by method MNDO. In: *Quantum chemical calculation of unique molecular system.* **2010**, *I.* Publisher VolSU, c. Volgograd, 93–95.
14. Babkin, V. A.; Dmitriev, V. Yu.; Zaikov, G. E. Quantum chemical calculation of molecule octene-1 by method MNDO. In: *Quantum chemical calculation of unique molecular system.* **2010**, *I.* Publisher VolSU, c. Volgograd, 103–105.

15. Babkin, V. A.; Dmitriev, V. Yu.; Zaikov, G. E. Quantum chemical calculation of molecule pentene-1 by method MNDO. In: *Quantum chemical calculation of unique molecular system.* **2010**, *I.* Publisher VolSU, c. Volgograd, 105–107.

16. Babkin, V. A.; Dmitriev, V. Yu.; Zaikov, G. E. Quantum chemical calculation of molecule propene-1 by method MNDO. In: *Quantum chemical calculation of unique molecular system. I.* Publisher VolSU, c. Volgograd, **2010**, 107–108.

17. Babkin, V. A.; Dmitriev, V. Yu.; Zaikov, G. E. Quantum chemical calculation of molecule ethylene-1 by method MNDO. In: *Quantum chemical calculation of unique molecular system. I.* Publisher VolSU, c. Volgograd, **2010**, 108–109.

18. Babkin, V. A.; Andreev, D. S. Quantum chemical calculation of molecule butadien-1, 3 by method MNDO. In: *Quantum chemical calculation of unique molecular system. I.* Publisher VolSU, c. Volgograd, **2010**, 235–236.

19. Babkin, V. A.; Andreev, D. S. Quantum chemical calculation of molecule 2-methylbutadien-1, 3 by method MNDO. In: *Quantum chemical calculation of unique molecular system. I.* Publisher VolSU, c. Volgograd, **2010**, 236–238.

20. Babkin, V. A.; Andreev, D. S. Quantum chemical calculation of molecule 2, 3-dimethylbutadien-1, 3 by method MNDO. In: *Quantum chemical calculation of unique molecular system. I.* Publisher VolSU, c. Volgograd, **2010**, 238–239.

21. Babkin, V. A.; Andreev, D. S. Quantum chemical calculation of molecule pentadien-1, 3 by method MNDO. In: *Quantum chemical calculation of unique molecular system. I.* Publisher VolSU, c. Volgograd, **2010**, 240–241.

22. Babkin, V. A.; Andreev, D. S. Quantum chemical calculation of molecule trans-trans-hexadien-2, 4 by method MNDO. In: *Quantum chemical calculation of unique molecular system. I.* Publisher VolSU, c. Volgograd, **2010**, 241–243.

23. Babkin, V. A.; Andreev, D. S. Quantum chemical calculation of molecule cis-trans-hexadien-2, 4 by method MNDO. In: *Quantum chemical calculation of unique molecular system. I.* Publisher VolSU, c. Volgograd, **2010**, 243–245.

24. Babkin, V. A.; Andreev, D. S. Quantum chemical calculation of molecule cis-cis-hexadien-2, 4 by method MNDO. In: *Quantum chemical calculation of unique molecular system. I.* Publisher VolSU, c. Volgograd, **2010**, 245–246.

25. Babkin, V. A.; Andreev, D. S. Quantum chemical calculation of molecule trans-2-methylpentadien-1, 3 by method MNDO. In: *Quantum chemical calculation of unique molecular system. I.* Publisher VolSU, c. Volgograd, **2010**, 247–248.

26. Babkin, V. A.; Andreev, D. S. Quantum chemical calculation of molecule trans-3-methylpentadien-1, 3 by method MNDO. In: *Quantum chemical calculation of unique molecular system. I.* Publisher VolSU, c. Volgograd, **2010**, 249–250.

27. Babkin, V. A.; Andreev, D. S. Quantum chemical calculation of molecule cis-3-methylpentadien-1, 3 by method MNDO. In: *Quantum chemical calculation of unique molecular system. I.* Publisher VolSU, c. Volgograd, **2010**, 251–252.

28. Babkin, V. A.; Andreev, D. S. Quantum chemical calculation of molecule 4-methylpentadien-1, 3 by method MNDO. In: *Quantum chemical calculation of unique molecular system. I.* Publisher VolSU, c. Volgograd, **2010**, 252–254.

29. Babkin, V. A.; Andreev, D. S. Quantum chemical calculation of molecule cis-3-methylpentadien-1, 3 by method MNDO. In: *Quantum chemical calculation of unique molecular system. I.* Publisher VolSU, c. Volgograd, **2010**, 254–256.

30. Babkin, V. A.; Andreev, D. S. Quantum chemical calculation of molecule 1, 1, 4, 4-tetramethylbutadien-1, 3 by method MNDO. In: *Quantum chemical calculation of unique molecular system. I.* Publisher VolSU, c. Volgograd, **2010**, 256–258.
31. Babkin, V. A.; Andreev, D. S. Quantum chemical calculation of molecule 2-phenylbutadien-1, 3 by method MNDO. In: *Quantum chemical calculation of unique molecular system. I.* Publisher VolSU, c. Volgograd, **2010**, 260–262.
32. Babkin, V. A.; Andreev, D. S. Quantum chemical calculation of molecule 1-phenyl-4-methylbutadien-1, 3 by method MNDO. In: *Quantum chemical calculation of unique molecular system. I.* Publisher VolSU, c. Volgograd, **2010**, 262–264.
33. Babkin, V. A.; Andreev, D. S. Quantum chemical calculation of molecule chloropren by method MNDO. In: *Quantum chemical calculation of unique molecular system. I.* Publisher VolSU, c. Volgograd, **2010**, 264–265.
34. Babkin, V. A.; Andreev, D. S. Quantum chemical calculation of molecule trans-hexathrien-1, 3, 5 by method MNDO. In: *Quantum chemical calculation of unique molecular system. I.* Publisher VolSU, c. Volgograd, **2010**, 266–267.

SECTION 6.4

1. Kennedy, J. *Cation polymerization of olefins*. The World. Moscow. **1978**, 430.
2. Shmidt, M. W.; Baldrosge, K. K.; Elbert, J. A.; Gordon, M. S.; Enseh, J. H.; Koseki, S.; Matsvnaga, N.; Nguyen, K. A.; Su, S. J.; et al. *J. Comput. CHEM. 14*, 1347–1363, **1993**.
3. Babkin, V. A.; Fedunov, R. G.; Minsker, K. S. et al. *Oxidation communication*, **2002**, *25(1)*, 21–47.
4. Bode, B. M. and Gordon, M. S. *J. Mol. Graphics Mod., 16*, **1998**, 133–138.
5. Babkin, V. A.; Dmitriev, V. Yu.; Zaikov, G. E. Quantum chemical calculation of molecule hexene-1 by method MNDO. In: *Quantum chemical calculation of unique molecular system. I.* Publisher VolSU, c. Volgograd, **2010**, 93–95.
6. Babkin, V. A.; Dmitriev, V. Yu.; Zaikov, G. E. Quantum chemical calculation of molecule heptene-1 by method MNDO. In: *Quantum chemical calculation of unique molecular system. I.* Publisher VolSU, c. Volgograd, **2010**, 95–97.
7. Babkin, V. A.; Dmitriev, V. Yu.; Zaikov, G. E. Quantum chemical calculation of molecule decene-1 by method MNDO. In: *Quantum chemical calculation of unique molecular system. I.* Publisher VolSU, c. Volgograd, **2010**, 97–99.
8. Babkin, V. A.; Dmitriev, V. Yu.; Zaikov, G. E. Quantum chemical calculation of molecule nonene-1 by method MNDO. In: *Quantum chemical calculation of unique molecular system. I.* Publisher VolSU, c. Volgograd, **2010**, 99–102.
9. Babkin, V. A.; Andreev, D. S. Quantum chemical calculation of molecule isobutylene by method MNDO. In: *Quantum chemical calculation of unique molecular system. I.* Publisher VolSU, c. Volgograd, **2010**, 176–177.
10. Babkin, V. A.; Andreev, D. S. Quantum chemical calculation of molecule 2-methylbutene-1 by method MNDO. In: *Quantum chemical calculation of unique molecular system. I.* Publisher VolSU, c. Volgograd, **2010**, 177–179.

11. Babkin, V. A.; Andreev, D. S. Quantum chemical calculation of molecule 2-methyl-butene-2 by method MNDO. In: *Quantum chemical calculation of unique molecular system. I.* Publisher VolSU, c. Volgograd, **2010**, 179–180.
12. Babkin, V. A.; Andreev, D. S. Quantum chemical calculation of molecule 2-methyl-pentene-1 by method MNDO. In: *Quantum chemical calculation of unique molecular system. I.* Publisher VolSU, c. Volgograd, **2010**, 181–182.
13. Babkin, V. A.; Dmitriev, V. Yu.; Zaikov, G. E. Quantum chemical calculation of molecule butene-1 by method MNDO. In: *Quantum chemical calculation of unique molecular system. I.* Publisher VolSU, c. Volgograd, **2010**, 89–90.
14. Babkin, V. A.; Dmitriev, V. Yu.; Zaikov, G. E. Quantum chemical calculation of molecule hexene-1 by method MNDO. In: *Quantum chemical calculation of unique molecular system. I.* Publisher VolSU, c. Volgograd, **2010**, 93–95.
15. Babkin, V. A.; Dmitriev, V. Yu.; Zaikov, G. E. Quantum chemical calculation of molecule octene-1 by method MNDO. In: *Quantum chemical calculation of unique molecular system. I.* Publisher VolSU, c. Volgograd, **2010**, 103–105.
16. Babkin, V. A.; Dmitriev, V. Yu.; Zaikov, G. E. Quantum chemical calculation of molecule pentene-1 by method MNDO. In: *Quantum chemical calculation of unique molecular system. I.* Publisher VolSU, c. Volgograd, **2010**, 105–107.
17. Babkin, V. A.; Dmitriev, V. Yu.; Zaikov, G. E. Quantum chemical calculation of molecule propene-1 by method MNDO. In: *Quantum chemical calculation of unique molecular system. I.* Publisher VolSU, c. Volgograd, **2010**, 107–108.
18. Babkin, V. A.; Dmitriev, V. Yu.; Zaikov, G. E. Quantum chemical calculation of molecule ethylene-1 by method MNDO. In: *Quantum chemical calculation of unique molecular system. I.* Publisher VolSU, c. Volgograd, **2010**, 108–109.
19. Babkin, V. A.; Andreev, D. S. Quantum chemical calculation of molecule butadien-1, 3 by method MNDO. In: *Quantum chemical calculation of unique molecular system. I.* Publisher VolSU, c. Volgograd, **2010**, 235–236.
20. Babkin, V. A.; Andreev, D. S. Quantum chemical calculation of molecule 2-methylbu-tadien-1, 3 by method MNDO. In: *Quantum chemical calculation of unique molecular system. I.* Publisher VolSU, c. Volgograd, **2010**, 236–238.
21. Babkin, V. A.; Andreev, D. S. Quantum chemical calculation of molecule 2, 3-dimeth-ylbutadien-1, 3 by method MNDO. In: *Quantum chemical calculation of unique molecular system. I.* Publisher VolSU, c. Volgograd, **2010**, 238–239.
22. Babkin, V. A.; Andreev, D. S. Quantum chemical calculation of molecule pentadien-1, 3 by method MNDO. In: *Quantum chemical calculation of unique molecular system. I.* Publisher VolSU, c. Volgograd, **2010**, 240–241.
23. Babkin, V. A.; Andreev, D. S. Quantum chemical calculation of molecule trans-trans-hexadien-2, 4 by method MNDO. In: *Quantum chemical calculation of unique molecular system. I.* Publisher VolSU, c. Volgograd, **2010**, 241–243.
24. Babkin, V. A.; Andreev, D. S. Quantum chemical calculation of molecule cis-trans-hexadien-2, 4 by method MNDO. In: *Quantum chemical calculation of unique molecular system. I.* Publisher VolSU, c. Volgograd, **2010**, 243–245.
25. Babkin, V. A.; Andreev, D. S. Quantum chemical calculation of molecule cis-cis-hexa-dien-2, 4 by method MNDO. In: *Quantum chemical calculation of unique molecular system. I.* Publisher VolSU, c. Volgograd, **2010**, 245–246.

26. Babkin, V. A.; Andreev, D. S. Quantum chemical calculation of molecule trans-2-methylpentadien-1, 3 by method MNDO. In: *Quantum chemical calculation of unique molecular system. I.* Publisher VolSU, c. Volgograd, **2010**, 247–248.
27. Babkin, V. A.; Andreev, D. S. Quantum chemical calculation of molecule trans-3-methylpentadien-1, 3 by method MNDO. In: *Quantum chemical calculation of unique molecular system. I.* Publisher VolSU, c. Volgograd, **2010**, 249–250.
28. Babkin, V. A., Andreev, D. S. Quantum chemical calculation of molecule cis-3-methylpentadien-1, 3 by method MNDO. In: *Quantum chemical calculation of unique molecular system. I.* Publisher VolSU, c. Volgograd, **2010**, 251–252.
29. Babkin, V. A.; Andreev, D. S. Quantum chemical calculation of molecule 4-methylpentadien-1, 3 by method MNDO. In: *Quantum chemical calculation of unique molecular system. I.* Publisher VolSU, c. Volgograd, **2010**, 252–254.
30. Babkin, V. A.; Andreev, D. S. Quantum chemical calculation of molecule cis-3-methylpentadien-1, 3 by method MNDO. In: *Quantum chemical calculation of unique molecular system. I.* Publisher VolSU, c. Volgograd, **2010**, 254–256.
31. Babkin, V. A.; Andreev, D. S. Quantum chemical calculation of molecule 1, 1, 4, 4-tetramethylbutadien-1, 3 by method MNDO. In: *Quantum chemical calculation of unique molecular system. I.* Publisher VolSU, c. Volgograd, **2010**, 256–258.
32. Babkin, V. A.; Andreev, D. S. Quantum chemical calculation of molecule 2-phenylbutadien-1, 3 by method MNDO. In: *Quantum chemical calculation of unique molecular system. I.* Publisher VolSU, c. Volgograd, **2010**, 260–262.
33. Babkin, V. A.; Andreev, D. S. Quantum chemical calculation of molecule 1-phenyl-4-methylbutadien-1, 3 by method MNDO. In: *Quantum chemical calculation of unique molecular system. I.* Publisher VolSU, c. Volgograd, **2010**, 262–264.
34. Babkin, V. A.; Andreev, D. S. Quantum chemical calculation of molecule chloropren by method MNDO. In: *Quantum chemical calculation of unique molecular system. I.* Publisher VolSU, c. Volgograd, **2010**, 264–265.
35. Babkin, V. A.; Andreev, D. S. Quantum chemical calculation of molecule trans-hexathrien-1, 3, 5 by method MNDO. In: *Quantum chemical calculation of unique molecular system. I.* Publisher VolSU, c. Volgograd, **2010**, 266–267.

CHAPTER 7

ENERGY OF A HOMOLYTIC CLEAVAGE OF COMMUNICATION OH IN 4-REPLACED 2,6-DI-TERT. BUTYLPHENOLS

A. A. VOLODKIN, G. E. ZAIKOV, N. M. EVTEEVA, and S. M. LOMAKIN

CONTENTS

7.1   INTRODUCTION

Quantum-chemical calculations by the method of Hartrii-Focks with parameters UHF and RHF 4-replaced 2,6-di-*tert.*butylphenol values energy homolytic dissociation O-H of communications – $D_{(OH)}$ are found Value energy formation and energy of homolytic cleavage $D_{(OH)}$ communications in molecules phenols depend on calculation approach (PM6 or PM3) and parameters of a method of Hartrii-Focks. There are dependences $D_{(OH)}$ from k_7 reactions 4-substituted investigated 2,6-di-*tert.*butylphenol with iso-propylbenzene peroxide compound. Results of calculations $D_{(OH)}$ of sterically hindered phenols in approach PM6 with parameter RHF are corrected to experimental data.

Energy of a homolytic cleavage of communication OH ($D_{(OH)}$) of hindered phenols is an efficacy quantity indicator in reactions inhibiting oxidations. Experimental results of definition $D_{(OH)}$) [1–3] are known. In work [2], 2,4,6-tree-*tert.*butylphenol used with $D_{(OH)} = 328.4$ kJ/mol⁻¹, in Ref. [3] — 2,4,6-tree-*tert.*butylphenol with $D_{(OH)} = 339$ kJ/mol⁻¹. Values $D_{(OH)}$ 2,4,6-tree- *tert.*butylphenol in the specified works differ among themselves on value $D_{(OH)} \sim 9.6$ kJ/mol⁻¹. Thereupon quantum-chemical calculations $D_{(OH)}$ of phenols are actual [4, 5].

In the present work approaching's PM3 and PM6 and operators of Hartrii-Focks in the conditions of the decision of the template equations the unlimited and circumscribed method in the program "Mopac2009" [6] used. In the conditions of the initiated oxidation at 50°C defined constants of speeds of reactions peroxy radical *iso*-propylbenzene with 4-Z-replased -2,6-di-*tert.*butylphenols. Results of the analysis of dependences of calculations $D_{(OH)}$ from k_7 have allowed to recommend approach PM6 with operators RHF for calculation $D_{(OH)}$ of sterically hindered phenols.

7.2   EXPERIMENTAL PART

Calculation of parameters of the received compounds made under the program "MOPAC2009, Version 8.288W," visualization of frames in to the program Chembio3D, mathematical processing of results in to the program "Origin 6.1."

Reaction constants peroxy radical of *iso*-propylbenzene with 4-Z-re-plased -2,6-di-*tert*.butylphenols (k_7) defined in conditions inhibiting oxidations *iso*-propylbenzene by oxygen at 50°C in presence azodiisobutyronitrile.

Compounds **1– 3** are products "Serva," on analysis findings by chromatography with the maintenance of the basic material of 98–99%. **1**, m.p. 35–36°C; **2**, m.p.70°C; **3**, m.p. 129–130°C; **4** it is synthesized on a method [7], m.p.98–99°C; **5** it is synthesized on a method [8], m.p. 80–8°C; **6** it is synthesized from **2** on a method [9], m.p.147°C; **7** it is synthesized from **2** on a method [10], m.p.187–188°C; **8** it is synthesized on a method [11], m.p.146–147°C; **9** it is synthesized on a method [12], m.p. 217–218°C; **10** – it is synthesized on a method [13], m.p.154–155°C.

7.3   DISCUSSION OF RESULTS

Calculation $D_{(OH)}$ is based on energy phenol formations, phenolic radical, atom of hydrogen and is expressed by the formula [14]:

$$D_{(OH)} = -H^\circ_{f\,(In.)} + H^\circ_{f\,(H)} - (-H^\circ_{f\,InH}),$$

where $H^\circ_{f\,(In.)}$ –energy of formation phenolic radical, $H^\circ_{f\,InH}$ -energy of phenol formation, $H^\circ_{f\,(H)}$ -energy of hydrogen formation ($H^\circ_{f\,(H)} \approx 219.0$ kJ/mol^{-1}).

Properties of homologues of phenols with two *tert*.butyl groups in an ortho-positions are bound to influence of the next atoms of hydrogen *tert*. butyl groups on reactivity of phenolic hydroxyl and geometric parameters of an aromatic cycle. This influence causes appearance of antioxidative properties which are characterized by value k_7 [15]. Operator Hartrii-Focks with an odd number of electrons considers a spin component that is used at calculation energy formations phenoxy radicals. Calculations of structure nonsubstituted phenol (**1**) are carried out and **2–10** 4-Z-replased -2,6-di-*tert*.butylphenols in approach PM3 and PM6 too (Eq. (1)).

$$1\ Z = Me\ (2),\ {}^tBu\ (3),\ Cl\ (4),\ Br\ (5),\ CN\ (6),\ CHO\ (7),$$
$$COCH_3\ (8),\ COOH\ (9),\ NO2\ (10). \tag{1}$$

Results of calculations energy formations $(-H_f^\circ)$ and $D_{(OH)}$ compounds **1–10** and phenoxy radicals by quantum-chemical method with parameters UHFи RHF are presented in Table 1.

TABLE 1   Energy of formation of phenols $(-H^\circ_{f\ inH})$, phenoxyl radicals $(-H^\circ_{f\ in})$ and energy of homolytic cleavage O-H of communication $D_{(OH)}$ in phenols (1)–(10).

	$-\mathbf{H}^\circ_{f\ inH}$ 4-Z-phenol	$-\mathbf{H}^\circ_{f\ in}$ 4-Z-phenoxyl		$\mathbf{D}_{(OH)}$
	KJ/mol⁻¹			
	UHF	UHF	RHF	UHF RHF
1	88.1	28.1	61.1	
2	317.6	225.1	194.1	310.5 341.5
3	377.8	284.1	252.2	311.7 343.5
4	323.9	224.7	192.5	317.1 349.4
5	275.3	172.0	138.9	320.9 354.4
6	154.8	47.3	12.6	325.5 360.2
7	400.4	289.5	255.2	328.4 363.2
8	463.2	351.5	317.3	329.3 363.9
9	661.1	545.6	511.5	333.5 367.6
10	316.1	195.5	160.7	314.4 373.4

From the data of calculation of phenol **1** and corresponding phenoxyl radical with parameter RHF follows that the calculated value $D_{(OH)}$ (367.7 kJ/mol⁻¹) is close enough to experimental data [16] (369.0 kJ/mol⁻¹). Hence, results with parameter RHF are comprehensible that proves to be true the data of correlations of estimated values $D_{(OH)}$ from k_7 among 4-Z-replaced -2,6-di-*tert.*butylphenols.

The season of inhibition of oxidation with antioxidant participation (τ) and coefficient of stopping of chain (f) at constant speed initiation ($W_i = 1.5 \times 10^{-8}$ mol l⁻¹ s⁻¹) defined a gasomet-ric method. The specified parameters are bound by expression: $f = \tau \cdot W_i \cdot [\text{InH}]^{-1}$, where InH – compounds **2 10**. Specific reaction rate value (k_7) defined from dependence $\Delta[O_2] / [\text{RH}] = -k_2/k_7 \cdot \ln(1 - t \cdot \tau^{-1})$, where k_2 – kinetic constant ROO· with *iso*-propylbenzene (according to [17] k_2 = 1.75 1 mol^{-1} s^{-1}), [RH] – concentration *iso*-propylbenzene (7.18 mol·l^{-1}), Δ [O$_2$] – quantity of absorbed oxygen. Values k_7 are used for comparison of results of calculations $D_{(OH)}$ depending on parameters of approach (PM6 and PM3) and method of Hartrii-Focks (UHF or RHF) a Fig. 1.

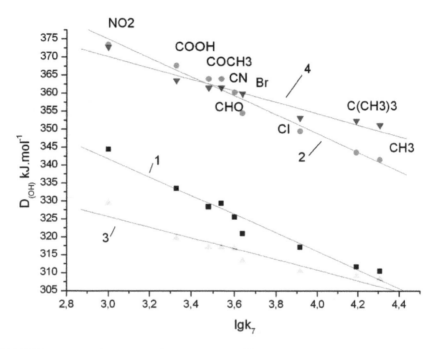

FIGURE 1 Dependences $D_{(OH)}$ from lgk$_7$ at inhibiting oxidation *iso*-propylbenzene in the presence of antioxidants **2–10**. 1 – $D_{(OH)}$ – calculation in approach PM6, UHF; 2 – $D_{(OH)}$ – calculation in approach PM6, RHF; 3 – $D_{(OH)}$ – calculation approach PM3, UHF; 4 – $D_{(OH)}$ – calculation approach PM3, RHF.

Comparison of known experimental data of definition $D_{(OH)}$ above-stated phenols a method of crossed parabolas [3, 18] with calculated in the present work specifies (Fig. 2) that the closest are results in a case of use of approach PM6 about parameter RHF.

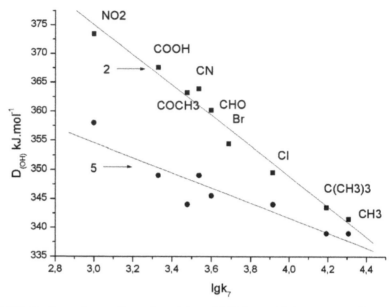

FIGURE 2 Dependences $D_{(OH)}$ from lgk$_7$ at inhibiting oxidation *iso*-propylbenzene in the presence of antioxidants **2–10**. 2 — $D_{(OH)}$ - calculation in approach PM6, RHF; a correlation coefficient r^2 =0.9875; 5 – the data from Ref. [3], a correlation coefficient r^2 = 0.9229.

The specified dependences are linear, but with different coefficients of correlations. Apostatis in values $D_{(OH)}$ in calculation and experiment on a method of crossed parabolas 4 phenols and corresponding phenoxyl radical increase in process of augmentation of electro-negativity of the substituent in position. In case of phenols **2** and **3** with alkyl group to sweep in position 4 coincidence of results of experiment and calculation within 2.5 ÷ 4.5 kJ.mol^{-1}. In process of augmentation of electronegativity of the substituent apostatis in calculation and experiment increases, and in compound **10** has 15.4 kJ.mol^{-1}. Substituent influence, is bound to changes of entropy of a transition state and geometry at transferring from aromatic structure of phenol to structure of phenoxyl radical. A role of entropy of a transition state in a method crossed parabolas it is considered in work [19]. The data of calculations of geometrical parameters of phenols **2–10** and phenoxyl radicals **2a–10a** is presented in Tables 2 and 3.

TABLE 2 Charges on atoms of oxygen, carbon, lengths of communication and angles in 4-replaced 2,6-di-*tert.*butylphenols **2–10**.

	Charge q		Length d/Å	Angle w / degree	
	O(6)	C (5)	O(6) – C (5)	C (3) – C (1) – C(9)	C (1) – C (5) – O(6)
2	−0.284	+0.175	1.422	17.7	114.5
3	−0.284	+0.173	1.383	17.9	114.6
4	−0.294	+0.132	1.379	17.3	114.3
5	−0.287	+0.149	1.378	17.4	114.4
6	−0.255	+0.211	1.374	17.5	114.5
7	−0.233	+0.233	1.373	18.1	114.0
8	−0.240	+0.231	1.374	17.6	114.7
9	−0.242	+0.224	1.371	16.1	114.5
10	−0.244	+0.216	1.367	17.4	114.6

TABLE 3 Charges on atoms of oxygen, carbon, lengths of communication and angles in structures 4-replaced 2,6-di-*tert.*butylphenoxyl radical **2a–10a**.

	Charge q		Length d /Å	Angle ω / degree	
	O (6)	C(5)	O (6) – C (5)	C (3) – C (1) – C(9)	C (1) – C (5) – O(6)
2a	−0.231	+0.377	1.227	16.8	121.1
3a	−0.253	+0.383	1.226	17.1	121.4
4a	−0.411	+0.372	1.224	16.6	121.0
5a	−0.439	+0.360	1.224	16.7	121.1
6a	−0.208	+0.375	1.223	16.7	121.1
7a	−0.200	+0.374	1.223	16.8	120.9
8a	−0.196	+0.372	1.226	16.9	121.0
9a	−0.198	+0.372	1.223	16.8	121.1
10a	−0.208	+0.374	1.221	16.6	121.1

Substituents affect on a charge of atom of oxygen (O6) in limens q = $-0.233 \div -0.294$, atom of carbon of an aromatic cycle (C5): $q = + 0.132 + \div 0.233$, and also length of communication (O6) – (C5): $\div 1.371 \; 1.422$ Å. The deflection of communication of carbon *tert*.butyl group from a plane of a hexatomic cycle equal $17.3 - 18.1°$, the angle between carbons of a hexatomic cycle and hydroxyl group oxygen is equal $114.0 \div 114.7°$.

Transferring of aromatic structure of phenol into structure phenoxyl radical is accompanied by changes of charges on atoms of oxygen (O6), carbon (C5), lengths of communications (O6) – (C5) and angles (C1) – (C5) – (O6) that affects on change of entropy in the course of transferring of phenol in phenoxyl radical.

Calculation in the program "Mopac2009" with parametre "Thermo" have calculated entropies of phenols and phenoxyl radicals (Table 4).

TABLE 4   Entropies of phenols **2–10** and phenoxyls radicals **2a–10, 2a–10a**.

Z	S° J/K/mol		ΔS° J/K/mol
	4-Z-phenol 2–10	4-Z-phenoxyl 2a–10a	
CH3	139.6	43.0	+3.4
^{t} B u Cl	157.1	160.2	+3.1
Cl	136.9	136.8	−0.1
Br	139.6	135.4	−4.2
CN	113.0	140.4	+27.4
COCH3	149.2	149.3	+0.1
CHO	141.1	143.4	+2.3
COOH	146.4	153.5	+7.1
NO2	144.1	133.6	−10.5

Results of calculations confirm entropy change at a transformation of phenoles in phenoxyles radicals. The nature to sweep in this process is ambiguous. In a case alkyls substituents ($Z = CH_3$, $(CH_3)_3$) entropy change compounds $\Delta S = + 3.4 \div +3.1$ Entropy change 4-nitro-2,6-di-*tert*.butyl-phenol leads to negative value ($\Delta S° = -10.5$). At transferring from 4-zian-

2,6-di-*tert.*butylphenol to radical $\Delta S^\circ = +27.4$. That can be bound with delocalization not coupled electron on atom of nitrogen. This assumption will be coordinated with a dimerization direction 4-zian-2,6-di-*tert.*butylphenol a radical [20] in which process diamagnetic compound with communication -N-N- is formed.

7.4 CONCLUSIONS

Results of quantum-chemical calculations of phenols and phenoxyls radicals a method of Hartrii-Focks with parameters UHF, RHF are used for calculations of energy of homolytic decomposition of communication OH bonds. Results of the analysis of dependences of calculations $D_{(OH)}$ from k_7 allow to recommend approach PM6 with parameter RHF for calculation $D_{(OH)}$ of sterically hindered phenols.

KEYWORDS

- apostatis
- Hartrii-Focks
- homolytic decomposition
- quantum-chemical calculations

REFERENCES

1. Beljakov, V. A.; Shanina, E. L.; Roginsky, V. A.; Miller, V. B. Izv. Akad. Nauk SSSR, *Ser. Chim.*, **1975**, *2685* (in Russian).
2. Vasserman, A. M.; Buchachenko, A. L.; Nikiforov, G. A.; Ershov, V. V.; Neumann, M. B. *J.Phys.Chim.*, **1967**, *41,* 705 (in Russian).
3. Denisov, E. T. *J.Phys.Chim.*, **1995**, *69(623)* (in Russian).
4. Djubchenko, O. I.; Nikulina, V. V.; Terah, E. I.; Prosenko, A. E.; Grigoriev, I. A. Izv. Akad. Nauk SSSR, *Ser. Chim*, **2007**, *1107* (in Russian).
5. Egorov, A. N.; Venus, M. B.; Zhuko, I. S.; Kozak, D.; Tsirelson, V. E. Izv. Akad. Nauk SSSR, *Ser. Chim*, **2008**, *1157* (in Russian).
6. Stewart, P. J. J. *Computational J. Mol. Mod.*, **2007**, *13,* 1173.

7. Ley, K.; Muller, E.; Mayer, R.; Scheffler, K. *Chem. Ber.*, **1958**, *91,* 2670.
8. Muller, E.; Ley, K.; Schlechte, G. *Chem. Ber.*, **1957**, *90,* 2660.
9. Cohen, L. *J. Organ. Chem.*, **1957**, *22,* 1333.
10. Coppinger, G.; Campbell, T. *J. Amer Chem. Soc.*, **1953**, *75,* 734.
11 Cook, C.; Gilmour, N. *J. Organ Chem.*, **1960**, *25,* 1429.
12 Coffield, T.; Filbey, A.; Ecke, G.; Kolka, A. *J. Amer Chem. Soc.*, **1957**, *79,* 5129.
13. Barness, T.; Hickinbottom, F. *J. Chem. Soc.*, **1961**, *953.*
14. Belezkaia, I. P. *Ions and ion Pairs in Organic Reactions*. Moscow ,.Mir, **1975**, *54* (in Russian).
15. Roginsky, V. A. Phenolic antioxidants. *Reactivity and Efficacy*. Moscow, Nauka, **1975**, *247* (in Russian).
16. Denisov, E. T.; Tumanov, V. E. *Successes of chemistry*, **2005**, *74,* 905 (in Russian).
17. Tsepalov, V. F. *Research of synthetic and connatural antioxidants in vivo and in vitro.* Moscow, Nauka, **1992**, *16* (in Russian).
18. Denisov, E. T.; Denisova, T. G. *Handbook of Antioxidants*, London-New York. CRC Press, **2000**, *88.*
19. Denisov, E. T. *J. Chim. Phys.* **2011**, *30(11)* (in Russian).
20. Muller, E.; Riecker, A.; Ley, K.; Mayer, R.; Schelfler, K. *Chim. Ber.*, **1959**, *92,* 2278.

CHAPTER 8

REACTION OF OZONE WITH SOME OXYGEN CONTAINING ORGANIC COMPOUNDS

S. RAKOVSKY, M. ANACHKOV, and G. E. ZAIKOV

CONTENTS

8.1   INTRODUCTION

This chapter, based on 92 references, is focused on degradation of organics by ozonation and it comprises various classes of oxygen-containing organic compounds – alcohols, ketones, ethers and hydroxybenzenes. The mechanisms of a multitude of ozone reactions with these compounds in organic solvents are discussed in details, presenting the respective reaction schemes and the corresponding kinetic parameters are given and some thermodynamic parameters are also listed. The dependences of the kinetics and the mechanism of the ozonation reactions on the structure of the compounds, on the medium and on the reaction conditions are revealed. The various possible applications of ozonolysis are specified and discussed. All these reactions have practical importance for the protection of the environment.

The ozonolysis of oxygen-containing compounds is a promising process that takes place under mild conditions and yields compounds of a higher oxidation state than that of the starting compounds. It may find various applications in chemical and pharmaceuticals industries, fine organic synthesis, etc. [1–2]. The widest application, in this respect, has been found for the oxidation of primary and secondary alcohols respectively into their corresponding aldehydes and ketones. For example in the cases of the oxidation of open-chain (simple) and cyclic secondary alcohols the yield of ketones is within the range 57% up to 83% [3]. Moreover, the considered interactions are extremely important from an ecological point of view for the utilization and purification of industrial wastewaters, originating from hydroxybenzene production, through their partial or complete oxidation [4–24]. The importance of this process for theory and practice gave us an impetus to carry out systematic investigations [25–35]. The aim of the present paper is the precise determination of the rate constants of ozonolysis of some more widely occurring representatives of the studied classes of organic compounds. Applying the activated complex method (ACT) [36] and collision theory (CT) [36] some theoretical investigations have also been carried out elucidating the structure of the transition state formed in the course of the reaction. On the basis of the correlation between the results from the experimental and theoretical studies some peculiarities in the mechanism of ozone reactions with the considered classes of oxygen containing compounds have been established.

8.2   EXPERIMENTAL PART

Ozone has been obtained from dry oxygen by means of a silent discharge of 5–8 kV at an oxygen flow rate of 0.1 l/min. The ozone concentration, 10^{-5}–10^{-3} M, has been measured spectrophotometrically in the wavelength region of 254–300 nm in a 5 cm quartz gas cell.

8.3   METHODS

The UV, IR, ESR spectra were registered on standard equipments, as well as HPLC, and GC analyses.

8.4   KINETIC MEASUREMENTS

8.4.1   STATIC METHOD

Pure reagent or reagent solution was injected into thermostatic 1 cm quartz cuvette, containing a solution of ozone in CCl_4, the time of mixing being less than 0.2 second. Ozone concentration was monitored spectrophotometrically in the region of 270–290 nm. At $[RH]_o/[O_3]_{lo} > 100$, the ozone pseudomonomolecular constant $k' = k[RH]_o$ was determined on the basis of the equation $\lg([O_3]_{lo}/[O_3]_{lt}) = k't$ where $[O_3]_{lo}$ and $[O_3]_{lt}$ are the initial and current concentrations of ozone in solution, respectively [27].

8.4.2.   DYNAMIC METHOD

The ozone was bubbled through a cylindrical glass reactor with inner diameter $\phi = 1.7$–3.7 cm and height 7–15 cm, supplied with porous glass grit-G2 at its bottom. The accuracy of maintaining constant temperature was ±0.1°C. Conventionally, gas flow rate was v=0.1 L/min; the solutions volume was V=10 ml; the ozone concentrations at the reactor inlet ($[O_3]_o$) varied from 10^{-6} to 10^{-3} M; the solvent was CCl_4; $[RH]_o = 10^{-4}$–10^1 M. The inlet and outlet ozone concentrations were measured in the gas phase in the 254–300 nm wavelength range. The determination of rate constants is based on the approach, which connects the balance of consumed ozone with the rate of the chemical reaction – Eq. (1):

$$\omega([O_3]_0.[O_3]_g) = k[O_3]_l[RH] \tag{1}$$

where ω is the relative flow rate of ozone-oxygen gas mixture (in litres per L of solution per sec); $[O_3]_0$ and $[O_3]_g$ are the ozone concentrations at the reactor inlet and outlet, respectively, $[O_3]_l$ is ozone concentration in the solution; [RH] is concentration of the reagent. This model is valid in all cases, when the rate of ozone absorption is considerably greater than the rate of the chemical reactions. If in the case of a bimolecular reaction, in accordance with Henry's Law, $[O_3]_l$ is substituted with $\alpha[O_3]_g$, where α is Henry's coefficient, Eq. (1) can be transformed into Eq. (2) (1):

$$k = \omega.\Delta[O_3]/([RH]. \alpha[O_3]_g) \tag{2}$$

One of the widely applied criteria with respect to the conditions of validity (applicability) of Henry's Law is the expression:

$$D_{O3}.k_1'/k_L^2 \ll 1$$

where D_{O3} is the diffusion coefficient of ozone in the solution; $k_L = D_{O3}/\delta$ is the coefficient of mass transfer in the liquid phase, and δ is thickness of the boundary layer in the hydrodynamic model of renovation surface; or $k_L = (D_{O3}.s)^{1/2}$, where s is the time interval of renovation. In the case of applying the bubbling method with small bubbles (diameters up to 2.5–3 mm) then $k_L = 0.31 \times (gv)^{1/3} \times (D_{O3}/v)^{2/3}$, where $v = \eta/\rho$, is the kinematic viscosity of the solvent, η is the viscosity of the solvent, ρ is the solvent density, g is the earth acceleration. Usually the k_L values are of the order of 0.1–0.05 cm/s. As $k_1' = k[RH]_0$ in most of the cases it is possible to select such values for $[RH]_0$, at which the criterion for applicability of Henry's Law is fulfilled [34].

In order to minimize the influence of the so-called "effect of delay in the response function"– $[O_3]g=f(\tau)$ upon calculating the values of k such sections of the kinetic curves are selected, which appear to be practically parallel or only slightly inclined with respect to the abscissa: $k_1'\alpha[O_3]_g \gg d[O_3]_g/d\tau$. The advantages and limitations of this method have been discussed in detail in [37, 38]. Despite some contradictory observations, the significant part of rate constants of ozone with organic compounds and polymers are obtained on the basis of Eq. (2) [1, 34].

8.4.3 ALCOHOLS

A number of authors have tried to elucidate the kinetics and mechanism of the ozonolysis of alcohols and their application to the selective preparation of ketones and aldehydes in high yields under mild conditions [3, 39–46]. The basic concepts on such reactions are discussed in several references (1, 2, 25, 34, 35, 39]. It has been established that the rate of ethanol ozonolysis does not change when the hydroxyl group is deuterated [43] and its value is 4.17 times higher than that when the methylene group is deuterated. On the basis of this fact, together with the data from the analysis of the product composition and the kinetics of their formation the authors have suggested that the H-atom abstraction by the ozone molecule is the rate-determining step of the reaction. This conclusion is confirmed by the data in Table 1 as reported by various authors [41–43]. The ratio between the relative reactivity of tertiary: secondary: primary alcohols are 1:12:241 according to reference [41], or 1:156:817 according to Ref. [7]. On the basis of the studies on the reaction of ozone with methyl, ethyl and 2-propyl alcohol, the following reaction mechanism is proposed (Scheme 1) [40]:

TABLE 1   Kinetic parameters of the ozone reaction with aliphatic alcohols at 25°C.

Alcohols	k, M^{-1}.s^{-1}	lg A	E, kcal/mol	Reference
t.-Butanol	0.01	4.6	9.0	[32]
t.-Butanol	0.05	–	–	[31]
Ethanol	0.35	6.7	9.8	[32]
Ethanol	0.25 (22°C)	6.7	9.8	[33]
n-Butanol	0.54	7.3	10.3	[32]
n-Butanol	0.39			[31]
Iso-Propanol	0.89	7.3	10.0	[32]
Cyclopentanol	1.35			[31]

$CH_3OH + O_3 \Leftrightarrow [HOC^•H_2...HO_3^•]^\# \rightarrow HOCH_2OOOH,$

and in the presence of a base:

$CH_3OH + O_3 + (B) \rightarrow HOCHO + {}^-O_2H + BH^+$

$CH_3CH_2OH + O_3 \Leftrightarrow [CH_3HC^•(OH)...HO_3^•]^\# \rightarrow CH_3HC(OOOH)OH \rightarrow$

$CH_3OOH + H_2O_2$ and/or $CH_3CHO + O_2 + H_2O$

$(CH_3)_2C(OH)H + O_3 \Leftrightarrow [(CH_3)_2C^•(OH)...HO_3^•]^\# \rightarrow (CH_3)_2C(OH)OOOH$

$\downarrow$-O_2

$CH_3C(OH)=CH_2 + H_2O + O_2 \ (CH_3)_2CO + H_2O$

$\downarrow$+O_3

$CH_3COOH + HCHO$

SCHEME 1.

It is supposed that an intermediate ion or pair of radicals is formed, whose recombination in the kinetic cage yields α-hydroxyl-hydrotrioxide. The latter leaves the cage and passes into the volume of solution.

The authors of Ref. [40] found that the product composition of 2-propanol ozonolysis depends on the method of propanol purification. Taking into account the above given consideration about the reaction pathway, in our opinion, the mechanism still remains unclear and new data should be provided for its elucidation. In this connection, we have carried out intensive experimental studies on the kinetics of ozonolysis of MeOH, *t*-BuOH, EtOH, *n*-PrOH, *n*-BuOH, *i*-PrOH, *s*-BuOH, *c*-HexOH both by the static and barbotage methods, the results of which are summarized in Figs. 1–3 and Tables 2–4.

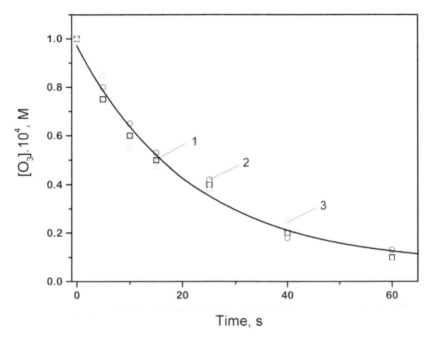

FIGURE 1   Kinetics of ozone consumption in the reaction of ozone with: 1 – methanol, 22°C, 0.74 M; 2 – ethanol, 3°C, 0.24 M; and 3 – isopropanol, 3.5°C, 0.037 M.

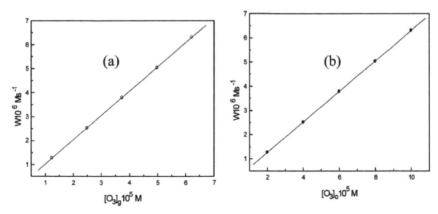

FIGURE 2   Dependence the rate of ozone consumption on the ozone concentrations (a) inlet $[O_3]_o$ and (b) outlet $[O_3]_g$, at MeOH concentration of 0.74 M.

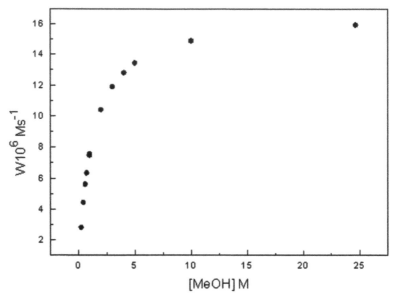

FIGURE 3 Dependence of the rate of ozone consumption on methanol concentration at $[O_3]_0 = 1.10^{-4}$ M (according to Eq. (1)).

TABLE 2 Kinetic parameters of ozone reaction with MeOH in carbon tetrachloride (CCl_4) and pure MeOH solutions: 22°C, $\omega = 0.167$ s^{-1}; $v = 1.67 \times 10^{-3}$ L/s, maximum rate of ozone inlet $- 1.67 \times 10^{5}$ M.s^{-1}

[MeOH], M	$[O_3]_0 \times 10^5$, M	$[O_3]_g \times 10^5$, M	$\Delta[O_3] \times 10^5$, M	$W \times 10^6$, M.s^{-1}	k, M^{-1}.s^{-1}
1	2	3	4	5	6
0	10	0	10	0	–
0.247	10	8.32	1.68	2.80	0.057
0.439	10	7.35	2.65	4.43	0.056
0.618	10	6.64	3.36	5.61	0.058
0.740	10	6.22	3.78	6.31	0.055
0.987	10	5.53	4.47	7.46	0.057
1.0	10	5.49	4.51	7.53	0.056
2.0	10	3.79	6.21	10.37	0.055

TABLE 2   *(Continued)*

[MeOH], M	$[O_3]_0 \times 10^5$, M	$[O_3]_g \times 10^5$, M	$\Delta[O_3] \times 10^5$, M	$W \times 10^6$, M.s^{-1}	k, M^{-1}.s^{-1}
3.0	10	2.89	7.11	11.87	0.057
4.0	10	2.34	7.66	12.79	0.058
5.0	10	1.96	8.04	13.42	0.054
10.0	10	1.09	8.91	14.88	0.056
24.7	10	0.47	9.53	15.91	0.058
0.740	8	4.98	3.02	5.04	0.057
0.740	6	3.73	2.27	3.79	0.055
0.740	4	2.49	1.51	2.52	0.056
0.740	2	1.24	0.76	1.27	0.058

TABLE 3   Dependence of k on the temperature in °C for ozone reaction with three types of alcohols

k, M^{-1}.s^{-1}	0°C	10°C	20°C	25°C	30°C
MeOH	0.008	0.021	0.049	0.072	0.108
t-BuOH	0.005	0.013	0.029	0.045	0.064
EtOH	0.14	0.28	0.54	0.74	1.10
n-PrOH	0.19	0.36	0.67	0.89	1.18
n-BuOH	0.15	0.30	0.56	0.76	1.10
i-PrOH	0.93	1.61	2.71	3.46	4.39
s-BuOH	0.88	1.54	2.58	3.29	4.18
c-HexOH	0.92	1.59	2.65	3.37	4.27

TABLE 4   Kinetic parameters of ozone reaction with some alcohols at 25°C.

Parameters	MeOH	t-BuOH	EtOH	n-PrOH	n-BuOH	i-PrOH	s-BuOH	c-HexOH
$k \times 10^2$, M^{-1}.s^{-1}	7.2	4.5	74	89	76	346	329	337

TABLE 4 *(Continued)*

Parameters	MeOH	t-BuOH	EtOH	n-PrOH	n-BuOH	i-PrOH	s-BuOH	c-HexOH
N, a-C–H	3	9 (b-C–H)	2	2	2	1	1	1
$k \times 10^2/n$, $M^{-1}.s^{-1}$	2.4	0.5	37	44	38	346	329	337
E_a, kcal/mol	13.9	13.7	10.9	10.1	10.4	8.5	8.5	8.4
$A \times 10^{-7}$, $M^{-1}.s^{-1}$	41	5.8	3.9	2.4	1.7	0.62	0.59	0.51

Figure 1 demonstrates the kinetic data of ozone consumption in solutions of methanol, ethanol and iso-propanol by means of the static method with time interval of mixing less than 0.2 seconds. It is seen that regardless of the type of alcohol being ozonized, the kinetic curves coincide with first-order rate law. With a view to more precise evaluation and comparison of the respective rate constants such concentrations of the respective alcohols have been selected, at which the rates of ozone consumption have close values. The kinetic curves of ozone reactions with MeOH, EtOH and i-PrOH gave the following values of the rate constants: 0.057, 0.17 and 1.13 M-1.s-1, respectively.

The same values were also obtained by carrying out the reactions in a bubble reactor. The values of k are calculated on the basis of Eq. (2) – the data are given in Table 2 and they are represented graphically in Figs. 2 and 3.

Processing the data from Table 2, columns 2, 3, and 5, based on Eq. (2), yield the linear dependencies shown in Figs. 2a and 2b. The dependence of the rate W on ΔO_3 has also been found out to be a linear one. The dependence of the rate W on the concentration of alcohol [ROH] (column 1) is a curve, which approaches a limit value at $W = 1.67 \times 10^{-5}$ M.s-1. The nonlinear character of this relationship is in accordance with the complexity of Eq. (2) (see also Fig. 3). In the case when [ROH]$\rightarrow \infty$, then we have $[O_3]_g \rightarrow 0$ and $\Delta O_3 \rightarrow [O_3]_0$. The values of k, calculated based on Eq. (2) at each one of the points on the curve (Fig. 3), are equal to 0.057.

Therefore the experimentally obtained linear dependences of W on $[O_3]_o$, $[O_3]_g$ and on $\Delta[O_3]$ as well as the profile of the curve in Fig. 3 correspond to the mathematical description of ozonolysis in a barbotage reactor – Eq. (2).

Judging from the analysis of the data obtained (Tables 3 and 4) it follows that the rate constant and the activation energies are strongly dependent on the alcohol structure. The interaction of ozone with MeOH possessing primary α-H atoms and with tert-BuOH having only primary C–H bonds has been found to be the slowest, and the value of k at 20°C per one α-H atom in the first case is 1.62×10^{-2} M^{-1}.s^{-1}, and in the second case, related to one primary H atom, is 3.22×10^{-3} M^{-1}.s^{-1}. The difference in the values of these constants is due to the fact that while the OH group in MeOH directly affects the α-H atom, in the case of the tert-BuOH, which does not possess any α-H atoms, the effect of the OH groups is transferred through one σ-bond and so it is considerably weaker.

As a result of this, the reactivity of tert-BuOH becomes similar to that of methane and neopentane. The interactions of ozone with EtOH, n-PrOH, n-BuOH take place at higher rates and the values of k per one α-H atom amount to: 0.27, 0.34 and 0.28, respectively. The higher rates of the ozone reactions with these alcohols are associated with the presence of secondary α-H atoms in their molecules, which have lower bond energies than the primary ones. The enhanced reactivity of n-PrOH compared with that of EtOH could be assigned to the donor effect of the second CH_3 group while the donor effect of the C_2H_5 group in n-BuOH is weaker than that of the methyl group and therefore k is lower. It has been found that the reactions of ozone with i-PrOH, s-BuOH and s-HexOH alcohols possessing tert-H atoms with the lowest bond energies is the fastest with k equal to 2.71, 2.58 and 2.65, respectively, i.e., in fact they are almost the same. The ratio between the reduced values of k at 25 for methanol: ethanol: sec. butanol are 1:15:137. Simultaneously, the values of E_a decrease with the decrease of the α-H atoms bond energy (D) and their ratio is 1:0.78:0.61 [12, 15]. This fact can be regarded as important evidence for the mechanism of α-C–H-atom abstraction by ozone.

The theoretical estimates of A were performed considering the two possible structures of the activated complex (AC): linear (LC) and cyclic (CC) (Fig. 4). The good agreement between the experimental data and theoretical estimates for A allows the determination of the AC structure and on this basis enables the selection of the most probable mechanism. Simultaneously, the value of the pre-exponential factor was calculated by the collision method and by comparing its value with the one calculated based on the AC theory the steric factor -p was determined Eq. (3) [48–52]:

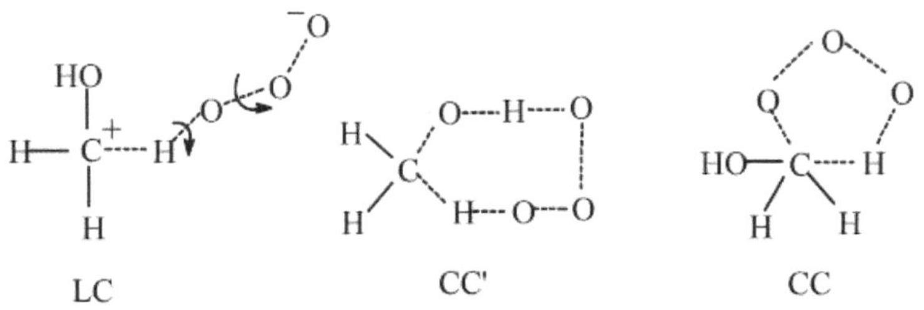

FIGURE 4 Structure of the probable activated complexes in the reaction of ozone with aliphatic alcohols: LC – linear with free fragment rotation and CC' and CC – cyclic complexes without free rotation.

$$k = p.Z_0.\exp(-E_a/RT), \qquad (3)$$

where: $Z_0 = \pi.(r_A+r_B)^2 \times (k_B T/\pi m^*)^{1/2}$; Z_0 is the collision factor; r_A and r_B are the van der Waals radii of the reagents; k_B is Boltzmann's constant; T is the absolute temperature; and m* is the reduced mass.

The pre-exponential factors for the reactions of ozone with alcohols, calculated according to the activated complex method (ACT) and collision theory (CT) are represented in Table 5.

TABLE 5. The values of A calculated by ACT with LC and CC and by CT, the sums (R_{fr}) and energy (E_{fr}) of inner rotation, steric factors (p), VDW radii of the molecules (r) and the ratio between the calculated and experimental values of pre-exponentials A^{LC}_{cal}/A_{obs}.

Parameters	MeOH	EtOH	n-PrOH	i-PrOH	n-BuOH	s-BuOH	t-BuOH	c-HexOH
$A^{CC} \times 10^{-4}$, $M^{-1}.s^{-1}$	52	8.3	3.7	3.5	2.4	2.2	6.3	1.6
$A^{LC} \times 10^{-4}$, $M^{-1}.s^{-1}$	56	15	6.7	6.3	4.4	4.0	4.9	2.9
$F_{fr} \times 10^{-2}$	7.1	9.8	12	11	13.8	12.6	12.9	12.6

TABLE 5   *(Continued)*

Parameters	MeOH	EtOH	n-PrOH	i-PrOH	n-BuOH	s-BuOH	t-BuOH	c-HexOH
$A^{LC}_{cal} \times 10^{-7}$, $M^{-1}.s^{-1}$	40	15	8.1	6.8	6.1	5.0	6.3	3.6
r, Å	2.24	2.57	2.91	2.88	3.13	3.23	3.13	3.23
$A^{CT} \times 10^{-11}$, $M^{-1}.s^{-1}$	2.6	2.6	2.7	2.7	2.8	2.8	2.8	2.7
$p \times 10^5$	22	5.8	2.5	2.3	1.6	1.4	1.8	4.1
A^{LC}_{cal}/A_{obs}	1.0	3.8	3.4	11	3.1	8.3	1.1	7.2
$E_{fr,}$ cal/mol	0	730	670	1368	701	1172	52	1081

Note: ALC is the pre-exponential factor calculated with LC without free fragment rotation (it is absent at $E_{fr} \geq 4$ kcal); A_{CC} – calculated with CC; ALC_{calc} = at free fragment rotation, i.e., when $\exp(-E_{fr}/RT) = 1$ or $E_{fr} = 0$ kcal; A_{ct} – calculated according to the collision theory at 300 K and r – calculated by the Alinger method (*PCMOD4* program) at the radius of the ozone molecule of 2A; E_{fr} – the calculated values of the rotation energy at ratio of $A^{LC}_{cal}/A_{obs} = 1$.

An interesting fact is the good agreement between the values of A^{LC} and A^{CC}. The values of A^{CC} with the five- and seven-member cyclic form of AC are practically similar as they differ by not more than 1%. However the free rotation produces an increase in A^{LC}_{cal} by three orders of magnitude compared with that of A^{CC}. The steric factor has a value, which is in agreement with a liquid phase reaction. The ratio between the calculated and experimental values of A^{LC}, is increasing in the sequence primary: secondary: tertiary alcohols is 1:3.4:8.8, i.e., the coincidence in this sequence is decreasing. This is connected with the fact that E_{fr} is not zero and it is increasing with the changes in the alcohol structure in the same sequence. It has been found that the values of A^{LC}_{calc} and A_{obs} coincide when the values of rotation energy are those given in line 10 in Table 5. We have obtained the same values by means of the *MOPAC 6* software package and therefore, the data on the free rotation energy (E_{fr}) in line 10 can also be considered as theoretically evaluated.

The good agreement between A^{LC}_{calc} and A_{obs} suggests that the rate-determining step of the ozone reaction with alcohols involves the formation of linear AC and abstraction of the α-H atom from the alcohol molecule.

The experimental and theoretical results conform well to Scheme (2).

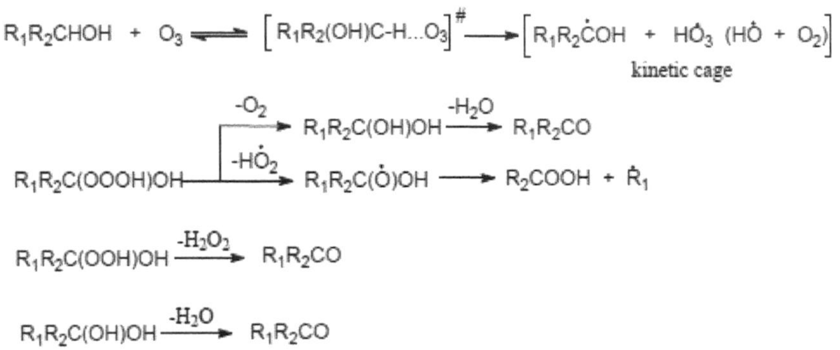

$$R_1R_2CHOH + O_3 \rightleftharpoons \left[R_1R_2(OH)C\text{-}H...O_3\right]^{\#} \longrightarrow \left[R_1R_2\dot{C}OH + H\dot{O}_3 \ (H\dot{O} + O_2)\right]$$

kinetic cage

$$R_1R_2C(OOOH)OH \xrightarrow[-H\dot{O}_2]{-O_2} \begin{array}{l} R_1R_2C(OH)OH \xrightarrow{-H_2O} R_1R_2CO \\ R_1R_2C(\dot{O})OH \longrightarrow R_2COOH + \dot{R}_1 \end{array}$$

$$R_1R_2C(OOH)OH \xrightarrow{-H_2O_2} R_1R_2CO$$

$$R_1R_2C(OH)OH \xrightarrow{-H_2O} R_1R_2CO$$

SCHEME 2

Ozone attacks the α-H atom, forming a LC, which further undergoes decomposition into a radical (or ion) pair in one kinetic cage. The α-hydroxy alcohol, α-hydroxyperoxy alcohol and α-hydroxytrioxy alcohol being unstable leave the cage and decompose rapidly to the corresponding aldehyde or ketone liberating water, hydroperoxide and oxygen or lead to the formation of hydroperoxy and alkoxy radicals. The latter species can further undergo monomolecular decomposition.

8.4.3.1   KETONES

The investigation of the reactions of ozone with ketones is of interest from the viewpoint of the theory of reactivity, the ozone chemistry, preparation of oxygen-containing compounds, and the degradation and stabilization of organic materials. In particular, the ozonation of methylethylketone (MEK) [53–55] yielded acetic acid (AcAc), diacetyl and hydroperoxide as the main reaction products in CCl_4. It has been assumed that the process is not a oxidation chain-radical reaction. Similar products are also formed when the reaction takes place in water. However, in this case ozone interacts with the enol form of MEK [53. The addition of nitric acid accelerates enolization, which is evidenced by the increase of the oxidation rate [55]. The decrease of the oxygen content in the ozone-oxygen mixture results in an increase of the oxidation rate, which is due to a change of the radical leading to oxidation, from $RO_2^•$ to $RO^•$. The ozonation of isomeric decanones indicates that the reactivity of ketones depends on the length and structure of the alkyl chain. In this case the main products are α-ketohydroperoxides, monocarboxylic acids, keto alcohols and alkylbutyric lactones. However, it should be noted that the conclusions in these studies were based on product analysis at very high degrees of conversion whereby the secondary reactions play an important role in the product formation. This gives us reasons to doubt whether the proposed mechanism describes adequately the initial steps of the reaction. The rate constant and the activation energy of interaction between cyclohexanone and ozone are 1.6×10^{-2} M-^{1}.s-^{1} and 11.2 kcal/mol, respectively.

The reactivity of ozonized methyl derivatives of cyclohexanone is considerably higher than that of the H form. An increase in the rate of interaction between the above derivatives and ozone is also observed depending on the location of substitution and it follows the sequence [56–58]: 3-methylcyclohexanone <4-methylcyclohexanone <2-methylcyclohexanone.

It is evident from the foregoing that so far no systematic investigations on the ozonolysis of ketones have been carried out. In our studies, on the ozonolysis of ketones we combined the kinetic approach and product analysis with theoretical methods. The results from the ozonolysis of some aliphatic ketones are shown in Table 6.

TABLE 6   Kinetic parameters of the reaction of ozone with some aliphatic ketones at 21°C.

Ketone	$k_0 \times 10^3$, M-^{1}.s-^{1}	$k_{st} \times 10^3$, M-^{1}.s-^{1}	k_{rel}	lg A	E_a, kcal/mol
CH_3COCH_3	3.5	7.7	0.2	10.823	18
$CH_3COC_2H_5$	12	24	0.7	10.63	17

TABLE 6 *(Continued)*

Ketone	$k_0 \times 10^3$, M^{-1}.s^{-1}	$k_{st} \times 10^3$, M^{-1}.s^{-1}	k_{rel}	lg A	E_a, kcal/mol
$CH_3COC_5H_{11}$	23	67	1	10.54	16.5
$CH_3COC_6H_{13}$	45	58	2.4	10.46	16
$CH_3COC_7H_{15}$	448	54	2.3	10.31	15.8
$CH_3COC_{10}H_{21}$	42	113	2.2	10.58	16.2
$C_2H_5COC_5H_{11}$	20	74	1	11.48	15.5
$CH_3COCH_2COCH_3$	29000	–	–	–	–
c-$C_7H_{12}O$	87	80	4.1	9.00	17
CH_3CO-2-Naph	22000	–	–	5.77	6
4-CH_3-c-C_6H_9O	117	101	6.2	8.97	12
4-t-Bu-c-C_6H_9O	166	123	8.1	7.93	11.8
Ketone	$k_0 \times 10^3$, M^{-1}.s^{-1}	$k_{st} \times 10^3$, M^{-1}.s^{-1}	k_{rel}	**lg A**	E_a, kcal/mol

Note: k_0 is calculated according to Eq. (2); k_{rel} is the relative rate constant determined by gas chromatography relative to ethylpentylketone.

The following six types of ketones have been the subjects of our investigations:

1. Acetone containing only α-C–H bonds,
2. Ketones with the general formula $CH_3CO(CH_2)_nCH_3$, where $n = 1$, 4, 5, 6 and 9, which contain secondary α, β, γ C–H bonds,
3. Ketones with the general formula $CH_3(CH_2)_m CO(CH_2)_nCH_3$, where $m = 1$, $n = 4$, which contain α, α'-secondary C–H bonds and cycloheptanone, which contains α, β and γ secondary C–H bonds, 4-4-methyl or 4-*tert*-butyl cyclohexanone with γ-*tert*-C–H bond,
4. 2-Naphthylmethylketone with various substituents, and
5. Acetylacetone containing α-C–H bond activated by two keto groups.

The rate constant of keto-enol tautomerism for the latter compound is considerably higher than those of the other ketones studied, because of appearance of the C+C-double bonds.

$$CH_3-CO-CH_2-CO-CH_3 \Leftrightarrow CH_3-CO-\mathbf{CH=C}(OH)-CH_3$$

The values of k vary, depending on the ketone structure, particularly on the content of primary, secondary or tertiary α-C–H-bonds, in a wide range. The ratio of their reactivities is $1 : 3.4 \div 24.8 : 33.4 \div 47.4$. In contrast to paraffins these values are more similar, which are associated to the activation of α-C–H-bonds by the keto group. Ozone is sensitive not only to the activation of the α-C–H bonds but also to the keto-enol equilibrium. The rate of its reaction with acetylacetone is by three orders of magnitude higher than that with monoketones. Ketones with different substituents, for example, 2-naphthylmethylketones also show higher reactivity than the aliphatic ketones. It is concluded that the reaction center in this case is the naphthyl ring and not the α-C–H bond in the methyl group.

Another interesting observation is the higher values of E_a compared with those of paraffins (with some 2–3 kcal/mol). As ozone is an electrophilic reagent and the activation of the α-C–H bonds from the keto group favours the nucleophilic attack, it is reasonable to expect an increase in E_a.

Figure 5 shows a typical kinetic curve illustrating the change of ozone concentration at the reactor outlet during ozonolysis of ketones in solution.

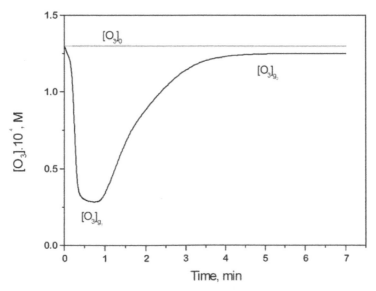

FIGURE 5   Ozonolysis of 0.712 mmol cycloheptylketone in 10 ml CCl_4, 21°C, v – 0.1 l/min.

The horizontal line describes the inlet ozone concentration and the curve corresponds to the change in its concentration at the reactor outlet. It is seen that the addition of ketone results in a sharp decrease in the ozone concentration, followed by a section with only slight changes of $[O_3]_{g1}$. After some time the ozone concentration rises up again to reach a region, where the respective curve is practically parallel to the abscissa ($[O_3]_{g2}$).

The shape of the kinetic curve of ozone consumption (Fig. 5) is complicated and this implies that ozone reacts with more than one compound [1, 26, 27]. The two horizontal sections of the curve are the result of the interaction of ozone with two compounds, the rate constants of which differ considerably. The areas over the curve correspond to the amount of absorbed ozone. Knowing the concentrations of the reacted compounds the stoichiometry coefficients and other kinetic parameters can be calculated.

This type of kinetic curve for the ketones is associated with the keto-enol equilibrium and with the ability of ozone to react with the C=C bonds 106 times faster than with the C–H bonds with almost zero activation energy [1].

The processes corresponding to the ozone-gas curve (Fig. 5) are the following:

1. The ozone starts its reaction with ketone (probably with the enolic form) and an abrupt fall of its concentration is registered;
2. When the rate of ozone supply becomes equal to the rate of the chemical reaction, the first step is formed;
3. The end of the step is connected with the consumption of the enolic form and the ozone concentration begins to rise up;
4. At this moment the ozone begins to react with the keto form and the second step is formed, whereby the rate of ozone supply becomes equal to the rate of the second reaction. The sharp transition between the two steady-state regions gives evidence that the rate of restoration of the keto-enolic equilibrium is considerably lower than the rate of ozone interaction with the keto form.

The concentration of the enolic form can be estimated from the area enclosed below the kinetic curve in Fig. 5 and then the equilibrium constant of enolization could be measured:

It is known that the stoichiometry of the reaction of ozone with C=C is 1:1 and thus the amount of the enolic form can be judged based on the amount of absorbed ozone, i.e., in this particular case the concentration of the enolic form is 0.134 mM. As the initial concentration of ketone is known, the equilibrium quantity will be the difference between the initial concentration and that of the enolic form. Then the equilibrium constant can be easily calculated using the formula:

$$K_e = [Enol]_e / [Keton]_e, \tag{4}$$

and it amounts to 1.9×10^{-2}.

Based on the curve in Fig. 5 one can also determine the rate constants of the ozone interaction with the two tautomeric forms. Taking into account the parameters of the first section of curve ($[O_3]_{g1}$) and by using Eq. (2), the value of the rate constant with the enolic form is found to be $k_E = 2.9 \times 10^2 \, M^{-1}.s^{-1}$ and therefore the second rate constant with the keto form is $k_K = 0.048 \, M^{-1}.s^{-1}$. In this way the possibility of application of the method of ozone titration to the measurement of the equilibrium constant and the rate constants of ozone interaction with the two tautomeric forms in one kinetic experiment has been demonstrated.

With a view to evaluate the possible influence of the processes of mass transfer on the values of the rate constants, determined by the barbotage method, beside the observation of the criteria, described in the experimental section, some additional experimentation has been carried out, connected with the changes in the size of the reactor, in the volume of the solution, and in the concentration of the ozone at the reactor inlet. The results, obtained upon varying the above-listed parameters within the range up to about 30–50%, did not show any significant deviations in the respective values of k, and therefore, there is noticeable effect of the conditions of mass transfer on these values.

It is evident that the rate constants of ozone interaction with ketone, measured by the static method (Fig. 6) through mixing of ozone and ketone solutions at $[K]_o >> [O_3]_o$ are higher than those found by the bubbling method (Table 6). This is not difficult to explain because in the former case the obtained values represent the total effective constant of interaction of both the enol and the keto form.

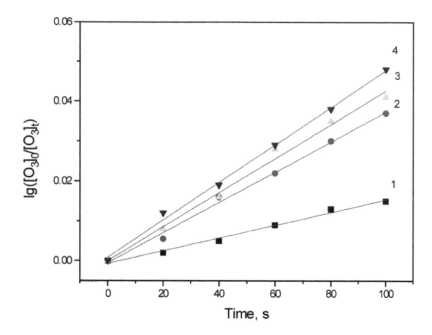

FIGURE 6 Semilogarithmic anamorphosis of the kinetic curve of ozonolysis of: 1 – methylethylketone; 2 – methylpentylketone; 3 – ethylpentylketone; and 4 – cycloheptylketone; 21 °C, ketone concentration 28 mM at $[O_3]_{lo}=1.10^{-3}$ M.

The results from the calculation of the equilibrium constants of keto-enol tautomerism for some aliphatic ketones in CCl_4 are given in Table 7. The data for acetylacetone and 2-naphthylmethylketone are not presented in the table because in the former case the rate of reaching the equilibrium is commensurable with the rate of ozone consumption and in the latter case the ozone reacts with the double bonds in the naphthyl ring. The equilibrium constants do not differ from those found within the temperature range of 21°C to 3°C and agrees with data from the literature [59–62].

TABLE 7 Equilibrium constants of keto-enol tautomerism of some ketones in CCl_4 solution at 21°C determined by titration with ozone.

Ketone	CH_3COCH_3	$CH_3COC_2H_5$	$CH_3COC_5H_{11}$	$CH_3COC_6H_{13}$	$CH_3COC_7H_{15}$	$CH_3COC_{10}H_{21}$	$c-C_7H_{12}O$
$K_e \times 10^4$	1.2	1.2	46	29	96	130	190

The kinetics of ozone reactions with ketones is also determined using gas chromatographic analyses. The relative rate constants shown in Table 6, column 4 demonstrate that only acetone and methylketone possess lower reactivity than the standard. It is seen that the rate constants, calculated from the relative values and the value of the standard constant correspond to those found by the bubbling method. The main products of ozone interaction with methylethylketone are 2-hydroxymethylethylketone, diacetyl, peroxides – alkyl and hydro, acetaldehyde and AcAc.

On the basis of the kinetic results obtained and the product analyses we suggest Eq. (3) for the proceeding of ketone ozonolysis:

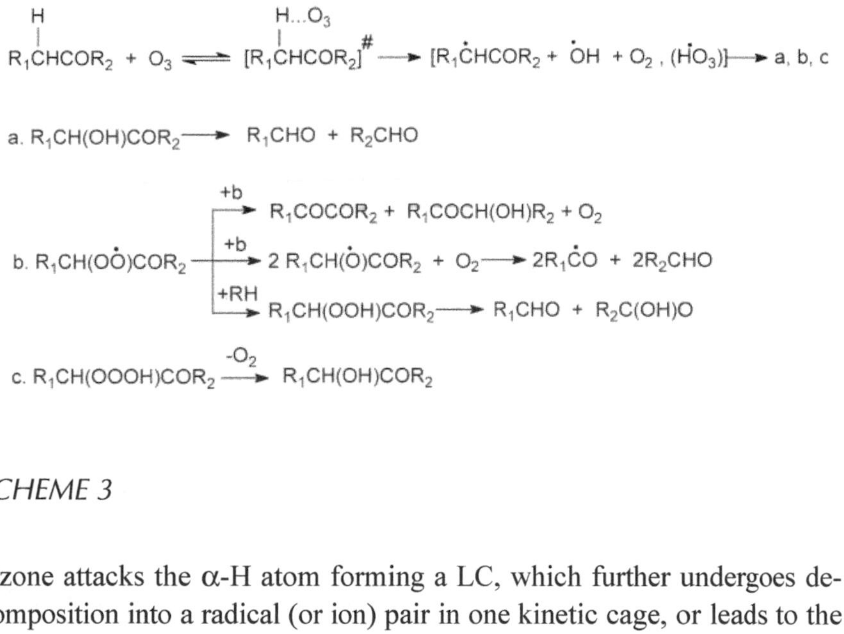

$$\overset{H}{\underset{|}{R_1CHCOR_2}} + O_3 \rightleftharpoons [R_1\overset{H...O_3}{\underset{|}{CHCOR_2}}]^{\#} \longrightarrow [R_1\dot{C}HCOR_2 + \dot{O}H + O_2, (\dot{H}O_3)] \longrightarrow a, b, c$$

a. $R_1CH(OH)COR_2 \longrightarrow R_1CHO + R_2CHO$

b. $R_1CH(O\dot{O})COR_2$
  $\xrightarrow{+b}$ $R_1COCOR_2 + R_1COCH(OH)R_2 + O_2$
  $\xrightarrow{+b}$ $2\,R_1CH(\dot{O})COR_2 + O_2 \longrightarrow 2R_1\dot{C}O + 2R_2CHO$
  $\xrightarrow{+RH}$ $R_1CH(OOH)COR_2 \longrightarrow R_1CHO + R_2C(OH)O$

c. $R_1CH(OOOH)COR_2 \xrightarrow{-O_2} R_1CH(OH)COR_2$

SCHEME 3

Ozone attacks the α-H atom forming a LC, which further undergoes decomposition into a radical (or ion) pair in one kinetic cage, or leads to the formation of hydroperoxy and alkoxy radicals. The latter can further undergo monomolecular decomposition.

The intermediate formation of LC is assumed in the first stage, followed by breaking of α-C–H bond leading to the formation of a radical (or ion) pair. Then α-hydroxyketone, α-peroxyketone radicals and α-hydroxytrioxyketone appear as a result of C–C bond breaking can decompose into two aldehydes.

The α-peroxyketone radical, reacting with the initial ketone, can be transformed into an α-hydroxyperoxyketone, which is decomposed through breaking of the C–C bond to aldehyde and acid or the two radicals recombine giving rise to diketone and α-hydroxyketone or two α-oxyketone radicals. The latter react with the initial ketone and are transformed into α-hydroxyketone or through a monomolecular decomposition and breaking of the C–C bond to an alkoxy radical and aldehyde. α-hydrotrioxyketone is converted into α-hydroxyketone and it is transferred in its turn into two aldehydes.

8.4.3.2   ETHERS

Investigations of ozone reactions with ethers began as early as the last century [63–67]. Results of these studies are summarized and reported by Bailley in his review [67]. The main products formed during ozonolysis of aliphatic ethers are alcohols, aldehydes, esters, acids, hydrotrioxides (HTO), hydrogen and organic peroxides, singlet oxygen and water. Price and co-workers [68] proposed the so-called 'insertion' mechanism, according to which ozone is inserted into the α-C–H bonds at the first step through a 1,3-dipolar addition, thus forming unstable HTO. This mechanism was also supported by the data of Erickson and Bailey [2, 67, 69] and Murray and co-workers [70], who measured the activation parameters of the decomposition of a series of HTO, obtained by the ozonolysis of some ethers. Giamalva and co-workers [71] summarized the possible mechanisms known today, i.e., interaction with the ether oxygen atom, 1,3-dipolar insertion of ozone into the α-C–H bonds, homolytic abstraction of the α-H atom and heterolytic abstraction with carboanion and carbocation formation. These authors clearly indicate the predominance of the one-step mechanisms with transfer of a hydrogen atom, hydride anion or cation.

As it was mentioned above, the low-temperature ozonolysis of ethers yields HTO, which however are stable at very low temperatures. At normal temperatures the composition of the products is found to be different from that after the decomposition of HTO. These two reasons gave us impetus to carry out extended studies on the ozone reaction with ethers in order to establish the real mechanism of these reactions at normal temperatures.

It was estimated, based on kinetic curve of ozone consumption (Fig. 7) and Eq. (2), that the kinetic constant of the ozonation of diethyl ether (DEE) at $t° = -5°C$ is 0.9 M^{-1}.s^{-1}. Taking into account the area, comprised between the straight line y=[O$_3$]$_0$ and the kinetic curves of DEE, similar to that in Fig. 5, which have been obtained at the lower concentrations of the reactant, the amount of ozone, consumed in the reaction has been determined. Juxtaposing this amount with the initial quantity of DEE in the reactor the stoichiometric coefficient of DEE ozonolysis has been estimated to be 1.

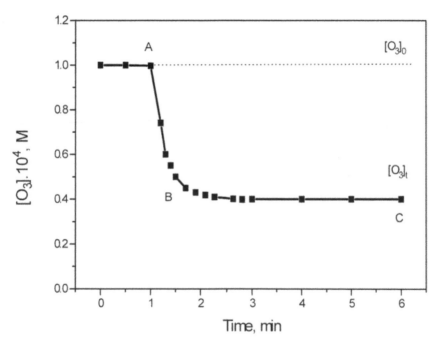

FIGURE 7   Kinetic curve of ozone concentration at the reactor outlet: 10 ml, 92 mM DEE, at −5°C.

The calculated stoichiometric coefficient of the ozonation reaction with other ethers was also unity. The kinetics of ozonolysis was studied with the example of n-dibutylether (DBE). The kinetic curves of the reaction product formation and the initial ether consumption are shown in Fig. 8.

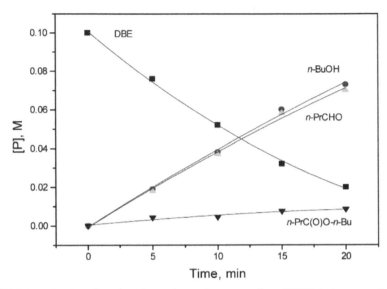

FIGURE 8   Kinetics of product formation and consumption of DBE during ozonolysis at ambient temperature, $[O_3]_0 = 2.34 \times 10^{-4}$ M.

The following products were identified by gas chromatography-mass spectrometry in the reaction mixture after 10 minutes (mass spectra of the obtained compounds are given below):

130 * 57 41 87 56 55 101 39 130 43 45 * n-Dibutylether,

74 * 56 41 43 42 55 39 57 45 40 41 * n-Butanol,

72 * 44 43 41 72 57 42 38 37 40 71 * Butanal,

88 * 60 73 41 42 43 45 39 55 61 88 * Butiryc acid;

144 * 71 89 56 43 41 57 60 73 55 42 * Butyl butyrate and

164 *& 247 * – Chloro-containing compounds like:

$CH_3(CH_2)_2CH(Cl)O(CH_2)_3CH_3$ (M_w 164.4) and

$CH_3(CH_2)_2CH(CCl_3)O(CH_2)_3CH_3$ (M_w 247)

The rate of DBE decomposition (7.4×10^{-5} M.s^{-1} determined on the basis of Fig.8) is almost equal to the sum of butanal (6.5×10^{-5} M.s^{-1}), butanol (6.71×10^{-5} M.s^{-1}) and butyl butyrate (1.0×10^{-5} M.s^{-1}) formation rates. The kinetic curves of butanol and butanal formation have almost the same slope and they start without any induction period. This could mean that they are formed in parallel reactions from a common precursor. The rate of butylbutyrate accumulation, as it is demonstrated by its kinetic curve, is approximately seven times lower. The latter indicates that their formation is occurring in a parallel reaction to butanol (butanal) formation, most probably from the same precursor. The ratio of product amount to initial DBE amount after 10 minutes was found to be 1:1. This means that the share of the auto-oxidation process is negligibly small, although we have identified some butyric acid by its IR spectrum (1765 cm^{-1} (monomer) and $\varepsilon = 1470$ M^{-1}.cm^{-1}) resulting from the butanal oxidation.

The possible existence of a common precursor raises the question about its nature. Such a precursor could be either α-hydroxyether (EOH) or α-hydrotrioxyether (EOOOH), which can produce further through intermolecular disproportionation simultaneously aldehyde and alcohol:

$$CH_3(CH_2)_2\,HC\!\!-\!\!O\!\cdot\!(CH_2)_3CH_3 \xrightarrow{\;-O_2\;} CH_3(CH_2)_2CHO \;+\; CH_3(CH_2)_2CH_2OH$$

(5')

$$CH_3(CH_2)_2\,HC\!\!-\!\!O\!\cdot\!(CH_2)_3CH_3 \longrightarrow CH_3(CH_2)_2CHO \;+\; CH_3(CH_2)_2CH_2OH$$

Probably, the intermolecular reaction of EOOOH disproportionation is preferable, because of the six-member-ring transition state, while EOH disproportionation occurs via a strained four-member-ring transition state, which is energetically unfavorable.

If EOH is assumed to be the precursor, then the butylbutyrate should be a major reaction product as a result of the rapid oxidation of EOH by ozone:

$$\text{[structure]} \xrightarrow{-H_2O,\ -O_2} CH_3(CH_2)_2C(O)O(CH_2)_3CH_3 \qquad (6)$$

which has not been observed experimentally.

When EOH is assumed to be the precursor, it is difficult to imagine its one-step transformation into an ester without any additional assumptions. Also the solvent, CCl_4, could not affect at all this transformation, bearing in mind its weak oxidizing properties and its non-specificity.

The formation of the ester from EOOOH in a single step can be easily presented through a four-member-ring transition state:

$$\text{[structure]} \xrightarrow{-H_2O_2} CH_3(CH_2)C(O)O(CH_2)_3CH_3 \qquad (6')$$

The formation of four-member transition state will be more unfavorable than the six-member transition state. This conclusion is in agreement with the kinetic data, namely, the rate of ester formation is approximately seven times lower than that of alcohol and aldehyde formation. Direct evidence for EOOOH formation was found only after prolonged ozonation for 24 h at $-78°C$ [72–75]. The NMR spectrum of the oxidate has a signal at $\delta = 13.52$ ppm, which is attributed to OOOH.

Figure 9 represents the dependences of the rate of ozonation of DBE on the concentrations of the ozone and that of DBE. Judging from the linear character of the dependences, in conformity with Eq. (5), it follows that the order of the reaction is unity with respect to each one of the two reactants and therefore rate law can be written in the form $W = k \times [DBE][O_3]$.

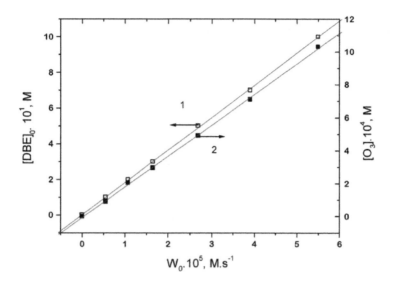

FIGURE 9   Dependence of the reaction rate of DBE ozonation (21°C) on the concentration of ozone (1) at [DBE] = 10 mM and DBE (2) at [O_3] = 0.1 mM.

Based on the results discussed above, a scheme of the ozone reaction with aliphatic ethers is proposed in Scheme 4.

Ether + $O_3 \rightarrow$ EOOOH (k_1)

EOOOH $\rightarrow$ Alcohol + Aldehyde + O_2 (k_2)

EOOOH $\rightarrow$ Ester + H_2O_2 (k_3)

SCHEME 4.

The formation of chlorine-containing compounds (1–2%) can be explained by the presence of radical intermediates in the reaction mixture. This means that EOOOH will also be decomposed via radical route:

EOOOH $\rightarrow$ EO$^\bullet$ + $^\bullet O_2H$ (k_4)

EO$^\bullet$ + EH $\rightarrow$ EOH + E$^\bullet$ (k_5)

E$^\bullet$ + $O_2 \rightarrow$ EO$_2^\bullet$ (k_p)

$E^{\bullet} + CCl_4 \rightarrow ECl + CCl_3^{\bullet}$ $(k_{p'})$

$EO_2^{\bullet} + EH \rightarrow EOOH + E^{\bullet}$ (k_p)

$E^{\bullet} + CCl_3^{\bullet} \rightarrow ECCl_3$ (k_t)

$2EO_2^{\bullet} \rightarrow$ non-radical products (k_t)

The validity of the mechanism, indicated above, was confirmed by the good agreement between the experimental points on the DBE decomposition curve and the product accumulation and by the theoretical curves, calculated according to the scheme and given in Fig. 8. For example, the curve describing the DBE consumption is obtained from Eq. (7):

$$[DBE]_t = [DBE]_0 . \exp(-k't) \tag{7}$$

where k' is the constant, measured by the stop-flow method – 1.26×10^{-3} s^{-1}, while the corresponding bimolecular constant is 6.3 M^{-1}.s^{-1}.

A comparison of the experimental values and calculated estimates of A, assuming linear (LC) and cyclic (CC) forms of the activated complex (AC) has been made. The kinetic constants of ozone reaction with diethylether (EtE), dichlorodiethylether (DClEtE), di-iso-propylether (i-PrE), di-n-butylether (n-ButE), di-iso-amylether (i-AmE) and di-n-amylether (n-AmE) in CCl$_4$ are given in Table 8.

TABLE 8 Rate constants of ethers ozonations at various temperatures (°C)

Ether	−15°C	−6°C	−5°C	3.5°C	4°C	10°C	13°C	21°C	24°C
EtE	0.5	0.9	1.0	1.2	1.6	2.2	2.4	3.1/3.0	3.6
DClEtE	–	–	0.009	0.016	–	–	0.028	0.035/0.04	–
i-PrE	–	–	1.6	1.9	–	–	3.4	5.4/5.1	–
n-ButE	–	–	1.9	2.1	–	–	4.3	6.3/6.3	–
i-AmE	–	–	1.9	2.5	–	–	4.1	7.8/7.0	–
n-AmE	–	–	2.2	2.8	–	–	4.2	7.8/7.5	–

Note: In column 9, after the slash the values of the constants determined by stop-flow method are given.

The rate constants for EtE, *n*-ButE, *i*-AmE and *n*-AmE have similar values. All these ethers have close α-C–H bond energies and similar electronic environment. The value of the rate constant for *i*-PrE ozonation is also close to those mentioned above, while the constant for DClEtE has 100 times lower value. In DIPE the presence of *tert*-C–H bonds, should contribute to their higher reactivity because of the lower energy of these bonds [34]. The presence of oxygen, however, and the more difficult stabilization of the transition state due to steric factors, makes this interaction slower. The very low rate constant of DClEtE ozonation can be attributed to the strong electron-accepting properties of the chlorine atom.

Arrhenius parameters, calculated on the basis of the data in Table 8, are summarized in Table 9.

TABLE 9   Arrhenius parameters of ozone reaction with some aliphatic ethers

Ether	k, $M^{-1}s^{-1}$	lg A	E_a, kcal/mol
EtE	3.1	6.4 (7.3)	7.9 (9.9)
DClEtE	0.035	4.9 (5.5)	8.5 (10.1)
i-PrE	5.4	6.3 (7.9)	7.6 (10.0)
n-ButE	6.3 (6.0)	6.6 (7.4)	7.8 (9.7)
i-AmE	6.8	6.6	7.8
n-AmE	7.8	6.5	7.6
THydP	(1.3)	(6.8)	(9.7)
THydF	(12.3)	(6.6)	(8.2)

Note: The values in brackets are taken from Ref. (43); THydP and THydF are tetrahydropyrane and tetrahydrofurane, respectively.

The comparison of the results, obtained with reference data, shows a good coincidence in regard to the values of the rate constants at 21°C. Only a slight difference in the activation energies was observed. Perhaps, it is the heat of ozone dissolution in CCl_4 that needs to be taken into account as the reason for these differences.

All reaction schemes, found in the current literature, describe the ozonation of ethers proceeding through two geometric forms of the activated complex – LC and CC:

The CC has a more compact structure, without any possibility for free rotation. It can be a transition state for the following reaction:

$$EH + O_3 \rightarrow EOOOH \rightarrow products \qquad (a)$$

where EOOOH is being formed in one step via 1,3-dipolar insertion.

The LC structure is an open one, allowing free rotation around H–O and O–O bonds. LC can be a transition state for the reactions:

$$EH + O_3 \rightarrow E^{\bullet} + HO_3{}^{\bullet} \rightarrow EOOOH \rightarrow products \qquad (b)$$

$$EH + O_3 \rightarrow E^{+} + HO_3{}^{-} \rightarrow EOOOH \rightarrow products \qquad (c)$$

in which an H atom or hydride anion abstraction occurs and EOOOH is formed during the second step.

The necessary parameters for estimating the A values are represented in Table 10.

TABLE 10 Heats of formation (ΔH) in kcal of initial, intermediate and final products, van der Waals radii of the ethers and free rotation energy (E_{fr}) around H–O bond.

Ether	r, [Å]	ΔH, EH	ΔH, EOOOH	ΔH, $E^{\bullet}$	ΔH, E^{+}	E_a, kcal/mol
EtE	3.18	−68	−66	−32	135	1.3
dClEtE	3.51	−74	−73	−41	132	2.2
i-PrE	3.48	−76	−74	−42	125	1.4
n-ButE	4.03	−81	−79	−47	120	1.3
i-AmE	4.31	−87	−86	−53	115	1.3
n-AmE	4.45	−86	−85	−51	116	1.3

The calculated values of A are shown in Table 11. The good agreement between the calculated estimates and experimental values for A confirms the conclusion that the geometry of the transition state is a linear (LC) and abstraction mechanisms (a) and (c) seems more probable.

TABLE 11   The calculated pre-exponents (A) and steric factors (p).

Ether	lg A, CT	lg A, CC	lg A, LC	p × 10⁶, CC	p × 10⁶, LC	A_{calc}^{LC}/A_{exp}
EtE	11.66	4.02	6.40	0.02	5.5	1
dClEtE	11.66	3.62	4.99	0.009	0.2	1.23
i-PrE	11.67	4.28	6.35	0.04	4.8	0.91
n-ButE	11.72	3.99	6.09	0.02	2.3	0.31
i-AmE	11.75	3.90	6.00	0.01	1.8	0.25
n-AmE	11.74	3.77	5.87	0.01	1.3	0.23

Note: the values of A in columns 3 and 4 are per one equivalent α-C–H atom.

In order to evaluate which one of the mechanisms is occurring we used the thermodynamic parameters shown in Table 10. The calculated heats of formation of the individual compounds fit well to the experimentally measured values [48, 49]. Using different expressions for the Hamiltonians, we have obtained the following values for ΔH: O_3 –33.3 kcal against the literature data of (34.1); HOOOH –17.8 (–17.1); HOOO –9.2 (–17.8); HOOO –20.3(–24.9); HO –5.2(–25.8).

$$\Delta H = -\Delta H\,(EH) - \Delta H\,(O_3) + \Delta H\,(EOOOH) \tag{a}$$

$$\Delta H = -\Delta H\,(EH) - \Delta H\,(O_3) + \Delta H\,(E^{\bullet}) + \Delta H\,(HO_3^{\bullet}) \tag{b}$$

$$\Delta H = -\Delta H\,(EH) - \Delta H\,(O_3) + \Delta H\,(\text{Å}^+) + \Delta H\,(HO_3^-) - e^2/r_{ip} - E_s \tag{c}$$

where both types of values: (1) for energy of interaction between two charged particles (e^2/r_{ip} = 5.4 eV = 2.06×10^{-16} kcal), and (2) for energy of solubility in tetrachloromethane solution (E_s = 17 kcal) were taken from reference [76]. The calculated values for ΔH are depicted in Table 12. From thermodynamic point of view, (a) is the preferable mechanism and out of (b) and (c), (b) is more plausible.

TABLE 12   Calculated heats of investigated reactions according to mechanisms a, b and c.

Ether	a, kcal/mol	b, kcal/mol	c, kcal/mol
EtE	−32	−16	0
dClEtE	−33	−19	6
i-PrE	−32	−18	1
n-ButE	−32	−18	1
i-AmE	−32	−18	2
n-AmE	−33	−17	2

8.4.3.3 HYDROXYBENZENES

The reactions of ozone with mono- and dihydroxybenzenes have provoked so far a particular interest (1, 2, 34, 73–75, 77–79), namely, because of their great importance for environment protection, chemical stabilization and the theory of reactivity. The ozonation of phenol, pyrocatechol, resorcinol and hydroquinone has been studied in different solvents – aqueous and organic, aimed at the deriving of the kinetic parameters and product composition [21–24, 80–86). The rate constants of phenol and resorcinol ozonation in water at room temperature are 1.3×10^3 M^{-1}.s^{-1} and $>3 \times 10^5$ M^{-1}.s^{-1}, respectively, whereas the rate constants of benzene, toluene and anisole ozonation in organic media are 2, 14 and 2.9×10^2 M^{-1}.s^{-1} [72–75].

Gurol and co-workers [77] found that the relative rates of pyrocatechol/phenol and resorcinol/phenol ozonation in water medium are 220 and 70, respectively. Provided that the rate constant of phenol ozonation is known [75], the calculated values of the rate constants of pyrocatechol and resorcinol ozonation are 2.86×10^5 $M^{-1}.s^{-1}$ and 9.1×10^4 $M^{-1}.s^{-1}$, respectively. However, in the case when the reaction is carried out in organic solvents the values are quite different. For example, in CCl_4 and at room temperature the following values have been obtained for: benzene – 0.06, ethylbenzene – 0.2, anisole – 1.1, phenol – 230 and pyrocatechol – 3.2×10^3 $M^{-1}.s^{-1}$ [70–73]. One of the possible explanations for the different values of the rate constants of hydroxyphenols ozonation obtained by the various researchers could be the great influence of the water on this reaction, for example for phenol they vary within the range 100–180 $M^{-1}.s^{-1}$, 500 $M^{-1}.s^{-1}$ for pyrocatechol and 300 $M^{-1}.s^{-1}$ for 3,6-di-*tert*-butylpyrocatechol. Pryor and co-workers [72] have reported that the rate constants of ozone reaction with α-tocopherol in CCl_4 and water are 5.5×10^3 $M^{-1}.s^{-1}$ and 1×10^6 $M^{-1}.s^{-1}$, with α-tocopherol acetate – 1.45×10^2 $M^{-1}.s^{-1}$ and for the reaction with α-tocopherolquinone it is 1.15×10^4 $M^{-1}.s^{-1}$.

Ozone is unstable when dissolved in water. The mechanism and the kinetics of the elementary reactions, involved in the ozone decomposition, have been investigated in numerous studies [87–92]. The stability of the ozone depends to a great extent on the water matrix, especially on its pH value. The pH of the water is important because the hydroxide ions initiate ozone decomposition, which involves the following reactions [87]:

$$O_3 + OH^- \rightarrow HO_2^- + O_2 \quad k = 70\ M^{-1}s^{-1} \tag{8}$$

$$O_3 + HO_2^- \rightarrow {}^\bullet OH + O_2^{\bullet -} + O_2 \quad k = 2.8 \rightarrow 10^6\ M^{-1}s^{-1} \tag{9}$$

$$O_3 + O_2^{\bullet -} \rightarrow O_3^{\bullet -} + O_2 \quad k = 1.6 \rightarrow 10^9\ M^{-1}s^{-1} \tag{10}$$

In aqueous solution the ozonide radical decomposes quickly into an OH radical, O_2 and OH anion [87, 88].

Additionally in the case of ozone reaction with electron-rich compounds, such as phenols (phenolates) and akoxylated aromatics in aqueous phase again OH radicals are formed, most probably via the mechanism of formation of the respective ozone adduct [88, 89]. For example it has been established that as a result of the electron transfer from phenol to ozone (reaction 11) at neutral pH the radical anion O_3^- is obtained with an yield of 22% [90].

$$PhO^- + O_3 \rightarrow PhO^{\bullet} + O_3^{\bullet -} \tag{11}$$

As a result the rate constant of one and the same compound (of phenols or alkoxylated aromatics type) in water, determined on the basis of its ozone consumption, varies within a wide range of values depending on the pH values, and on the presence of hydroxyl radical scavengers minimizing the interfering radical chain reactions. This rate constant differs from that one, determined on the basis of the compound consumption [91]. However, this aspect requires special studies and discussion, which are beyond the scope of the present investigation.

Product analysis of pyrocatechol ozonation in aqueous medium shows that 3 moles of ozone are readily absorbed to give CO_2 (24.8%), CO (6%), formic acid (32.5%) and glyoxal (4.2%). As CO_2 and formic acid are the main reaction products, it seems very likely that most of the pyrocatechol undergoes anomalous ozonolysis [82]. Radical formation has been observed during 2,6-di-*ter*-butylphenol ozonolysis. When the ratio of absorbed ozone to phenol is 0.8, 50% viscous yellow oil and 3,5, 2,6-di-*tert*-butyl-*o*-quinone have been identified in the products [85]. In 3,6-di-*tert*-butyl-pyrocatechol ozonation the corresponding quinone was found to be the main product after complete consumption of the initial substrate [86]. Side products such as 2,5-di-*tert*-butyl-, 3-hydroxy-*p*-quinone and 3,6-di-*tert*-butyl-1,2-phenylacetal-5-hydroxy-3,6-di-*tert*-butyl-*p*-benzophenone have also been found. The rate constant of this reaction amounts to 3×10^2 $M^{-1}.s^{-1}$. The same constant with the corresponding pyrocatechol with O-hydroxy-acetylated groups has been estimated to have a two orders of magnitude lower value. A mechanism has been supposed to involve the formation of either a 1,3-cyclic activated complex between two hydrogen atoms from two OH groups and one ozone molecule or π- or σ-complexes, formed with the benzene ring.

The different values of the literature constants and the various mechanisms proposed for this reaction imposes the necessity of further research of its kinetics and reaction pathway. In this connection we have studied the ozonolysis of the following hydroxybenzenes (Scheme 5).

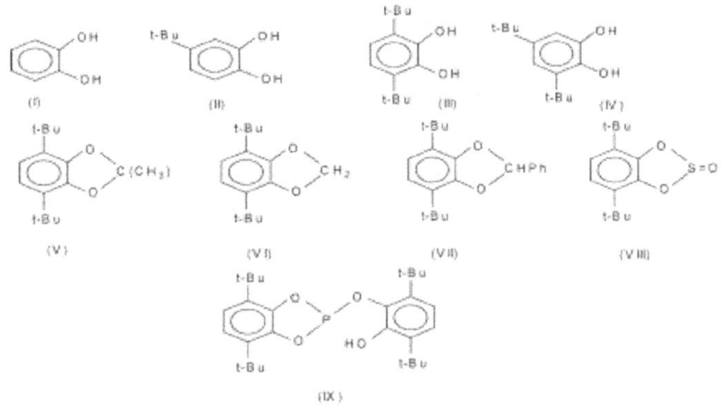

SCHEME 5.

The probable mechanisms of ozone interaction with dihydroxybenzenes are represented in Scheme 6.

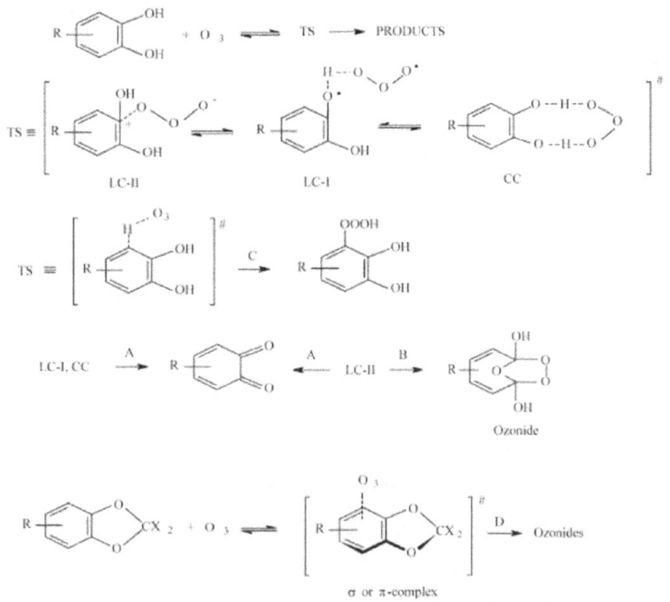

SCHEME 6.

The mechanism A with a cyclic complex (CC) formation in the transition state was supposed by Razumovskii and co-workers [86]. The mechanism B with linear complex LC–II in the transition state was put forward and discussed by Bailey [2]. The interaction of ozone with C–H bonds with the formation of trioxide [71, 73] as a possible parallel reaction is indicated in mechanism C. The acetylated forms of dihydroxybenzenes can react only via attack on the benzene ring according to mechanism D or C. Formally, mechanism D can be regarded as an extended version of mechanism B, involving the formation of TS similar to $\pi-$ or σ-complexes. We propose a new mechanism, an extended version of the Razumovskii mechanism, whereby the transition state is linear with LC–I structure.

Upon ozonation of any of the investigated catechols directly inside the electron spin resonance cell or after freezing the reaction products in liquid nitrogen no signals have been detected.

The kinetic curve of the changes in the ozone concentration at the bubbling reactor outlet (Fig. 10) is characterized by three different regions: AB – fast ozone consumption after the addition of pyrocatechol, BC – steady-state part, when the rate of the chemical reaction becomes equal to the rate of ozone supply, and CD – the ozone concentration begins to rise up due to the pyrocatechol consumption. The BC part of the curve allows calculation of the rate constant, and based on the area below the curve ABCD – evaluation of the stoichiometry of the reaction. The straight line designated $[O_3]_0$ is the ozone concentration at the reactor inlet. Curve 2 presents the o-quinone formation in the course of the reaction time. Its profile suggests the intermediate formation of o-quinone.

The kinetic curves of the product formation in pyrocatechol ozonolysis, its consumption and o-quinone consumption during its ozonation in a separate experiment are given in Fig. 11. The rate constants of pyrocatechol and o-quinone consumption, calculated based on the kinetic curves in Fig. 11, were 3.2×10^4 $M^{-1}.s^{-1}$ and 7.1×10^4 $M^{-1}.s^{-1}$, respectively. The initial rate of the o-quinone formation had almost the same value as that of pyrocatechol consumption. The small variation of the constants is due to the participation of pyrocatechol in parallel reactions. Actually, during the reaction small amounts of open-chain products have been identified.

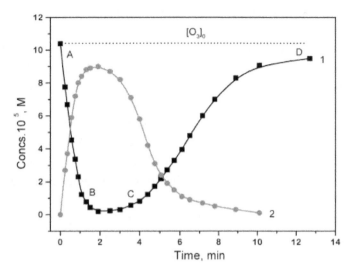

FIGURE 10   Kinetic curves of ozone absorption at reactor outlet [1] and *o*-quinone accumulation [2] in the course of pyrocatechol ozonation at ambient temperature, [PC] = 0.227 mM in CCl$_4$ (10 ml) during bubbling of ozone with 0.1 l/min flow rate.

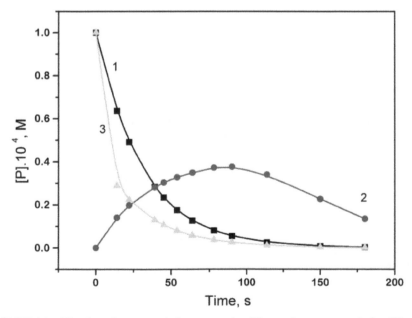

FIGURE 11   Kinetics of pyrocatechol consumption [1], *o*-quinone accumulation [2] and *o*-quinine consumption [3], at ambient temperature, [O$_3$] = 1 × 10^{-5} M.

Pyrocatechol ozonation at 15% conversion degree gave the following yields: o-quinone – 85%, pyrogallol – 3%, ozonide – 10%, muconic acid –2%, maleic acid and fumaric acids and the polymeric products – 1%. The ratio between the amount of absorbed ozone and the consumed pyrocatechol was calculated to be 6. Similar ratio values have also been obtained for the other hydroxybenzenes with free hydroxy groups.

It was found that with the increase of the conversion degree from 0 to 100%, the amount of open-chain products is continuously increasing, while that of the remaining products passes through a maximum. The individual compounds identified by ^{13}C–NMR are: muconic semialdehyde and acid – 20%, maleic and fumaric semialdehyde and acid – 40%, glyoxal, formic acid, oxalic acid, carbon dioxide and polymeric products – 40%. The reaction rate of dihydroxybenzene ozonation follows first-order kinetics in relation to each reagent (Figs. 12a and 12b).

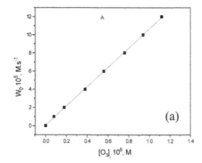 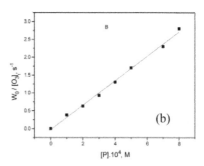

FIGURE 12   Dependence of the rate of pyrocatechol ozonation on the concentration of ozone (a) and on the concentration of pyrocatechol (b).

Table 13 represents the obtained experimental kinetic data on the ozonation of the investigated dihydroxybenzenes at various temperatures.

TABLE 13   Kinetic data of dihydroxybenzene ozonolysis in CCl_4, 20°C.

Substrate	k, $M^{-1}.s^{-1}$	lg A	E_a, kcal/mol
I	3230	5.35	2.5
II	3100	4.91	1.9
III	3100	5.07	2.1

TABLE 13   *(Continued)*

Substrate	k, $M^{-1}.s^{-1}$	lg A	E_a, kcal/mol
IV	3200	4.80	1.7
V	500	7.14	6.0
VI	749	7.17	5.8
VII	828	7.06	5.6
VIII	598	6.84	5.5
IX	111	6.81	6.4
Benzene	0.06	8.32	12.8
Toluene	0.40	7.05	10.3
Anisole	10	–	–
Phenol	160	7.31	7.0

A drastic difference between the values of the kinetic parameters of catechols I–IV and V–IX, in which the OH groups are acetylated, has been observed. The rate constants of catechols I–IV manifested 4–28 times higher values than those of catechols of V–IX types. On the other hand the pre-exponential factors demonstrated with about two orders of magnitude lower values with the former compounds. The activation energies are 2.05 ± 0.5 and 5.9 ± 0.5 kcal/mol, respectively, or the acetylation of the HO groups leads to an increase in E_a by about 4 kcal/mol. The ozone probably reacts predominantly with the hydrogen atoms of the HO groups and, only to a very small extent, with the benzene ring. The ratio of these ozone interactions varies from 94:4 to 80:20 depending on the hydroxybenzene nature. The lower activation energies of catechols I–IV with respect to those of V–IX are more consistent with the formation of an activated complex of contact type in the transition state, i.e., with structure LC–I (Scheme 7). This assumption could also be confirmed by the analysis of the kinetic parameters of the ozonation of anizol, benzenes, phenol and toluene. The reaction of ozone with benzene proceeds at a low specific rate of 0.06 $M^{-1}.s^{-1}$ and a relatively high activation energy equal to 12.8 kcal/mol. The methyl derivatives of benzene, i.e., toluene, reacts 6.7 times faster and the E_a is decreased by 2.5 kcal/mol, and the rate of anizol ozonation is even higher – 10 $M^{-1}.s^{-1}$. The exchange of ketyl group by a hydroxyl one leads to significant acceleration of the ozonation reaction rate and it approaches values of 160 $M^{-1}.s^{-1}$, and the activation energy is reduced by 5.8 kcal/mol. Probably, in this case the mechanism of the reaction is changed and ozone interacts predominantly with the hydrogen atoms of the OH groups. The decrease in the reactivity of anizol compared with that of phenol (~16 times) supports this assumption.

The analysis of the kinetic parameters of catechols in acetylated form and those of benzene, toluene, anizol and phenol shows that the rate constants increase and the activation energies go down upon increasing the number of the electron-donating substituents (Table 13). The high values of the pre-exponential factors could be associated with the high values of the energy of free rotation in the activated complex as a result of the steric hindrances caused by the presence of the bulky *tert*-butyl groups. This means that in the case when the hydroxyl groups are acetylated, the activated complex should have an LC–II like structure and the ozonation would lead only to the formation of ozonides and open ring products.

The kinetic parameters of some dihydroxybenzenes have been determined in the case of ozonation in aqueous medium (for those ones, which are water soluble – I–IV, VIII and IX). However, we have found that because of the diffusion limitations in the bubbling reactor we have not been able to measure constants higher than 1×10^4 $M^{-1}.s^{-1}$. In fact, the literature values for some of these constants are of this order, but they have been measured by a modified stop-flow technique [32].

The analyses made so far show that the reaction path of the ozone reactions with dihydroxybenzenes depends strongly on their nature and they precede via transition states with activated complex – LC–I or LC–II. The parameters needed for the calculations are represented in Table 14.

TABLE 14 Symmetry numbers (σ), VdW radii (r), heats of formation (ΔH) of dihydroxybenzenes and the corresponding quinones and ozonides

Substrate	σ	r, Å	ΔH, kcal/mol	$\Delta H^{quinone}$, kcal/mol	$\Delta H^{ozonide}$, kcal/mol
I	2	3.02	−62 (−85)	−39 (−43)	−73
II	1	3.82	−88	−66	−99
III	2	4.46	−109	−86	−125
IV	1	4.39	−110	−90	−123
V	2	5.07	−116		−125
VI	2	4.56	−90		−103
VII	1	5.37	−63		−86
VIII	1	4.77	−73		−77
IX	1	5.30	−135		−140
Benzene	6	2.86	19 (13)		21
Toluene	1	3.37	12 (3)		11
Anisole	1	3.45	−16 (−26)		−20
Phenol	1	2.96	−21 (−38)		−24

Note: The experimental values are in parentheses [92].

In addition, we have calculated the heats of formation of pyrogallol – –117.6 kcal/mol (–129 kcal/ mol [(92)]), 3-trihydroxy pyrocatechol – –59 kcal/mol, H_2O_3 – –21.2 kcal/mol (–17.7 kcal/mol [(92)]) and ozone – 34.1 kcal/mol.

Thus, the heats of the ozonation reaction according to the different mechanisms have been calculated and they amount to the following values: A – 33.1; B and D – 46.1 and C – 32.1 kcal/mol. All the mechanisms are exothermic and, therefore thermodynamically favorable. In this case only the magnitude of the activation energy and the entropy benefits will determine the reactivities of these compounds and the reaction pathway.

Calculated pre-exponential factors (Table 15) were compared with those obtained experimentally (Table 13) and it is seen that A have lower values (~10 times) for the cyclic form of the activated complex, if compared with the experimental values. At the same time the values of A for LC are about 200 times higher. The values of A for CC are the highest as predicted by the theory and they are lower than the experimental values. This supposes that the reaction takes place via a cyclic complex. The values of A calculated for LC in Table 15 have been obtained without taking into account the energy of free rotation. The latter was calculated as a sum of the rotation around H–O and O–O axes (by *mopac6*) and it amounts to 3.1 kcal. This means that the real values in column 4 are about 200 times smaller. The comparison between the A values corrected in this way and the experimental data for the compounds I–IV reveals complete agreement.

TABLE 15 Calculated pre-exponential values for ozone reaction with dihydroxybenzenes.

Substrate	lg A TC	lg A, CC	lg A, LC	lg A, calc.	A_{calc}/A_{obs}	lg p
1	2	3	4	5	6	7
I	11.318	4.346	7.69	5.39	1.10	5.928
II	11.422	3.905	7.25	4.95	1.10	6.472
III	11.500	4.070	7.42	5.12	1.12	6.380
IV	11.490	3.780	7.12	4.82	1.05	6.670
V	11.572	4.017	7.36	7.36	1.66	4.212
VI	11.511	4.056	7.40	7.40	1.70	4.102
VII	11.801	3.934	7.28	7.28	1.66	4.521
VIII	11.533	3.717	7.06	7.06	1.66	4.473
IX	11.575	3.689	6.88	6.88	1.10	4.695
Benzene	11.319	5.124	8.47	8.47–	1.41	2.849
Phenol	11.322	4.196	7.54	7.54	1.70	3.782

Note: p is the collision factor.

The reaction pathway of the ozonation reactions of compounds V–IX is quite different as the transition state includes the formation of a σ– or π–complex. For these compounds the formation of AC in the transition state is impeded due to the breaking of their aromatic character but simultaneously the free rotation is facilitated, whose energy may be even zero, due to the action of the principle of the lowest energy. If the energy of free rotation were assumed to be very low then the values of A calculated for LC–II would coincide with the experimentally found ones. Such a coincidence is observed for compound V in column 5 (Table 15).

8.6 CONCLUSION

The basic conclusion from the analysis of the results obtained for this reaction is that the kinetics and mechanism of the ozonation reaction of dihydroxybenzenes depend strongly on their structure and the type of the reaction medium.

KEYWORDS

- alcohols
- ethers
- hydroxybenzenes
- ketones

REFERENCES

1. Razumovskii, S. D.; Rakovsky, S. K.; Shopov, D. M.; Zaikov, G. E. Publishing House of the Bulgarian Academy of Sciences: Sofia, Bulgaria, **1983**.
2. Bailey, P. S. In *Ozonation in Organic Chemistry, Volume 2 – Nonolefinic Compounds*; Academic Press: New York, NY, USA, **1982**, 255.
3. Waters, W. L.; Rollin, A. J.; Bardwell, C. M.; Schneider, J. A.; Aanerud, T. W. *J. Org. Chem.* **1976**, *41*, 889.

4. Otal, E.; Mantzavinos, D.; Delga, M. V.; Hellenbrand, R.; Lebrato, J.; Metcalfe, I. S.; Livingston, A. G. *Journal of Chemical Technology and Biotechnology,* **1997**, *70(2),* 147.

5. Benitez, F. J.; Beltran-Heredia, J.; Acero, J. L.; Pinilla, A. *Journal of Chemical Technology and Biotechnology,* **1997**, *70(3),* 253.

6. Amalric, L.; Guillard, C.; Pichat, P. *Research on Chemical Intermediates,* **1995**, *21(1),* 33.

7. Heinzle, E.; Stockinger, H.; Stern, M.; Fahmy, M.; Kut, O.M. *Journal of Chemical Technology and Biotechnology,* **1995**, *62(3),* 241.

8. Beltran, F. J.; Alvarez, P. M.; Legube, B.; Allemane, H. *Journal of Chemical Technology and Biotechnology,* **1995**, *62(3),* 272.

9. Lee, M. J.; Arai, H.; Miyata, T. *Chemistry Letters,* **1994**, *6,* 1069.

10. Benko, E. M. ; Kovaleva, V. V.; Mamleeva, N. A.; Mitrofanova, A. N.; Voblikova, V. A.; Pryakhin, A. N.; Lunin, V. V. *Zhurnal Fisicheskoi Khimii,* **1994**, *68(11),* 1964.

11. Artemev, Y. M.; Artemeva, M. A.; Vinogradov, M. G.; Ilika, T. I. *Russian Journal of Applied Chemistry,* **1994**, *67(9* (Part 2), 1354.

12. Scheuer, C.; Wimmer, B.; Bischof, H.; Nguyen, L.; Maguhn, J.; Spitzauer, P.; Kettrup, A.; Wabner, D. *Journal of Chromatography A,* **1995**, *706(1-2),* 253.

13. Hoger, B.; Gilbert, E.; Eberle, S. H. *Vom Wasser,* **1993**, *80,* 187.

14. Ruck, W.; *Vom Wasser,* **1993**, *80,* 253.

15. Gab, S.; Turner, W. V.; Wolff, S.; Becker, K. H.; Ruppert, L.; Brockmann, K. J. *Atmospheric Environment,* **1995**, *29(18),* 2401.

16. Takeuchi, K.; Fushimi, C.; Nakamura, K.; Ibusuki, T. *Bunseki Kagaku,* **1993**, *42(10),* 625.

17. Pan, G. Y.; Chen, C. L.; Gratzl, J. S.; Chang, H. M. *Research on Chemical Intermediates,* **1995**, *21(3-5),* 205.

18. Wada, H.; Naoi, T.; Kuroda, Y. *Nippon Kagaku Kaishi,* **1995**, *4,* 306.

19. Marco, A.; Chamarro, E.; Esplugas, S. *Afinidad,* **1994**, *51(452),* 265.

20. Starek, J.; Zukal, A.; Rathousky, J. *Carbon,* **1994**, *32(2),* 207.

21. Mvula, E.; von Sonntag, C. *Organic & Biomolecular Chemistry,* **2003**, *1(10),* 1749.

22. Ben'ko, E. M.; Bokova, M. N.; Pryakhin, A. N.; Lunin, V. V. *Russian Journal of Physical Chemistry,* **2003**, *77(5),* 739.

23. Ramseier, M. K.; von Gunten, U. *Ozone-Science & Engineering,* **2009**, *31(3),* 201.

24. Valsania, M. C.; Fasano, F.; Richardson, S. D.; Vincenti, M. *Water research* **2012**, *46,* 2795.

25. Rakovsky, S. K.; Cherneva, D. R. *International Journal of Chemical Kinetics,* **1990**, *22(4),* 321.

26. Rakovsky, S. K.; Cherneva, D. R.; Shopov, D. M.; Razumovskiy, S. D. *Communications of the Department of Chemistry,* Bulgarian Academy of Sciences, **1976**, *9(4),* 711.

27. Rakovsky, S. K.; Cherneva, D. R. *Oxidation Communications,* **1989**, *12(3),* 108.

28. Rakovsky, S. K.; Kulak, L. G.; Kuramshin, E. M.; Zlotsky, S. S.; Rakhmankulov, D. L. *Communications of the Department of Chemistry,* Bulgarian Academy of Sciences, **1989**, *22(3),* 722.

29. Rakhmankulov, D. L.; Zlotsky, S. S.; Rudnik, L. Z.; Teregulova, G. T.; Rakovsky, S. K. *Communications of the Department of Chemistry*, Bulgarian Academy of Sciences, **1989**, *22(3)*, 652.

30. Anachkov, M. P.; Rakovsky, S. K.; Stoyanov, A. K.; Fotty, R. K., *Thermochemica Acta*, **1994**, *237*, 213.

31. Rakovsky, S. K.; Cherneva, D. R.; Deneva, M., *International Journal of Chemical Kinetics*, **1995**, *27(2)*, 153.

32. Rakovsky, S. K.; Cherneva, D. R.; Deneva, M.; Ershov, V. V. *Oxidation Communications*, **1997**, *20(2)*, 169.

33. Rakovsky, S. K.; Sheldon, R. A.; Rantwijk, F. V. *Oxidation Communications*, **1996**, *19(4)*, 482.

34. Zaikov, G. E.; Rakovsky, S. K., *Ozonation of Organic & Polymer Compounds, iSmithers, Smithers Rapra, Shawbury, Shreswsbury, Shropshire*, SY4 4NR, UK, **2009**.

35. Popova, D.; Rakovsky, S.; Anachkov, M. *Oxidation Communications*, **2007**, *30(3)*, 529.

36. Eyring, H.; Lin S. H.; Lin S. M. *BASIC Chemical Kinetics*, Moskow, Mir, **1983** (In Russian).

37. Rozovskii, A. *Ja. Heterogeneous Chemical Reactions (Kinetics and Macrokinetics)*, Nauka: Moscow, **1980**, (In Russian).

38. Rozenberg, M. M.; Brun, E. B. *Teor. Osn. Khim. Teknol.* (In Russian), **1990**, *24(2)*, 198.

39. Denisov, E. T.; Mitskevich, N. I.; Agabekov, V. E. *Mechanism of Liquid-Phase Oxidation of Oxygen Containing Compounds*, Science and Technique Publishing House, Minsk, **1975**.

40. Whiting, M.; Bolt, A.; Parish, J. *Oxidation of Organic Compounds – III*, Advances in Chemistry Series, **1968**, *77*, ACS, Washington, DC, USA, p 4.

41. Williamson, D.; Cvetanovich, R. J. *Journal of American Chemical Society*, **1970**, *92*, 2949.

42. Gerchikov, A. Ya.; Kuznetzov, E. R.; Denisov, E. T. *Kinetika i Kataliz*, **1970**, *15*, 109.

43. Shereshovetz, V. V.; Shafikov, N.; Komissarov, V. D. *Kinetika i Kataliz*, **1980**, *21*, 1596.

44. Shafikov, N. Ya.; Gusmanov, A. A.; Zimin, Yu. S.; Komissarov, V. D. *Kinetics and Catalysis*, **2002**, *43(6)*, 799.

45. Gerchikov, A. Ya.; Zimin, Yu. S.; Trukhanova, N. V.; Evgrafov, V. N. *React.Kinet. Catal.Lett.* **1999**, *68(2)*, 257.

46. Borodin, A. A.; Razumovskii, S. D. *Kinetics and Catalysis*, **2009**, *50*, 3385.

47. Rakovsky, S. K. *Kinetics and mechanism of the ozone reactions with paraffins in liquid phase*, Moscow, Institute of Chemical Physics, Academy of Science of the USSR, **1975**. (PhD Thesis).

48. Kondratiev, V. N. (Chief Editor), *Bond Energy, Ionization Potential and Electron Affinity*, [in Russian], Nauka Publishers: Moscow, **1974**, 83.

49. Gordon, A. J.; Ford, R. A. *The Chemists Companion*, Mir Publishing House: Moscow, **1976**.

50. Eiring, G.; Lin, S. G.; Lin, S. M. *The Principles of Chemical Kinetics*, Mir Publishing House: Moscow, **1983**.

51. Karadakov, K.; Rakovsky, S. K. Calculation of Pre-exponents ver.1.0, (Software Copyright, 1987, based on the J. Howbert, Software: The Molecular Animator, vers. 1–3, Copyright, **1983**.

52. Thiel, W.; Software, IBM, Program MNDOC, QCPE ver., Fachbereich Physikalische Chemie der Philipps-Universitat, auf den Zahbergen, D-3550, Marburg, Germany, based on M.Y.S. Dewar and W. Thiel, *Journal of the American Chemical Society,* **1977**, *99*, 4899.

53. Gerchikov, A. Ya.; Komissarov, V. D.; Denisov, E. T.; Kochemassov, G. B., *Kinetika i Kataliz,* **1972**, *13(1126)*; *Kinetika Kataliz* **1974**, *15*, 509.

54. Gerchikov, A. Ya.; Komissarov, V. D.; Denisov, E. T. *Kinetika i Kataliz,* **1974**, *15*, 230.

55. Gerchikov, A. Ya.; Komissarov, V. D.; Galimova, L. G.; Denisov, E. T. *Doklady Akademii Nauk SSSR*, **1973**, *213*, 881.

56. Korotkova, N. P.; Syroezhko, A. M.; Proskuryakov, V. A. *Zhurnal Prikladnoi Khimii,* **1981**, *54*, 885.

57. Komissarov, V. D.; Galimova, L. G.; Denisov, E. T. *Kinetika i Kataliz,* **1974**, *15*, 1063.

58. Syroezhko, A. M.; Korotkova, N. P.; Vihorev, A. A.; Proskuryakov, V. A. *Zhurnal Prikladnoi Khimii,* **1978**, *15*, 2562.

59. Barton, D.; Li, V. D., Eds., *General Organic Chemistry*, **1982**, *2*, 577, Khimiya Publishers: Moscow.

60. House, H. O.; Benjamin, W. A., Eds., *Modern Synthetic Reactions*; 2nd Edition, Menlo Park, CA, USA, **1972**.

61. Zabicky, J., Ed., *The Chemistry of the Carbonyl Group*, **1970**, *2* Interscience Publishers, London, UK.

62. Alcais, P.; Brouillard, R. *Journal of the Chemical Society – Perkin II,* **1972**, 1214.

63. Von Balo, L., *Annalen*, **1866**, *140*, 348.

64. Berthlot, M. *Comptes Rendus,* **1881**, *92*, 895.

65. Harries, V. *Analen*, **1905**, *343*, 311.

66. Fischer, F. G. *Analen der. Chemie*, **1929**, *476*, 233.

67. Bailey, P. S. *Chemical Reviews,* **1958**, *58*, 925.

68. Price, C. C.; Tumolo, A.L. *Journal of the American Chemical Society*, **1964**, *86*, 4691.

69. Ericson, R. E.; Bakalik, D.; Richards, C.; Seanlon, M. *Journal of Organic Chemistry,* **1966**, *31*, 461.

70. Stary, F. E.; Emge, D. E.; Murray, R. W. *Journal of American Chemical Society,* **1976**, *98*, 1880.

71. Giamalva, D. H.; Church, D. F.; Pryor, W. A. *Journal of American Chemical Society,* **1986**, *108*, 7678.

72. Pryor, W. A.; Ohto, N.; Church, D. F. *Journal of American Chemical Society,* **1983**, *105*, 3614.

73. Hoigne, J.; Bader, H. *Water Research,* **1983**, *17(2)*, 173.

74. Hoigne, J.; Bader, H. *Water Research,* **1983**, *17(2)*, 185.

75. Hoigne, J.; Bader, H.; Haag, W. R.; Staehlin, J. *Water Research,* **1985**, *19(8)*, 993.

76. Kuramshin, E. M.; Ivashov, U. B.; Zlotskii S. S.; Rahmankulov D. L. Izvestia Visshikh Uchebnih Zavedenii, *Khimia i Khimicheskaya Tekhnologia,* **1984**, *27*, 13.

77. Gurol, M. D.; Nekouinaini, S. *Industrial and Engineering Chemistry Fundamentals,* **1984**, *23(1)*, 54.

78. Bernatek, E.; Frangen, C. *Acta Chemica Scandanavica,* **1961**, *15*, 471.

79. Bernatek, E.; Moskeland, J.; Valen, K. *Acta Chemica Scandanavica,* **1961**, *15*, 1454.
80. Bernatek, E.; Frangen, C. *Acta Chemica Scandanavica,* **1962**, *16*, 2421.
81. Bernatek, E.; Vincze, A. *Acta Chemica Scandanavica,* **1965**, *19*, 2007.
82. Jarret, M.; Bermond, A.; Ducauze, C. *Analysis,* **1983**, *14*, 185.
83. Razumovskii, S. D.; Nikiforov, G. A.; Globenko, G. M.; Kefely, A. A.; Gurvich, Ya. A.; Karelin, N. A.; Zaikov, G. E. *Neftekhimia,* **1972**, *12*, 376.
84. Konstantinova, M. L.; Razumovskii, S. D.; Zaikov, G. E. *Izvestiya Akademii Nauk SSSR,* Seriya Khimicheskaya; **1991**, 324.
85. Razumovskii, S. D.; Konstantinova, M. L.; Zaikov, G. E. *Izvestiya Akademii Nauk SSSR,* Seriya Khimicheskaya; **1992**, 1203.
86. Konstantinova, M. L.; Vol'eva, V. B.; Razumovskii, S. D., *Izvestiya Akademii Nauk SSSR,* Seriya Khimicheskaya; **1992**, 1443.
87. Hoigne, J. In: *The Handbook of Environmental Chemistry*; Hrubec, J., Ed.; Springer-Verlag: New York, **1998**, *5*, 83.
88. von Gunten, U. *Water Research,* **2003**, *37*, 1443.
89. Pi, Yu.; Schumacher, Jo.; Jekel, M. *Water Research,* **2005**, *39*, 83.
90. Buffle, M. O.; von Gunten U. *Environ. Sci. Technol.* **2006**, *40*, 3057.
91. Yao, C. C. D.; Haag, W. R. *Water Research,* **1991**, *25(7)*, 761.
92. *Handbook of Chemistry and Physics*, 66th Edition, **1985–1986**, D278, CPC Press: Boca Raton, FL, USA.

CHAPTER 9

THE KINETICS AND MECHANISM OF THE SELECTIVE ETHYLBENZENE OXIDATION

L. I. MATIENKO, V. I. BINYUKOV, L. A. MOSOLOVA, and
G. E. ZAIKOV

CONTENTS

9.1   AIM AND BACKGROUND

The role of the intra- and intermolecular H-bonds in mechanisms of catalysis with triple heterobinuclear hetero ligand complexes $Ni^{II}(acac)_2 \times NaSt$(or LiSt)$\times$PhOH, including nickel and redox-inactive metal Na (Li), in the selective ethylbenzene oxidation by dioxygen into α-phenyl ethyl hydroperoxide is discussed. The AFM method has been used for research of possibility of the stable supramolecular nanostructures $\{Ni^{II}(acac)_2 \times NaSt$(or LiSt)$\times$PhOH$\}_n$ formation, due to intermolecular H-bonds.

The problem of selective oxidation of alkylarenes to hydroperoxides is economically sound. Hydroperoxides are used as intermediates in the large-scale production of important monomers. For instance, propylene oxide and styrene are synthesized from α-phenyl ethyl hydroperoxide, and cumyl hydroperoxide is the precursor in the synthesis of phenol and acetone [1]. The method of modifying the Ni^{II} and $Fe^{II,III}$ complexes used in the selective oxidation of alkylarenes (ethylbenzene and cumene) with molecular oxygen to afford the corresponding hydroperoxides aimed at increasing their selectivities has been first proposed by Matienko [2, 3]. This method consists of introducing additional mono- or multidentate modifying ligands into catalytic metal complexes. The mechanism of action of such modifying ligands was elucidated. New efficient catalysts of selective oxidation of ethylbenzene to α-phenyl ethyl hydroperoxide were developed [2, 3].

The phenomenon of a substantial increase in the selectivity (S) and conversion (C) of the ethylbenzene oxidation to the to a-phenyl ethyl hydroperoxide upon addition of PhOH together with alkali metal stearate M'St (M' = Li, Na) as ligands to metal complexes NiII(acac)$_2$ was discovered in our works [1, 2, 4].

The observed values of C [$C > 35\%$ at $(S_{PEH})_{max} = 85-87\%$], [ROOH]$_{max}$ (1.6–1.8 mol l^{-1} far exceeded those obtained with other ternary catalytic systems $\{Ni^{II}(acac)_2 + L^2 + PhOH\}$ (L^2 is N-metyl–2-pyrrolidone (MP), hexamethylphosphorotriamide (HMPA),) and the majority of active binary systems. These results by Matienko and Mosolova are protected by the Russian Federation patent (2004).

The advantage of these ternary systems is the long-term activity of the in situ formed complexes $Ni^{II}(acac)_2 \cdot L^2 \times PhOH$. The best results were obtained in the case of catalysis with system $\{Ni^{II}(acac)_2 + NaSt + PhOH\}$ (Fig. 1).

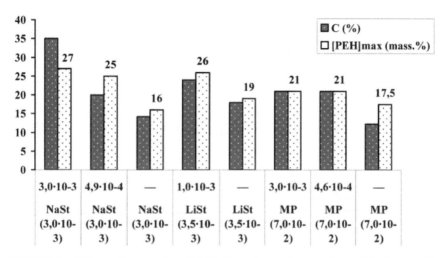

FIGURE 1 Values of conversion C (%) (I row), maximum values of hydroperoxide concentrations $[PEH]_{max}$ (mass.%) (II row) in reactions of ethylbenzene oxidation in the presence of triple catalytic systems $\{Ni(II)(acac)_2 + L^2 + PhOH\}$ (L^2 = NaSt, LiSt, MP: [PhOH] (mol/l)—on an axis of abscises (the top number), $[L^2]$ (mol/l)—on an axis of abscises (the bottom number)). $[Ni^{II}(acac)_2] = 3.0 \times 10^{-3}$ mol/l, 120°C.

The high efficiency of three-component systems $\{NiII(acac)2 + MSt + PhOH\}$ (M = Na, Li) in the reaction of selective oxidation of ethylbenzene to a-phenyl ethyl hydroperoxide was associated with the formation of extremely stable binuclear heteroligand complexes $Ni(acac)_2 \times MSt \times PhOH$. The stability of complexes $Ni^{II}(acac)_2 \times MSt \times PhOH$ can be associated with the formation of intermolecular H-bonds [5].

Nanostructure science and supramolecular chemistry are fast evolving fields that are concerned with manipulation of materials that have important structural features of nanometer size (1 nm to 1 μm) [6, 7]. The self-assembled systems and self-organized structures mediated by transition metals are considered in connection with increasing research interest in chemical transformations with use of these systems [8].

Earlier we have applied AFM technique for evidence of the possibility of stable supramolecular structures formation in the process of ethylbenzene oxidation to the to α-phenyl ethyl hydroperoxide in the presence of systems {Ni^{II}(acac)$_2$+ L^2} [9].

In this paper we have applied AFM technique for research the possibility of supramolecular structures formation in the process of ethylbenzene oxidation to the to α-phenyl ethyl hydroperoxide at catalysis with three-component systems {Ni^{II}(acac)$_2$+NaSt(or LiSt)+PhOH}:

{Ni^{II}(acac)$_2$+NaSt(or LiSt)+PhOH}→Ni^{II}(acac)$_2$×NaSt(or LiSt)×PhOH→

→{NiII(acac)$_2$×NaSt(or LiSt)×PhOH}$_n$

9.2   INTRODUCTION

Often metals of constant valency compounds are used in combination with redox-active transition-metal complexes to promote a variety of reactions involving the transfer of electrons [10]. This effect is typified in metalloproteins such as the copper zinc superoxide dismutase, in which both metal ions have been proposed to be functionally active [11].

The adducts of salts of metals of constant valency with b-diketonates and N,N'-ethylene bis (salycylideniminates) of CoII, NiII, CuII are known [12, 13]. Thus the coordination of salts of metals of constant valency M¢L (M¢ – metal of constant valency) with complexes of transition metals ML1 is carried out through ligand (complexes of nickel and copper with the Shiff bases, copper acetylacetonate: ML1—M¢L (bonds L1—M¢) (I type). In the other cases – anions of salts of metals of constant valency are coordinated with ions of transitional metal (acetylacetonates of nickel and cobalt: L1M—L M¢ (bonds M—L) (II type)). For example, in complexes [15-crown-5ÉCa(II)–(µ-OH)–MnIIIMST]+ calcium ions are bound with [MST]3– -ligand of Mn complex ([MST]3—tripodal ligand N,N′,N″-[2,2′,2″-nitrilotris(ethane-2, 1-diyl)]tris(2, 4, 6-trimethylbenzenesulfonamido) [14].

Earlier, we have received kinetic and spectrophotometry (UV spectra) proofs in favor formation of complexes of bis(acetylacetonate)nickel, $Ni^{II}(acac)_2$, with NaSt of II type, M–L, namely, $(acac)_2$ Ni–St(Na) [2, 3].

We discovered that the introduction of phenol together with the $\{Ni(acac)_2 + L^2\}$ catalyst in the reaction system in the initial stage of ethylbenzene oxidation is one of the most efficient methods of designing catalytic systems for the ethylbenzene selective oxidation to α-phenyl ethyl hydroperoxide. The high efficiency of three-component systems $\{Ni(acac)_2+MSt+PhOH\}$ (M = Na, Li) in the reaction of selective oxidation of ethylbenzene to α-phenyl ethyl hydroperoxide was associated with the formation of extremely stable heteroligand complexes $Ni(acac)_2 \times MSt \times PhOH$ [4]. The stability of these triple complexes throughout oxidation process allowed us to assume the formation of stable supramolecular structures due to H-bonding [5].

9.3 EXPERIMENTAL

AFM SOLVER P47/SMENA with Silicon Cantilevers NSG11S (NT MDT) with curvature radius 10 nm, tip height: 10–15 μm and cone angle ≤ 22° in taping mode on resonant frequency 150 KHz was used.

As substrate the polished Silicone surface special chemically modified was used.

Waterproof modified Silicone surface was exploit for the self-assembly-driven growth due to H-bonding of complexes $Ni^{II}(acac)_2 \times NaSt$(or LiSt)×PhOH with Silicone surface. The saturated chloroform ($CHCl_3$) solution of complex $Ni^{II}(acac)_2 \times NaSt$(or LiSt)×PhOH (1:1:1) was put on a surface, maintained some time, and then solvent was deleted from a surface by means of special method – spin-coating process.

In the course of scanning of investigated samples it has been found, that the structures are fixed on a surface strongly enough due to H-bonding. The self-assembly-driven growth of the supramolecular structures on the basis of complexes $Ni^{II}(acac)_2 \times NaSt(or\ LiSt) \times PhOH$ due to H-bonds and perhaps the other non covalent interactions was observed on Silicone surface. One can watch these structures on the basis of $Ni^{II}(acac)_2 \times NaSt \times PhOH$ with big height and volume. In check experiments it has been shown that for binary systems $\{Ni^{II}(acac)_2 + NaSt\}$, and $\{Ni^{II}(acac)_2 + PhOH\}$, the formation of the similar structures (exceeding on height of 2–10 nanometers) is not observed. In the case of $\{Ni^{II}(acac)_2 + LiSt\}$ binary system the observed structures had interesting unusual form with height about 8–10 nm, unlike those which we received in the case of $Ni^{II}(acac)_2 \times LiSt \times PhOH$ complexes. The particles, observed in the case binary systems $\{LiSt(NaSt) + PhOH\}$ had smaller height about 4–6 nm.

Ethylbenzene (RH) was oxidized with dioxygen at 120°C in glass bubbling-type reactor in the presence of three-component systems $\{Ni(II)(acac)_2 + L^2 + PhOH\}$ ($L^2 = NaSt, LiSt$) [4].

9.3.1 ANALYSIS OF OXIDATION PRODUCTS

α-Phenylethylhydroperoxide (PEH) was analyzed by iodometry. By-products, including methylphenylcarbinol (MPC), acetophenone (AP), and phenol (PhOH) as well as the RH content in the oxidation process were examined by GLC [4].

The order in which PEH, AP, and MPC formed was determined from the time dependence of product accumulation rate rations at $t \rightarrow 0$. The variation of these rations with time was evaluated by graphic differentiation [4] (Figs. 2, and 3). Experimental data processing was done using special computer programs Mathcad and Graph2Digit.

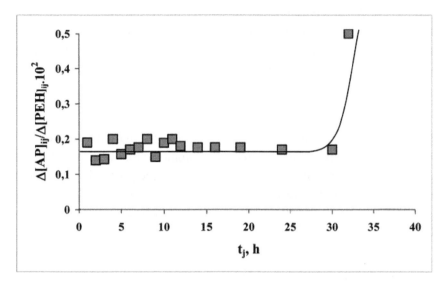

FIGURE 2 Dependence $\Delta[AP]_{ij}/\Delta[PEH]_{ij} \times 10^2$ on time t_j in the course of ethylbenzene oxidation, catalyzed with complexes $Ni(II)(acac)_2 \times NaSt \times PhOH$ (1:1:1), 120°C.

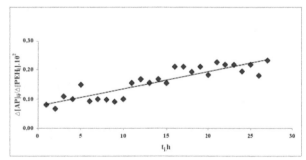

FIGURE 3 Dependence $\Delta[AP]_{ij}/\Delta[PEH]_{ij} \times 10^2$ on time t_j in the course of ethylbenzene oxidation, catalyzed with complexes $Ni(II)(acac)_2 \times LiSt \times PhOH$ (1:1:1), 120°C.

9.4 RESULTS AND DISCUSSION

Earlier we have established, that the increase in the initial rate of the ethylbenzene oxidation with dioxygen, catalyzed with $Ni^{II}(acac)_2$ in the presence of additives of metalloligand MSt (M=Li, Na, K), is due to higher activity of formed complexes $Ni^{II}(acac)_2 \cdot MSt$ in the micro stages of chain initiation and/ or decomposition of PEH with free radical formation. As it was established the increase in rates of chain initiation and/ or radical decomposition of PEH (with free radical formation) seems to be due to the increase in DN of MSt as ligand-modifiers in row:

$$LiSt < NaSt < KSt.$$

At that the participation of catalyst $Ni^{II}(acac)_2 \times MSt$ in micro steps of chain propagation and, probably, in chain termination must also be taken into account. [2]. The results found in [14] illustrate the possibility that redox-inactive metal ions can be used to facilitate the activation of dioxygen (see the abstract scheme of [14], above), which will be coordinated with our data.

At catalysis with triple complexes $Ni(acac)_2 \times L^2 \times PhOH$ ($L^2 = NaSt$, LiSt) the parallel formation of α-phenyl ethyl hydroperoxide, acetophenone and MPC was observed.

($w_{AP(MPC)}/w_{PEH} \neq 0$ at $t \to 0$, $w_{AP}/w_{MPC} \neq 0$ at $t \to 0$) throughout the reaction of ethylbenzene oxidation) (e.g., Figs. 2 and 3).

A more considerable increase in the selectivity (SPEH) at the catalysis by $Ni(acac)_2 \cdot L_2 \cdot PhOH$ ($L_2 = NaSt$, LiSt) complexes as compared with non-catalytic oxidation was associated with the change in the route of acetophenone and methylphenylcarbinol formation (AP and MPC form in parallel with PEH rather than as a result of its decomposition as observed in the non-catalytic ethylbenzene oxidation) and with the inhibition of the PEH heterolytic decomposition.

Thus we had shown that the triple complexes $Ni(II)(acac)_2 \times NaSt$ (or LiSt)$\times PhOH$ unlike binary complexes $Ni(II)(acac)_2 \times NaSt$ (or LiSt) seem to be not active in the reaction with hydroperoxide, but active in the reactions of chain initiation (O_2 activation) and the chain propagation ($Cat+RO_2 \cdot \to$) (and, probably, in chain termination). In these systems dioxygen activation may be promoted through the formation of intramolecular H-bonds [2]. The role of intramolecular H-bonds are established by us in mechanism of formation of triple catalytic complexes $\{Ni(II)(acac)_2 \cdot L^2 \times PhOH\}$ ($L^2 =$ N-methylpirrolidon-2) in the process of ethylbenzene oxidation with molecular oxygen [4].

As mentioned above, the advantage of ternary systems {Ni(acac)$_2$+MSt+PhOH} (M = Na, Li) is the long-term activity of the *in situ* formed complexes Ni(acac)$_2$ × L^2 × PhOH (activity in chain initiation (O$_2$ activation) and the chain propagation (Cat+RO$_2$·→)); the acac ligand does not undergo transformations in the course of ethylbenzene oxidation. So the ternary systems {Ni(acac)$_2$+MSt+PhOH} (M = Na, Li) are the high efficient as catalysts of the selective ethylbenzene oxidation to a-phenyl ethyl hydroperoxide that impressed in the considerable increase in the degree of conversion of ethylbenzene to PEH at selectivity S_{PEH}=90% and in the yield of α-phenyl ethyl hydroperoxide. The high stability and long-term activity of heterobinuclear heteroligand complexes Ni^{II}(acac)$_2$· MSt · PhOH seems to be due to formation of active supramolecular structures {Ni^{II}(acac)$_2$×NaSt(or LiSt)×PhOH}$_n$ as a result of intermolecular (phenol–carboxylate) H-bonds and, possibly, of the other non-covalent interactions [5].

The possibility of association of triple complexes Ni^{II}(acac)$_2$×NaSt(or LiSt)×PhOH to supramolecular structures due to H-bonding is followed from analysis of AFM data, received by us in this work. Results are presented on the next Figs. 4–9 and Table 1.

TABLE 1 The mean values of area, volume, height, length, width of the AFM nanoparticle images on the basis of Ni^{II}(acac)$_2$×NaSt×PhOH formed on the surface of modified silicone.

Variable	Mean values	Confidence −95.000%	Confidence +95.000%
Area (μm²)	0.13211	0.11489	0.14933
Volume (μm³)	14.11354	11.60499	16.62210
Z (Height) (nm)	80.56714	73.23940	87.89489
Length (μm)	0.58154	0.53758	0.62549
Width (μm)	0.19047	0.17987	0.20107

Figures 4 and 5 demonstrated three-dimensional and two-dimensional AFM image (30×30 and 10×10 μm) of the structures on the basis of triple complexes Ni^{II}(acac)$_2$×NaSt×PhOH formed at drawing of a uterine solution on a surface of modified silicone.

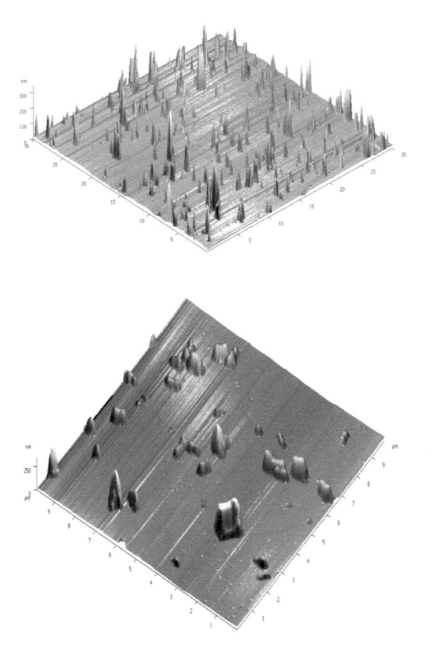

FIGURE 4   The AFM three-dimensional image ((a) 30×30 μm and (b) 10×10 μm) of the structures formed on a surface of modified silicone on the basis of triple complexes $Ni^{II}(acac)_2 \times NaSt \times PhOH$.

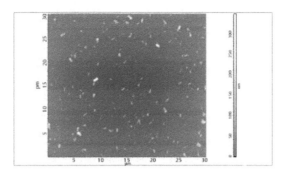

FIGURE 5   The AFM two-dimensional image (30×30 μm) of nanoparticles on the basis Ni^{II}(acac)$_2$×NaSt×PhOH formed on the surface of modified silicone.

In the Fig. 6, the histogram of mean height of nanoparticles on basis of Ni^{II}(acac)$_2$×NaSt×PhOH is presented. As can see, structures are various on heights from the 25 nm to ~ 250–300 nm for maximal values. The distribution histogram shows that the greatest number of particles—is particles of the mean size of 50–100 nm on height.

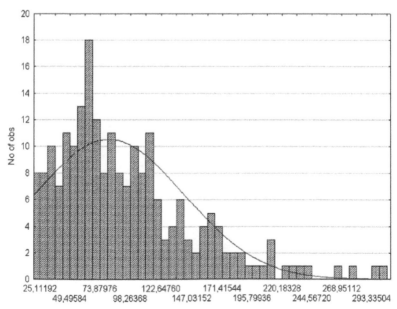

FIGURE 6   Histogram of mean values of height (nm) of the AFM images of nano structures based on Ni^{II}(acac)$_2$×NaSt×PhOH, formed on the surface of modified silicone.

Table 1 shows the mean values of area, volume, height, width, length of nanoscale structures on basis of triple complexes $Ni^{II}(acac)_2 \times NaSt \times PhOH$ formed on the surface of modified silicone.

We have revealed an interesting fact that the length of the formed nanoparticles in the XY plane exceeds the width of the nanoparticles about three times (Table 1).

On the next pictures (Figs. 7–9) the nanoparticle image on the basis of $Ni^{II}(acac)_2 \times LiSt \times PhOH$ and $Ni^{II}(acac)_2 \times LiSt$ complexes is demonstrated.

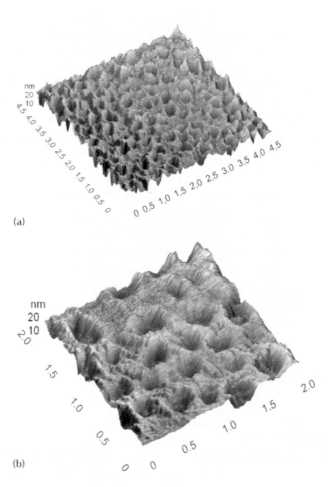

FIGURE 7   The AFM three-dimensional image (4.5×4.5 (a) and 2×2 (b) (μm)) of the structures formed on a surface of modified silicone on the basis of triple complexes NiII(acac)2×LiSt×PhOH.

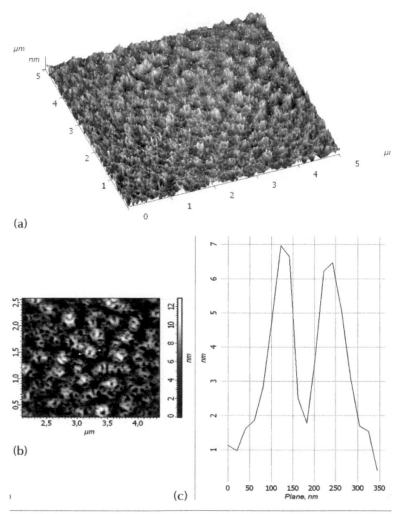

FIGURE 8   The AFM three- and two-dimensional image (5×5 (a) and 2.5×2.5 (b) (μm)) of the structures with more simple form received on a surface of modified silicone on the basis of triple complexes $Ni^{II}(acac)_2 \times LiSt \times PhOH$ and profile of one of these structures (c).

(a)

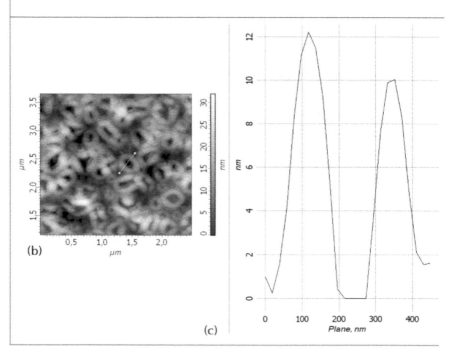

(b)

(c)

FIGURE 9   The AFM three-dimensional image (2.5×2.5 (μm)) (a), two-dimensional image (2.5×2.5 (μm)) and profile (b) of the nanostructures formed on a surface of modified silicone on the basis of $Ni^{II}(acac)_2×LiSt$ complexes.

As one can see, nano structures on the basis of triple complexes $Ni^{II}(acac)_2 \times LiSt \times PhOH$ have the interesting cell form with cell height ~ 10 nm and cell width ~ 0.5 μm.

On the Fig.8 three- and two-dimensional image (5×5 and 2×2 (μm)) and profile of one of the nanostructures on the basis of triple complexes $Ni^{II}(acac)_2 \times LiSt \times PhOH$ with more simple form (cell height ~ 7–12 nm and cell width ~ 60 nm), which we observed also on a surface of modified silicone, are presented.

As can see, the nanostructures on the basis of $Ni^{II}(acac)_2 \times LiSt \times PhOH$, presented with Fig. 7 and 8, are different from nanostructures on the basis of triple complexes $Ni^{II}(acac)_2 \times NaSt \times PhOH$ by form and have lesser height (~ 10–12 nm unlike ~ 80 nm for nanostructures on the basis of triple complexes $Ni^{II}(acac)_2 \times NaSt \times PhOH$) (Fig. 6, Table 1).

In the case of binary complexes, $\{Ni^{II}(acac)_2 \cdot LiSt\}$ and $LiSt \cdot PhOH$, nanostructure were observed also, but had or lesser height (~ 4–6 nm in the case of $LiSt \cdot PhOH$) or less neatly regular structure ($Ni^{II}(acac)_2 \cdot LiSt$, see, for example, Fig. 8) in comparison with nanostructure on the basis of triple complexes $Ni^{II}(acac)_2 \times LiSt \times PhOH$.

So, it can be concluded that the high the degrees of conversion of ethylbenzene to PEH and the yields of α-phenyl ethyl hydroperoxide in the case of catalysis of three-component systems $\{Ni(acac)_2 + MSt + PhOH\}$ (M = Na, Li) in the reaction of selective oxidation of ethylbenzene to α-phenyl ethyl hydroperoxide, as one of the reasons, may be associated with the formation of heterobinuclear heteroligand complexes $Ni^{II}(acac)_2 \times MSt \times PhOH$, which are self organized during ethylbenzene oxidation to extremely stable supramolecular structures $\{Ni^{II}(acac)_2 \times NaSt(or\ LiSt) \times PhOH\}_n$ as a result of intermolecular (phenol–carboxylate) H-bonds and, possible, the other non-covalent interactions [5]. The more effectivity of heterobinuclear heteroligand complexes $Ni^{II}(acac)_2 \times NaSt \times PhOH$, including metalloligand NaSt, as selective catalysts as compared with $Ni^{II}(acac)_2 \times LiSt \times PhOH$, seems to be due to formation of more stable supramolecular structures $\{Ni^{II}(acac)_2 \times NaSt \times PhOH\}_n$. Data, presented on Figs. 4–9, testify in favor of this supposition.

9.5. CONCLUSION

Thus, in the present work we applied AFM method in the analytical purposes to research the possibility of the formation of supramolecular structures on basis of heterobinuclear heteroligand triple complexes Ni(II) (acac)$_2$×NaSt(or LiSt)×PhOH.

We have shown what the self-assembly-driven growth seems to be due to H-bonding of triple complexes Ni(II)(acac)$_2$×NaSt(or LiSt)×PhOH with a surface of modified silicone, and further formation supramolecular nanostructures {Ni(II)(acac)$_2$×NaSt(or LiSt)×PhOH}$_n$ due to directional intermolecular (phenol–carboxylate) H-bonds [5], and, possibly, other non-covalent interactions (van Der Waals-attractions and π-bonding).

These data support the very probable supramolecular structures appearance on the basis of heterobinuclear heteroligand triple complexes Ni(II)(acac)$_2$×NaSt(or LiSt)×PhOH in the course of the ethylbenzene oxidation with dioxygen, catalyzed by three-component catalytic system {Ni(II)(acac)$_2$+NaSt(or LiSt)+PhOH} and this can be one of the explanations of the high values of conversion of the ethylbenzene oxidation into α-phenyl ethyl hydroperoxide at selectivity S_{PEH} preservation at level not below S_{PEH}=90% in this process. The more effectivity of heterobinuclear heteroligand complexes Ni^{II}(acac)$_2$×NaSt×PhOH, including metalloligand NaSt, as selective catalysts as compared with Ni^{II}(acac)$_2$×LiSt×PhOH, seems to be due to formation of more stable supramolecular structures {Ni^{II}(acac)$_2$×NaSt×PhOH}$_n$.

KEYWORDS

- AFM data
- ethylbenzene oxidation
- heterobinuclear heteroligand complexes
- hexamethylphosphorotriamide
- metalloligand

REFERENCES

1. Weissermel, K.; Arpe, H.-J. *Industrial Organic Chemistry*, 3nd ed., transl. by Lindley, C. R. VCH: New York, USA, **1997**.
2. Matienko, L. I. In: *Reactions and Properties of Monomers and Polymers*, D'Amore, A. & Zaikov, G. (Eds.), Chapter 2, Nova Sience Publ., Inc.: New York, **2007**, *21*.
3. Matienko, L. I.; Mosolova, L. A.; Zaikov, G. E. *Selective Catalytic Hydrocarbons Oxidation*. New Perspectives, Nova Science Publ., Inc.: New York, USA, **2010**.
4. Matienko, L. I.; Mosolova, L. A. *Oxidation Commun.* **2007**, *30*, 282.
5. Dubey, M.; Koner, R. R.; Ray, M. *Inorg. Chem.*, **2009**, *48*, 9294.
6. Leninger, St.; Olenyuk, B.; Stang, P. *J. Chem. Rev.*, **2000**, *100*, 853.
7. Stang, P. J.; Olenyuk, B. *Acc. Chem. Res.*, **1997**, *30*, 502.
8. Beletskaya, I.; Tyurin, V. S.; Tsivadze, A. Yu.; Guilard, R. *Stem, Ch. Chem. Rev.*, **2009**, *109*, 1659.
9. Matienko, L. I.; Binyukov, V. I.; Mosolova, L. A.; and Mil, E. M. *J. of Characterization and Development of Novel Materials.* **2011**, *4(1)*, 54 (Nova Publ., Canada).
10. Fukuzumi, S.; Ohkubo, K. *Coord. Chem.Rev.* **2010**, *254*, 372.
11. Valentine, J. S.; Doucette, P. A.; Potter, S. Z. *Annu. Rev. Biochem.* **2005**, *74*, 63.
12. Dani, C. M.; Das, A. K. *Indian J. Chem. Sect. A.*, **1984**, *23*, 1041.
13. Aminabhovi, T. M.; Biradar, N. S.; Divaka, M. S. *Inorg. Chim. Acta*, **1984**, *92*, 99.
14. Park, Y. J.; Ziller, J. W.; Borovik, A. S. *J. Am. Chem. Soc.*, **2011**, *133*, 9258.

CHAPTER 10

COMPOSITE MATERIALS ON THE BASIS OF EPOXYCONTAINIG ORGANOSILICON COMPOUNDS

E. MARKARASHVILI, T. TATRISHVILI, N. KOIAVA, A. BERLIN, G. ZAIKOV, J. ANELI, and O. MUKBANIANI

CONTENTS

10.1.   AIMS AND BACKGROUND

There are some ways of synthesis of epoxyorganosiloxanes. From these methods it is significant to denote oxidative epoxidation of unsaturated bond containing organosilanes and siloxanes [1, 2], hydrosilylation reactions of ≡Si-H bond containing silanes and siloxanes to allyl glycidyl ether in the presence of catalysts [3, 4] and the reactions of oganosilanoles or organo siloxanoles with epichlorohydrin [5, 6].

Some oxidation reactions of alkenes give cyclic ethers, in which both carbons of a double bond become bonded to the same oxygen atom [1]. An important method for preparing epoxides is the reaction with peracids, RCO_3H. The oxygen-oxygen bond of such peroxide derivatives is not only weak (ca. 35 kcal/mol), but is polarized in this case so that the acyloxy group is negative and the hydroxyl group is positive. If we assume electrophilic character for the OH moiety, the following equation may be written (Scheme 1):

SCHEME 1. OXIDATIVE EPOXIDATION OF ALKENES IN THE PRESENCE OF PERACIDS.

It is unlikely that a dipolar intermediate, as shown above, is actually formed. The epoxidation reaction is believed to occur in a single step with a transition state incorporating all of the bonding events shown in the equation. Consequently, epoxidation by peracids always have syn-stereoselectivity, and seldom give structural rearrangement. Presumably the electron shifts induce a charge separation that is immediately neutralized by electron shifts.

Authors of works made also rather essential supervision connected to rather smaller or greater oxidability of double bonds, caused by influence of the neighboring substitutions, (oxymethylene groups, aldehyde and ketone groups) which is testified by publications of different years [7–9]. From peracids for oxidation of olefins apply peracetic acid [7, 10], pertrifluoroacetic acid [11], m-chloroperbenzoic acid [10, 12] and perbenzoic acid [4, 13].

Now the opinion on the greatest activity of catalysts is standard on the basis of molybdenum, vanadium, tungsten and some other metals of variable valency. These metals have such properties, as presence of a high positive charge, ability to connection of electronic pair due to vacant electronic orbits. Essential feature of these metals is tendency to formation of complexes of a different degree of stability with olefins, oxygen containing compounds and other substances. To number complexing compounds, undoubtedly, hydro peroxides concern also. Consequence complexation is heterolytic decomposition of O–O bonds.

In oxidative epoxidation reactions besides of oxidizing agent often various catalysts systems are used [14]. It was established that the rate of heterolytic decomposition of O–O bonds in tertiary-butyl peroxide in the presence of catalysts such as $Mo(CO)_6$ proceeds in result of complexation between metal and hydroperoxide. By the authors suggested the probable schemes transition condition, without discussion of valent state of metal (Scheme 2).

SCHEME 2. PROBABLE SCHEMES OF TRANSITION CONDITIONS.

It was shown, that the rate of decomposition of tertiary-butyl hydro peroxide in toluene solution at 87°C in the presence of $Mo(CO)_6$ is much lower than the rate of epoxidation. It allows conducting calculations, not considering the decomposition of hydro peroxide, which is not resulting in formation of oxide.

The analysis of the literature allows counting, that is already developed highly effective molybdenum containing catalysts of epoxidation, convenient not only for laboratory practice, but also perspective for industrial application. Most of such derivatives of molybdenum, as $MoCl_5$, $Mo(CO)_6$ and molybdenum naphthenate are frequently used in oxidative epoxidation reaction. In addition during oxidative epoxidation by peroxides and hydro peroxides solvent renders essential influence on the rate of reaction [15].

The atom of silicon connected with unsaturated aliphatic group, should facilitate its oxidation by peracids. In the same conditions vinyl derivative of silicon reacts with peracetic acid [4], perbenzoic acid [12] much more slowly than allyl-derivative. In siliconorganic compounds where double bond is removed more than on two carbon atoms from silicon, influence of atom of silicon on oxidation of unsaturated group [4] does not affect.

Organosilicon compounds, containing two unsaturated groups, are oxidized faster, than compounds containing one unsaturated bond [16].

The rate of oxidation reaction by perbenzoic of silanes, containing one unsaturated group, decreases in a line: trimethyl–> tripropyl–> triethyl–> tribenzyl–> triphenylalkenylsilane [17].

By epoxidation of unsaturated organosilicon compounds by perbenzoic, perphthalic, 40% peracetic acid [4] and peroxytrifluoroacetic acid [4] was shown, that the best epoxidation agent is peroxytrifluoroacetic acid. During epoxidation of 1.1.3.3-tetramethyldivinyldisiloxane and 1.1.3.3.5.5.7.7-octaphenyl–1.7-divinyltetrasiloxane by perbenzoic acid corresponding diepoxyderivatives has been obtained with 60–65% yield.

Epoxidation of vinylsilanes by 40% solution of peracetic acid in acetic acid proceeds very slowly with the yield less than 20%. With an increase of temperature and the reaction time the secondary decomposition reactions takes place and the yield of epoxide products decreaces.

The oxidative epoxidation reactions of unsaturated siliconorganic compounds by perbenzoic acid in chloroform and tetrachloromethane proceeds significantly rapidly than another solvents, though by using less active solvents (ether, tetrahydrofurane) is possible exception of secondary reactions. In addition it was shown [18] that oxidative epoxidation of vinyl containing siliconorganic compounds by tertiary-butyl alcohol in the presence of $Mo(CO)_6$ in benzene solution at 40°C epoxy derevatives is abot 60–70%.

Novolac resins have been widely used in many applications such as adhesives, coatings, construction, and composite matrices in the aerospace and electronic industries. This is due to their high strength, excellent flameretardant, low cost, good chemical and corrosion resistance and good dimensional stability [19–21]. Elastomeric and thermoplastic modifiers have been successfully incorporated into epoxy networks to improve their fracture energies and impact strengths [22].

Functionalized polysiloxanes have been investigated as elastomers for this purpose [23, 24]. Polysiloxanes, widely known as silicone, are of particular interest due to their extremely low glass transition temperatures and flexibility, their hydrophobic surface properties, good thermal stability and excellent flame retardant [25].

By the authors [26] the novolac-epoxy networks was modified with reactive epoxy functional polysiloxane oligomers containing various levels of polar nitrile pendent groups was investtigated. It was shown that the using with high content of polysiloxanes as an elastomeric modifiers component in epoxy-novolac networks exhibited micro phase separation.

10.1.1 OXIDATIVE EPOXIDATION OF UNSATURATED BOND CONTAINING ORGANOSILANES AND SILOXANES

Now an actual direction is carrying out of researches on development of methods of synthesis of organosilicon compounds containing in molecules reactionable carbofunctional groups.

The basic methods of obtaining of oligoepoxysiloxanes are the hydrosilylation reactions of hydrideorganosilanes and siloxanes with allyl glycidyl ether, oxidative epoxidation of siliconorganic compounds containing in the structure unsaturated vinyl and allyl groups, and also co-oligomerization reactions short-chain epoxy siloxanes with organocyclosiloxanes.

In this chapter, the oxidative epoxidation reactions of 1.1.3.3-tetramethyl–1.3-divinyldisiloxane and 1.1.3.3.5.5-hexamethyl–1.5-divinyltrisiloxane in the presence of various oxidizers and catalyst is investigated. Apparently from the literary review there are many data on application as oxidizers of olefins perbenzoic and peracetic acids where the yield of diepoxy-derivatives is about 60%.

For oxidative epoxidation of divinyldimethylsiloxanes as oxidizers, tertiary-butyl hydro peroxide and hydrogen peroxide have been used by us for the first time. The reaction was carried out in the presence of molybdenum containing catalysts at the ratio of siloxane: peroxide 1:1, 1:2 and 1:2.5. The oxidative epoxidation of divinylsiloxanes in the presence of tertiary-butyl hydro peroxide proceeds according to the following (Scheme 3) [27]:

SCHEME 3. OXIDATIVE EPOXIDATION REACTIONS OF DIVINYLDIMETHYLSILOXANES IN THE PRESENCE OF TERTIARY-BUTYL HYDROPEROXIDE.

where, n=1 (I), n=2 (II).

It is necessary to note, that during epoxidation of divinyltetramethyld-isiloxane by the specified oxidizers at 1:1 molar ratio of siloxane: oxidizer mono epoxy derivatives are obtained; and at 1:2.5 molar ratio the basic product – dioxirine are obtained. The reaction products easily are divided during its distillation on vacuum. The structure and composition of synthesized diepoxy derivatives were determined by means of elementary and functional analyzes, by finding of molecular masses, by FTIR, ^{1}H NMR spectra data.

In the FTIR spectra one can observe absorption bands at 1020 cm^{-1} characteristic for asymmetric valence oscillation of linear $\equiv$Si-O–Si$\equiv$ bonds, also there are new absorption bands at 820–840, 917 cm^{-1} characteristic for epoxy rings. Absorption band at 1150 cm^{-1} corresponds to asymmetric valence oscillation for the C–O–C simple ether bonds. In the ^{1}H NMR spectra of compounds I and II one can observe singlet signals for methyl protons attached at silicon atom with chemical shift δ=0.1 ppm, multiplet signals for protons of methylene and methine groups in oxirane ring with center of chemical shifts δ=2.7– 2.4 ppm and δ=2.5 ppm accordingly.

Influence of the solvents, catalysts, reaction temperatures and comparative oxidizing ability of tertiary-butyl hydro peroxide, hydrogen peroxide on the yield of compounds I and II has been investigated and obtained results are presented in Tables 1 and 2.

TABLE 1 Influence of various factor on the oxidative epoxidation reactions of 1.1.3.3-tetramethyl–1.3-divinyldisiloxane (molar ratio of siloxane : oxidizer is 1:2,5).

No.	Reaction temperature T°C	Yield, %	Catalyst	Solvent	Oxidizer
1	90	72	$(NH_4)_6Mo_7O_{24} \cdot 10H_2O$	Toluene	TBH
2	90	75	$Mo(CO)_6$	Toluene	TBH
3	70	67	$(NH_4)_6Mo_7O_{24} \cdot 10H_2O$	Benzene	TBH
4	70	70	$Mo(CO)_6$	Benzene	TBH
5	120	73	$(NH_4)_6Mo_7O_{24} \cdot 10H_2O$	Chlorobenzene	TBH

TABLE 1   *(Continued)*

No.	Reaction temperature T°C	Yield, %	Catalyst	Solvent	Oxidizer
6	120	83	$Mo(CO)_6$	Chlorobenzene	TBH
7	70	84	$(NH_4)_6Mo_7O_{24} \cdot 10H_2O$	CCl_4	TBH
8	70	86	$Mo(CO)_6$	CCl_4	TBH
9	90	63	$(NH_4)_6Mo_7O_{24} \cdot 10H_2O$	Toluene	H_2O_2
10	90	66	$Mo(CO)_6$	Toluene	H_2O_2
11	70	54	$(NH_4)_6Mo_7O_{24} \cdot 10H_2O$	Benzene	H_2O_2
12	70	56	$Mo(CO)_6$	Benzene	H_2O_2
13	120	62	$(NH_4)_6Mo_7O_{24} \cdot 10H_2O$	Chlorobenzene	H_2O_2
14	120	73	$Mo(CO)_6$	Chlorobenzene	H_2O_2
15	70	80	$(NH_4)_6Mo_7O_{24} \cdot 10H_2O$	CCl_4	H_2O_2
16	70	83	$Mo(CO)_6$	CCl_4	H_2O_2

For compound I calculated values: MM=218, containing epoxy groups – 39.45%, reaction duration – 36 hours; founded values: MM=219, containing epoxy groups – 39.2%.

TABLE 2 Influence of various factor on the oxidative epoxidation reactions of 1.1.3.3.5.5-hexamethyl–1.5 divinyldisiloxane (molar ratio of siloxane : oxidizer is 1:2,5).

No.	Reaction temperature T°C	Yield, %	Catalyst	Solvent	Oxidizer
1	90	72	$(NH_4)_6Mo_7O_{24} \cdot 10H_2O$	Toluene	TBH
2	90	75	$Mo(CO)_6$	Toluene	TBH
3	70	60	$(NH_4)_6Mo_7O_{24} \cdot 10H_2O$	Benzene	TBH
4	70	68	$Mo(CO)_6$	Benzene	TBH
5	120	69	$(NH_4)_6Mo_7O_{24} \cdot 10H_2O$	Chlorobenzene	TBH

TABLE 2   *(Continued)*

No.	Reaction temperature T°C	Yield, %	Catalyst	Solvent	Oxidizer
6	120	73	$Mo(CO)_6$	Chlorobenzene	TBH
7	70	80	$(NH_4)_6Mo_7O_{24} \cdot 10H_2O$	CCl_4	TBH
8	70	85	$Mo(CO)_6$	CCl_4	TBH
9	90	70	$(NH_4)_6Mo_7O_{24} \cdot 10H_2O$	Toluene	H_2O_2
10	90	75	$Mo(CO)_6$	Toluene	H_2O_2
11	70	62	$(NH_4)_6Mo_7O_{24} \cdot 10H_2O$	Benzene	H_2O_2
12	70	60	$Mo(CO)_6$	Benzene	H_2O_2
13	120	60	$(NH_4)_6Mo_7O_{24} \cdot 10H_2O$	Chlorobenzene	H_2O_2
14	120	72	$Mo(CO)_6$	Chlorobenzene	H_2O_2
15	70	80	$(NH_4)_6Mo_7O_{24} \cdot 10H_2O$	CCl_4	H_2O_2
16	70	85	$Mo(CO)_6$	CCl_4	H_2O_2

For compound II calculated values: MM=292, containing epoxy groups – 29.45%, reaction duration – 36 hours; founded values: MM= 290, containing epoxy groups – 28%.

The reaction was carried out at various temperatures in benzene, toluene, chlorobenzene or carbon tetrachloride at 70–120°C (depending on boiling temperature of solvent) temperature. Apparently from the data of Table 1, 2 and Fig. 1, during epoxidation reaction in identical conditions, the highest yield of dioxirine are achieved in case of carrying out of reactions in carbon tetrachloride and the least – in benzene.

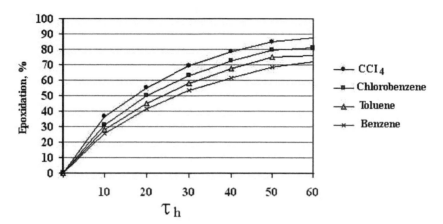

FIGURE 1   Dependence of the yield of compound I on the time in oxidative epoxidation reactions of tetramethyldivinyldisiloxane by tertiary-butyl hydroperoxide in various solvents, at 1:2.5 molar ratio of siloxane : oxidizer.

At comparison of oxidative epoxidation abilities of tertiary-butyl hydro peroxide and hydrogen peroxide (at the presence of the same catalyst) appeared, that oxidizing ability of tertiary-butyl hydro peroxide is higher, than hydrogen peroxide (see the yields of diepoxy derivatives).

The type of the used catalyst during oxidative epoxidation tetramethyldivinyldisiloxane also influences on the yield of diepoxide. For oxidation of divinylorganosiloxanes molybdenum containing catalysts, such as: $Mo(CO)_6$ – hexacarbonyl molybdenum, $(NH_4)_6Mo_7O_{24} \cdot 10H_2O$ – ammonium paramolybdate have been used by us for the first time. By investigation of catalytic activity of the specified catalysts it was established, that hexacarbonyl molybdenum $Mo(CO)_6$ shows big catalytic activity in comparison with ammonium paramolybdate.

In the literary data there is information on an opportunity of immobilization of the catalyst on a polymeric matrix that can lead to simplification of epoxidation technology [28]. We have investigated influence of structure of the catalyst on the yield of dioxirine during oxidation divinyldiorganosiloxanes. For this purpose heterogeneous catalysts, immobilized on polymeric matrixes have been obtained. As a matrix ion exchange resin «Amberlyst IRA–45», high basic anionite AB–16 with active =NH, ≡N, –N(R)$_3$ groups and polyvinyl alcohol have been chosen.

Sufficiently stable catalyst managed to be received sedimentation ammonium paramolybdate on Amberlyst resin (70 h works) and polyvinyl alcohol (100 h works). Immobilization of hexacarbonyl molybdenum on ion exchange resin the authors [28] have presented by the following (Scheme 4):

SCHEME. 4. IMMOBILIZATION OF HEXACARBONYL MOLYBDENUM ON ION EXCHANGE RESIN.

Apparently from the resulted scheme molybdenum hexacarbonyl preliminary has been affected by UV-irradiation for reception of active molybdenum, which in turn immobilized on a polymeric matrix. To receive diepoxide compounds with the high yield with the use of immobilized catalyst systems was not possible. Structure of catalyst supporter as it was marked in the literature, does not especially influence on the yields of epoxy compounds.

Epoxy resins of various structure and composite materials on their basis have found wide application in various areas of techniques. Among epoxy resins the greatest application has found dian and epoxy-novolac resins. On their basis the polymeric materials possessing high physical and chemical, physical-mechanical and dielectric parameters are received.

However, specified epoxy resins at room temperature represent as a high viscous liquids or solid substances with 40–100°C softening temperature, which complicates their processing and application. At their use as a basis of glues, coatings, impregnating and potting compounds various purpose, for decreasing of viscosity apply active thinners, compounds with low viscosity, however in most cases they essentially reduce such important physical and chemical parameters as heat-resistance, water resistance, and also the some strength parameters of obtained polymeric materials.

In the given part of the work, properties of compositions and polymers on a basis epoxy-dian resin ED-20 and epoxy-novolac resin UP–643 (Russian products) which has been diluted by epoxy group containing silicon-organic compounds I and II have been considered.

For definition of possible fields of application of obtained epoxides in consisting of polymeric composite materials process of structurization of compounds I and II various hardener agents was investigated. Appeared, that in view of low ability to interaction with aliphatic and aromatic amines, and also with anhydrides of di-carbonic acids it is not possible to obtain on their basis qualitative samples of polymers in the conditions usually used for organic epoxides [29].

By mixing of the synthesized compounds with high-viscous and solid epoxy resins has allowed to reduce their viscosity that testifies to diluting ability of epoxy organosilicon compounds in 3–6 times. Mixtures of epoxy-methylsiloxane with dian and novolac resins were hardened with iso– methyltetrahydrophthalic anhydride and polyethylene polyamine.

The analysis of the experimental data has shown, that introduction of compounds I and II in 20–30% amount in epoxy-dian resins allows their processing at room temperature. At higher temperatures of processing about 10–20% are possible to reduce amount of an active thinner entered into a composition.

Such high viscous resin as UP–643 in a mix with compounds I and II can be processed at 50–60°C, and optimum viscosity of oligomer mixes is reached at introduction of a thinner in 20–35% amount.

In Table 3 some physical-mechanical and dielectric properties of modified and unmodified epoxide polymers are presented. From the data resulted in Table 3 it is visible, that by using of epoxy siliconorganic compounds I and II, as active thinners of epoxy compositions, their strength properties and dielectric parameters practically do not undergo essential changes.

TABLE 3   Strength and dielectric properties of hardened compositions on a basis of epoxy-dian resins ED-20 and epoxy-novolac resins UP–643 with active epoxy siliconorganic thinners (compounds I and II) and without them

The name of a parameter	Properties of a composition without a thinner			Properties of a composition with thinner			
	1	2	3	4	5	6	7
Ultimate strength, MPa							
At compression	112	143	110	109	140	101	107
At a stretching	74	65	67	76	74	73	69

TABLE 3   *(Continued)*

The name of a parameter	Properties of a composition without a thinner			Properties of a composition with thinner			
	1	2	3	4	5	6	7
Ultimate strength, MPa							
At a bending	120	85	117	115	92	110	113
Relative lengthening at break, %	2,5	2,2	2,0	2,1	2,1	2,5	2,3
Water absorption for 24 hours at 25°C, %	0,122	0,094	0,099	0,105	0,093	0,089	0,082
Specific volumetric electric resistance, Omxcm	$6,3 \times 10^{15}$	$9,4 \times 10^{15}$	$7,1 \times 10^{13}$	$7,7 \times 10^{14}$	$5,7 \times 10^{15}$	$2,4 \times 10^{13}$	$3,5 \times 10^{13}$
Dielectric permeability at 10^6 Hz	3,3	3,2	3,9	3,9	3,7	4,2	4,0
Tangent of a corner of dielectric losses about 10^6 Hz	0,033	0,0124	0,035	0,039	0,017	0,036	0,037

The note to the table:
- Compositions 1 and 2 on the basis of ED-20 and UP–643 accordingly, hardened with mix of *iso*-methyltetrahydrophthalic anhydride and UP–606/2;
- Compositions 3 on the basis of ED-20, hardener polyethylene-polyamine "B' type;
- Compositions 4, 5 on the basis of ED-20 and UP–643 according to addition 30 mass part. Compounds II, hardened with mix of *iso*-methyltetrahydrophthalic anhydride and UP–606/2;

- Compositions 6, 7 on the basis of ED-20 with addition 20 mass part. Compounds I and II, hardened by polyethylene polyamine.

It was shown that using of epoxy group containing dimethylsiloxane it is possible to modify epoxy-dian and epoxy-novolac resins, reducing their viscosity in 3–6 times, receiving homogeneous compositions. Thus process of reprocessing specified epoxy resins is facilitated, their hardening processes proceeds at lower temperatures with preservation strength and dielectric properties. Especially it is necessary to notice, that the obtained composite materials on the parameters outbalance known organic epoxide compositions at their long exploitation in damp conditions. So, their specific volumetric electric resistance after humidifying on the order is higher, in comparison with composite materials on a basis epoxy-dian polymer.

So the oxidative epoxidation of 1.1.3.3-tetramethyl–1,3-divinyldisiloxane and 1.1.3.3.5.5-hexamethyl–1.5-divinyltrisiloxane by tertiary-butyl hydro peroxide and hydrogen hydro peroxide at the presence of molybdenum catalysts in the medium of various organic solvents and ratio of initial compounds has been investigated and optimal conditions of the reaction has been established.

Possible fields of application the properties of compositions and polymers on a basis of dian-epoxy resin ED-20 and epoxy-novolac resin UP–643 which have been diluted by epoxy group containing siliconorganic compounds have been considered.

Another way of synthesis of linear organosilicon diepoxide is the condensation Reactions of $\alpha\omega$-dihydroxediorganosiloxanes with an epichlorohydrin.

For the purpose of reception of diglycidoxidesilicoorganic oligomers the reaction of condensation of an epichlorohydrin (ECH) with α-dihydroxediorganosiloxanes at presence 25% solutions of a sodium hydroxide have been investigated, at different temperatures (20/120°C), according to the literature data [30]. The reaction proceeds under the following general (Scheme 5) [31–34]:

SCHEME 5. CONDENSATION REACTION OF α,Ω-DIHYDROXEDIORGANOSILOXANES WITH AN EPICHLOROHYDRIN.

where, $R'=R''=CH_3$, n=9 (III); $R'=R''=CH_3$, n=14 (IV); $R'=CH_3$, $R''=C_6H_5$, n=60 (V); $R'=R''=C_6H_5$, n=3 (VI); $R'=CH_3$, $R''=Th$, n=2 (VII), n=10 (VIII).

As a result of reaction gum-like organosilicon diglycidoxide ethers (III–VIII) have been received.

For determination of the reactions optimal conditions, the polycondensation reaction at various temperatures, various amounts of the catalyst and molar ratios of initial compounds have been studied. Apparently from the data resulted in a Table 4 and Figs. 2 and 3, it was shown that the best yields of an epoxy group containing linear organosilicon oligomer is reached at temperature 110–120°C at presence 25% solutions of a sodium hydroxide and at molar ratios of ω,αdihydroxediorganosiloxanes: epichlorohydrin 1:2, both at simultaneous entering all components, and at mixing α,ωdihydroxydiorganosiloxane with an epichlorohydrin and constant addition of a water solution of a sodium hydroxide proceeds with education of oligomers III–VIII which yield is about 90%.

TABLE 4 Change of a yield diglycidoxipolymethylsiloxane (IV) depending on temperature of a reactionary mix and concentration of a solution of a sodium hydroxide.

№	Reaction time	Temperature of a reactionary mix	Concentration NaOH, %	Yield, %
1	10	20	70	15
2	20	30	65	30
3	30	40	60	30–35
4	40	50	55	55
5	55	60	50	65
6	70	70	45	65
7	85	80	40	75
8	100	90	35	85
9	115	100	30	85
10	130	110–120	25	90–95

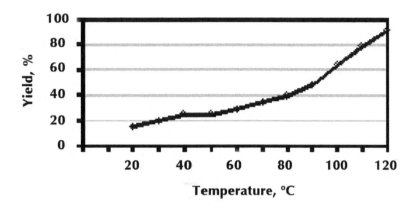

FIGURE 2 Dependence of a yield diglycidoxypolymethylsiloxane (IV) from temperature of a reactionary mix at a parity of initial components 1:2 (siloxane: ECH).

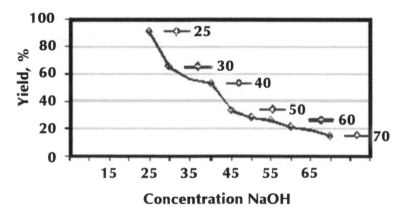

FIGURE 3 Dependence of a yield diglycidoxypolymethylsiloxane (IV) from concentration of a sodium hydroxide.

The composition and structure of oligomers has been established by the functional and elemental analysis, definition of molecular masses and by FTIR spectra data. In FTIR spectra of the synthesized compounds strips of absorption, characteristic for epoxy-groups are observed at 850 cm^{-1} and 4545 cm^{-1}. Some physical characteristics, reaction temperature and the yields properties and results of the element and functional analysis of synthesized oligomers III–VIII are resulted in Table 5.

TABLE 5   Yields and results of the element and functional analysis of synthesized oligomers III–VIII.

№	Yield, %	M mass*	Epoxy group, %	Elementary analysis**, %		
				C	H	Si
III	97	796	10.8	36.18	8.04	31.66
		772	10.48	35.11	7.80	30.70
IV	95	1166	7.38	35.00	8.06	33.62
		1108	7.01	33.25	7.66	31.94
V	96	8290	1.03	61.66	5.91	20.27
		8320	1.04	62.26	5.97	20.46
VI	90	724	11.88	69.61	4.42	4.42
		709	11.64	66.83	4.33	4.33
VII	95	416	20.67	46.15	5.77	13.46
		404	20.05	44.77	5.65	13.06
VIII	95	1532	5.62	43.86	3.39	18.28
		1484	5.33	41.60	3.22	17.34

*Molecular weights are certain on trailer epoxy groups.
**Where in numerator – the calculated values and in a denominator-found.

On the basis of oligomers III–VIII synthesized by us the composite materials K-1, K-2, K-3, K-4, K-5 which hardened different hardeners as amine type and anhydride type (methyltetrahydrophthalic anhydride) have been received: D-1 methylphenyldiamine (MPDA), D-2 4,4'-diaminodiphenylsulphone (4,4' DADPS), D-3 4,4'-diaminodiphenylmethane (4,4' DADPM), D-4 4,4'-diaminotriphenyloxide (4,4' DATPO) and D-5 methyltetrahydrophthalic anhydride (MTHPA). The constitution of composites, modes of their hardening are presented in Table 6.

TABLE 6 Constitution and modes of a hardening of composites on the basis of epoxy pitch the ED-22 and oligomers III–VIII.

No	Name of composites	Constitution of composites	Amount mass part	Hardening agent		T°C	Hardening, time, h
				Name	Mass part		
1	K-1	ED-22 III	100 40–45	D-1	10–15	110	5
2	K-1a	ED-22 III	100 40–45	D-2	10–15	120	4
3	K-1б	ED-22 III	100 40–45	D-3	10–15	120	5
4	K-1в	ED-22 III	100 40–45	D-4	10–15	150	5
5	K-2	ED-22 IV	100 40–45	D-1	10–15	110	4
6	K-2a	ED-22 IV	100 40–45	D-2	10–15	130	4
7	K-2б	ED-22 IV	100 40–45	D-3	10–15	120	5
8	K-2в	ED-22 IV	100 40–45	D-4	10–15	120	5
9	K-3	ED-22 V	100 40–45	D-1	10–15	120	5
10	K-3a	ED-22 V	100 40–45	D-2	10–15	120	5
11	K-3б	ED-22 V	100 40–45	D-3	10–15	120	5
12	K-3в	ED-22 V	100 40–45	D-4	10–15	130	5
13	K-4	ED-22 VI	100 40–45	D-1	10–15	130	5

TABLE 6   *(Continued)*

No	Name of composites	Constitution of composites	Amount mass part	Hardening agent		T°C	Hardening, time, h
				Name	Mass part		
14	K-4a	ED-22 VI	100 40–45	D-2	10–15	130	5
15	K-4б	ED-22 VI	100 40–45	D-3	10–15	130	5
16	K-4в	ED-22 VI	100 40–45	D-4	10–15	130	5
17	K-5	ED-22 VI	100 40–45	D-5	10–15	110	5

The kinetics of a hardening of compounds K-1, K-2, K-3, K-4 by allocation gel-fractions within 4–5 hours has been studied. Apparently from Fig. 4 hardening of composites K-1 and K-2 with a hardener D-5 begins at temperature 100°C, and the full hardening comes above 120°C temperature.

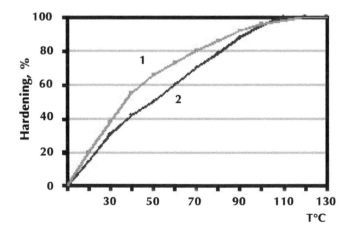

FIGURE 4   The hardening curves of composites K-1 (1) and K-2 (2) depending on temperature of a hardening with a binder MTHPA (D-5) ratio 100:15 a mass part (during 45 min).

Chemical stability of compounds K-1, K-2, K-3 and K-4 solutions to different chemical reagents during 90 day has been studied. For this purpose prepared films (thickness 200–300 microns) were lowered to the solution. After certain interval of time the films were took from a solution and checked external changes of sheets. The data of Table 7 show some swelling of samples after their stay in solutions: 0.1% of potassium permanganate and 10% of nitric acid in which samples changed external parameters.

TABLE 7   Chemical stability of the films based on compound K-1.

No.	Reagent	Hardening agent	Chemical stability		
			After 30 days	After 60 days	After 90 days
1	Distilled water	D-1	Without change	Without change	Changes in color
		D-2	—"—	—"—	—»—
		D-3	—"—	—"—	—"—
		D-4	—"—	—"—	—"—
		D-5	—"—	—"—	—"—
2	30% solution of hydrochloric acid	D-1	Without change	Without change	Changes in color
		D-2	—"—	—"—	—»—
		D-3	—"—	—"—	—"—
		D-4	—"—	—"—	—"—
		D-5	—"—	—"—	—"—
3	30% solution of sulfuric acid	D-1	Without change	Without change	Without change, exfoliates from glass
		D-2	—"—	—"—	
		D-3	—"—	—"—	—»—
		D-4	—"—	—"—	—"—
		D-5	—"—	—"—	—"—
4	10% solution of nitric acid	D-1	Cracks on the film surface	Darkening of films	The film has bulked up and has turned yellow
		D-2	—"—	—»—	—»—
		D-3	—"—	—"—	—"—
		D-4	—"—	—"—	—"—
		D-5	—"—	—"—	—"—

TABLE 7   *(Continued)*

No.	Reagent	Hardening agent	Chemical stability		
			After 30 days	**After 60 days**	**After 90 days**
5	0.1% solution of potassium permanganat	D-1 D-2 D-3 D-4 D-5	Covering with spots —"— —"— —"— —"—	Darkening of film and covering with spots —»— —"— —"— —»—	The film has bulked up —»— —"— —"— —"—

The thermal-oxidative destruction of the received epoxy compounds as basis products (III–VIII) have been investigated. For saving of stability of obtained results the thermal-oxidative destruction of composites was studied on samples with approximately identical degree of dispersivity, at identical rate of heating (5°/min) and weight of samples (50 mg). It was appeared, that composites were destructed at ~200–250°C. The picture of composites destruction is individual for each case and strongly varies at change of a parity of weight of siliconorganic epoxy compound and a hardener. From a Table 7 it is obvious, that the thermal-oxidative destruction of the considered samples begins at rather low temperatures, which may be explained by saving of organic fragments in structure of a copolymer. The observable result is in accordance with well known opinion that thermal stability of a copolymer is connect with thermal stability of corresponding homo polymers and copolymer. In the considered number of epoxy composites, in structure of system a considerable role-play the functional or pseudo-functional groups in polymer, which assist to destruction processes.

Establishment of the thermal stability of polymers From a Table 7 is difficult, however a conclusion in this respect a combination of products III and IV in compounds K-1 and K-2 with a hardener D-2 give optimum constitution.

The investigation of dependence of thermal-oxidative stability of the specified composites from a parity of weight of the synthesized oligomer and a hardener shows, that the most stable systems it turns out at observance of equivalence of epoxy and amine groups a lot of hardener ≈8–10 of %. Thermogravimetric curves (Figs. 5 and 6) allow us on thermal-oxidative stability to distribute the received composites in a following scheme:

K-1	K-2	K-2	K-2	K-3
		> > ≈ >		
D-2	D-2	D-1	D-3	D-2

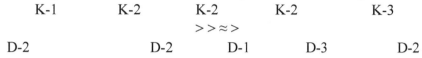

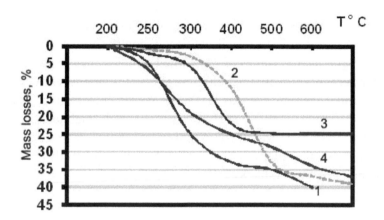

FIGURE 5   Thermogravimetric curves of a thermal-oxidative destruction of composite K-1 hardened by a hardener D-1 (4), D-2 (3), D-3 (2), D-4 (1).

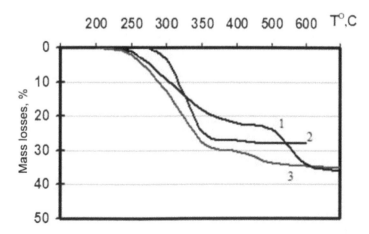

FIGURE 6   Thermogravimetric curves of a thermal-oxidative destruction of composites K-2 (1), K-3 (2), K-4 (3), hardened by a hardener D-1.

At comparison of thermal-oxidative stability for hardeners it is possible to conclude that compounds containing oligomers XI–XIV and the specified hardeners settle down in a following row:

D-2> D-1> D-3> D-4

It is shown, that the most heat-resistant composites are obtained at use of hardeners D-2, D-1, and as an established fact, equivalent parity of epoxy oligomers: the hardener (~8–10% surplus of a hardener) creates favorable conditions for increase of thermal stability of composites.

Thermogravimetric curves of the studied samples have rather complex character because of overlaying different processes against each other, as the differential thermal analysis (DTA) is poorly informative. The carbon rest of samples approximately on 30% exceed "maintenance" SiO_2 in samples. Most likely the thermal thermal-oxidative destruction of the polymers, accompanied also allocation of low-molecular siliceous products mainly proceeds (Table 8).

TABLE 8   Loss of weight of composites depending on temperature.

Hardening agent	Temperature °C	Compounds and loss of weight, %			
		K-1	K-2	K-3	K-4
	150	6	5	4	5
D-1	200	11	13	14	12
	230	18	20	21	26
	150	4	4	3	5
D-2	200	12	16	10	20
	230	19	21	25	21
	150	5	6	8	8
D-3	200	14	20	16	20
	230	22	28	23	26
	150	8	10	10	8
D-4	200	20	22	20	16
	230	21	26	26	35

Apparently from Table 10 a full hardening of composite K-1 with a hardener D-2 is reached at temperature 150°C during 4–5 hours (Table 9).

TABLE 9   Yield of gel-fractions of composite K-1 with a hardener D-1 depending on temperature of hardening.

No.	Hardening time, min	Hardening temperature, °C	Yield of gel-fractions, %
1	20	18	0
2	60	45	0.1
3	120	60	0.2
4	160	80	2.8
5	180	110	10.0
6	240	120	40.0
7	270	130	60,0
8	300	140	80.0
9	360	150	100.0

At obtaining of the combined composites of epoxyorganosiliconc oligomers (III, VI) and epoxydiane resin ED-22 chemical interaction between them, and also reduction of functional epoxy groups is not revealed. However at a hardening they behave differently. Presence of functional, epoxy groups, with high reactionary ability already at presence of hardeners of amine type leads to chemical interaction and gradual reduction of quantity of functional groups.

The monomers containing organosilicon fragments and epoxy groups in a molecule affirm at presence of amine hardeners and give three-dimensional cross-linked polymers which deformation temperature constancy reaches 250–300°C. The combined systems of different types on a miscellaneous behave at a hardening. At a cold hardening of composites with a polyethylene polyamine (PEPA), the time of a hardening increases up to 3 days.

On Figs. 7 and 8, the curves of kinetics of hardening processes of composites K-D and pure epoxy-dian pitch ED-22 with hardeners the PEPA and MTHPA are presented.

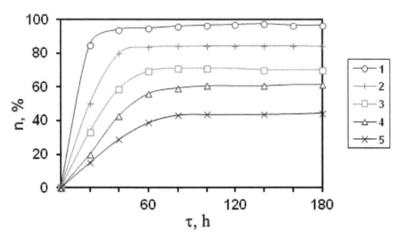

FIGURE 7   The kinetics of a hardening of epoxy resin ED-22 curve 1, composites K-1 curve 2, K-2 curve 3, K-3 curve 4 and K-4 curve 5, hardened by PEPA on a cold, where n-the maintenance of insoluble fractions.

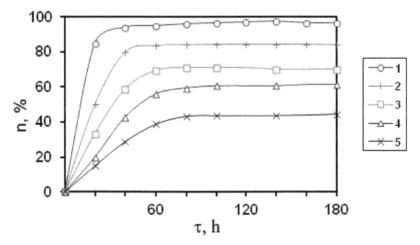

FIGURE 8   The kinetics of a hardening of epoxy resin ED-22 (1), composites K-1 (2), K-2 (3), K-3 (4) and K-4 (5), hardened MTHPA at temperature 120°C, where n is the maintenance of insoluble fractions.

At heating the same compositions the rate and completeness of hardening increases, however on the rate of a hardening epoxy pitch surpasses all siliceous composite materials. It once again proves that the modified composites, in comparison with epoxy pitch, are characterized by smaller reactionary ability and greater viability that enables their applications as caulking materials. Thus, at a hardening of the epoxy-dian pitch modified by epoxyorganosilicon oligomers (III, VI), it is possible to allocate two stages.

On the first stage of a hardening there is a process of slow gradual formation of linear polymers and increasing of a polymeric grid due to interaction of functional groups and a hardener, on what is spent up to 30% of their quantity therefore there is a hardening of a composition. The temperature of reactionary weight at the first stage of a hardening should not exceed 60°C. The second stage of process proceeds at heating a composition up to 120–150°C during 4–5 hours. Under such circumstances there is a formation of insoluble and not fusible polymer due to an expenditure of a great bulk of functional groups, and also formations of new cross-section bonds due to interaction of secondary hydroxyl groups.

Thus, apparently from results of discussion, polymeric materials on the basis of epoxy-dian resin ED-22, the modified epoxyorganosilicon oligomers, possess a complex of valuable operational properties: high thermal stability, stability to influence of solutions of acids, alkalis, salts, high dielectric and physical-mechanical properties that offer their perspective for use in various branches.

10.2 EPOXYCONTAINING POLYPHENYLSILSESQUIOXANES AS A MODIFICATOR OF NOVOLAC RESINS

Novolac resins have been widely used in many applications such as adhesives, coatings, construction, and composite matrices in the aerospace and electronic industries. This is due to their high strength, excellent flame-retardant, low cost, good chemical and corrosion resistance and good dimensional stability [35, 36]. Hexamethylenetetramine (HMTA) is normally used as a curing agent to produce high crosslink density networks [37, 38]. However, this reaction can generate volatile co– products such as ammonia which lead to voids in the networks.

For this reason, HMTA cured novolac networks are generally brittle in nature and have low fracture energies and impact strengths, probably due to the voids combined with the high crosslink densities. Curing novolac resins with epoxy resins can produce networks without volatile by products. These can be reinforced with glass or carbon fibers to yield tough, void-free composites, which also retain much of the flame-retardant properties of phenolics [39].

Materials based on epoxy resins have high application in different branches of industry. However, besides of their many positive properties (good chemical and corrosion resistance, good flame-retardant and dimensional stability) some exploitation properties of these materials not yet corresponds to modern technical requirements (e.g. impact strength, thermal stability). In the early works [40–42] the thermal stability of cured epoxides was increased at hardening of the blends of epoxy and novolac phenol-formaldehyde resins at 180– 200°C. With using in these reactions of aminocyclotriphosphasens the block oligomers have been produced [42]. These materials are characterized by high heat resistance in comparison with hardened pure epoxy resins and epoxy novolac compound. The oligomers of such types may be used as the basis for production of other oligomer materials with properties differed from ordinary epoxy novolac copolymers. Functionalized polysiloxanes are of particular interest due to their very low glass transition temperatures and flexibility, good thermal stability and flame retardant properties [43–46]. In combination with some elastomeric and thermoplastic modifiers polysiloxanes have been successfully incorporated into epoxy networks to improve their fracture energies and impact strength. High thermal firmness is inherent to the polymer compounds based on epoxy-novolac blocks epoxy including component of which is silicon-organic epoxy oligomer [43, 47].

For improving of dielectric properties of modified filler compounds the tetraepoxy organosilicon oligomers with cyclotetrasiloxane and organosilsesquioxane fragments in the chain have been obtained. For synthesis of 2.4.6.8-tetraepoxy–2.4.6.8-tetraorganocyclotetrasiloxanescis–2.4.6.8-tetrahydroxy–2.4.6.8-tetraphenylcyclotetrasiloxane   and   2.4.6.8-tetrahydroxy–2.4.6.8-tetrathienylcyclotetrasiloxane were used. The condensation reaction of epichlorohydrin with tetroles was conducted at surplus amount of epichlorohydrin with 25% solution of sodium hydroxide in diethyl-ester at room temperature. The reaction proceeds according to the Scheme 6 [48, 49]:

SCHEME 6. CONDENSATION REACTION OF TETRAHYDROXYTERAORGANOCYCLOTETRASILOXANES WITH AN EPICHLOROHYDRIN.

where, R=Ph (IX), Th (X).

In result of reaction the transparent– yellow viscous compounds well soluble in the usually organic solvents were obtained. Composition and structure of obtained compounds were proved on the basis of elemental analysis, definition of number of epoxy groups and molecular masses and IR spectra. Some physical and chemical data of obtained compounds are presented on the Table 10.

In the IR spectra of obtained compounds in the range of asymmetric and symmetric valence oscillations ≡Si-O–Si≡ bond the bifurcation of the strips with maximums v_{as} –1045 cm^{-1}, 1145 cm^{-1} и v_{as} – 455, 480 cm^{-1} are observed. Bifurcation of the signals for thienyl containing cyclotetrasiloxane fragment is saved. In the case of phenylcyclotetrasiloxane ring condensation at presence of nucleophilic sodium may be occur the opening of cyclotetrasiloxane ring, which leads to formation of the structure different from *cis*-configuration (see Figs. 9 and 10). This opinion is in accordance with known data [50, 51].

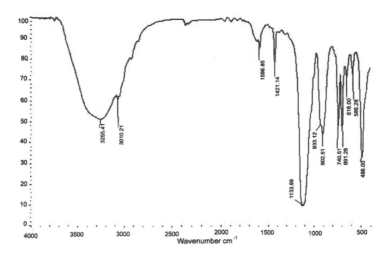

FIGURE 9 FTIR spectrum of cis–2.4.6.8-tetrahydroxy–2.4.6.8-tetraphenylcyclotetrasiloxane.

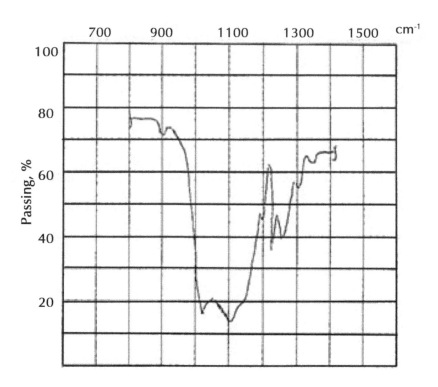

FIGURE 10   FTIR spectrum of 2.4.6.8-glycidoxy–2.4.6.8-tetraphenylcyclotetrasiloxane.

With the aim of definition of the effect of organocyclosiloxane frag-ment length in the tetraepoxy compounds on the properties of a flood composition materials based on them we have investigated the reactions analogically reaction (Eq. (6)) the condensation reactions of tetrahydroxy-oligoorganosilsesquioxanes with bifurcated epichlorohydrin at present of 25% solution of sodium hydroxide at room temperature were investigat-ed. Tetrahydroxyphenylsilsesquioxanes with *cis*-anti-*cis*-tactic structure as initial compounds obtained by thermal condensation of cis–2.4.6.8-tetrahydroxy–2.4.6.8-tetraphenylcyclotetrasiloxane were used. At interac-tion of the fragments of T_4 with cis-location of functional groups poly-phenylsilsesquioxanes (PPSSO) with *cis*-isotactic configuration may be formed if the atoms of silicon in the tetrole molecules are in one and same plane and in *cis*-anti-*cis*-tactic configuration, if the T_4 cycles are in the parallel planes (Fig. 11).

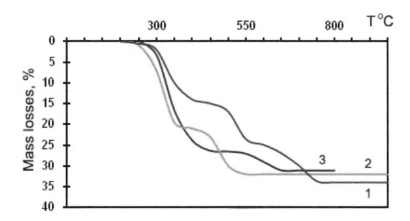

FIGURE 11 Thermogravimetric curves for the compounds G-1(1), G-2 (2), G-3 (3) hardened by hardener D-1.

However the realization of cis-isotactic structure is associated with several steric hindrances and characterized by a short length of the molecules. During the tetrole polycondensation without initiator of the basis type, which allows the obtaining of the polymer without break of siloxane bond in the organocyclosiloxane, the conditions of synthesis ensures the perfect cyclolinear ladder structure and the macromolecules with the Kuhn segment with length about 50 Å are formatted.

The spectroscopic investigations [50, 51] give the fundament for propose, than the structure of macromolecules of PPSSO, obtained by condensation of tetrole T_4, and differs from structure of one, obtained from phenyltrichlorosilane by anion polymerization at present of the initiator of basis type. The conducted experimental investigations allow suggest that the macromolecules with kun segment about 50 Å have the structure near for *cis*-anti-*cis*-tactic one [50] (Scheme 7):

SCHEME. 7. CIS-ANTI-CIS-TACTIC AND CIS-ISOTACTIC CONFIGURATION OF PPSSO.

A combination of the T_4 fragments with trans-displacement of the functional-groups the *cis*-sindiotactic structure of chain may be formatted *cis*-syndiotactic configuration (Scheme 8).

SCHEME 8. CIS-SYNDIOTACTIC CONFIGURATION OF PPSSO.

Such structure of the chain is satirically more profitable and is characterized by high rigidity of the chain. Therefore, the dual structure of the molecules of ladder fragments lay under hydrolytic condensation organotriclorosilane and products of it partial hydrolysis and condensation. Tetrahydroxythienyl(chlortienyl)silsesquioxanes have the structure near to *cis*-sindiotactic structure [50]. So the condensation reaction of tetrahydroxyoligoorganosilsesquioxanes with epichlorohydrin proceeds according to the following Scheme 9 [48, 49]:

SCHEME 9. THE CONDENSATION REACTION OF TETRAHYDROXYOLIGOORGANOSILSESQUIOXANES WITH EPICHLOROHYDRIN.

where, R= C_6H_5 – n=3 (XI), n=10 (XII); R= Th, n=3 (XIII), R=–ThCl, n=3 (XIV), n=5 (XV).

The obtained compounds are amber color viscous products well soluble in the acetone, methylethylketone and ethyl acetate. The yield, number of epoxy-groups, molecular masses and elemental analysis of obtained silicon-organic oligomers XI – XV are presented in the Table 10.

TABLE 10   Yields, number of epoxy groups, molecular masses and elemental analysis of tetraepoxyorganosilicon oligomers IX– XV.

No.	Yield, %	Number of epoxy groups, %	M*
IX	96	24.16 24.05	712 709
X	95	21.83 20.95	788 756
XI	95	9.86 10.00	1744 1769
XII	96	3.23 3.18	5324 5242

TABLE 10 *(Continued)*

No.	Yield, %	Number of epoxy groups, %	M*
XIII	94	9.41	1828
		9.36	1818
XIV	95	7.62	2254
		7.64	2260
XV	94	4.74	3626
		4.80	3672

*Molecular masses were determined by end epoxy group analyses.

Numerator is calculated in values, and the denominator is experimental.

In the IR spectra of obtained compounds XI–XV in the region of asymmetrical and symmetrical valence oscillation of $\equiv$Si–O–Si$\equiv$ bonds absorption bonds with maximums v_{as} –1045 cm^{-1}, 1145 cm^{-1} and v_{as} – 455, 480 cm^{-1} is observed. Bifurcation of the signals of $\equiv$Si–O–Si$\equiv$ bonds for thienyl and chlorothienyl containing cyclotetrasiloxane fragment is saved. As for phenylcyclotetrasiloxane ring in the ladder fragment at the present of nucleophilic sodium hydroxide the opening of cyclotetrasiloxane ring takes place as in case of ionic polymerization of 2.4.6.8-tetrahydroxy–2.4.6.8-tetraphenylcyclotetrasiloxane, in result of which the structure differed from *cis*-configuration is obtained. This opinion corresponds to known data [50].

Epoxyorganosilicon compounds IX–XV were used as modifiers for epoxy-dian resin ED-22. The compounds were prepared on the basis of ED-22 with 20–24% epoxy-groups (100 mass parts)at different ratio of modifier and hardener: D-1 – methylphenyldiamine (MPDA),D-2 – 4,4' – diaminodiphenylsulfone (4,4' -DADPS), D-3 – 4,4'diaminodiphenylmethane (4,4'-dADPM), D-4 – 4,4'-diaminotriphenyloxide (4,4'-DATPO) and D-5 – metatetrahydrophtalanhydride (MTHFA). The hardening of composites was conducted both at room and high temperatures. Composition and the hardening regime of obtained composites were conducted also both at room and high temperatures. The composition and hardening regime of obtained compounds are presented on the Table 11 and physical-mechanical and electric properties – on the Table 12. The obtained compounds have the high dielectric, mechanic characteristics and thermal-oxidative stability. The results of investigation of Dependence of physical-mechanical and dielectric properties on the number of the introduced modifier are presented on the Table 13. The study of change of compounds viscosity in the process of hardening on temperature shows that compound differ from analogous by high viability in the range 80–90°C, however at temperature 150–160°C its viability sharply decreases, which is connected with quick destruction of the functional groups and its consequent hardening.

TABLE 11 Composition and hardening regime of the composites based on tetraepoxypolyphenylsilsesquioxanes and epoxy-resin ED-22.

Composite	Composite content	Mass part	Composite Color	Hardening tem-re, T°C	Hardening time, h
G-1	Epoxy pitch ED-22 Oligomer IX Binder D-1	100 40–45 13	Light yellow viscous mass	160	4
G-2	Epoxy pitch ED-22 Oligomer XI Binder D-2	100 40–45 13	Light yellow, transparent mass	155	4
G-3	Epoxy pitch ED-22 Oligomer XII Binder D-4	100 40–45 10–15	Dark-yellow viscous mass	155	5

TABLE 12 Mechanical and electric properties of compounds based on tetraepoxypolyphenylsilsesquioxanes and epoxy resin ED-22.

No.	Characteristic	Compound			Unit
		G-1	G-2	G-3	
1.	Electrical strength	24,5	22	23,8	кВ/mm
2.	Specific surface electrical resistivity at 20°C	1×10^6	$1,2 \times 10^6$	$0,6 \times 10^{17}$	Om
3.	Specific volumetric electric resistivity at 20°C	$1,5 \times 10^{13}$	2×10^{14}	5×10^{13}	Om/cm
4.	Strength at bending	111,0	113,5	120,0	MPA
5.	Heat resistance according to Martens	77,8	77,5	87,5	°C

TABLE 13 Physical-mechanical properties of epoxy resin ED-22, modified by epoxyorganosilicon oligomer XII with hardener MTGPA (D-5).

No.	Containing of modi-fier, mass parts (at 100 mass parts of the resin)	Destruction stress at stretching, MPa	Relative Elongation at rupture,%	Hardness by Brinnel, MPa	Electric durability, kV/mm	Tangent of dielectric losses angle at 10^3Gz	Dielectric penetration at 10^3Gz	Specific volumetric electric resistance, $\times 10^{13}$ Ohmxcm
1.	10	7.15	10	1.0	61	0.022	5.92	4.3
2	15	8.1	12	1.0	68	0.029	6.06	4.3
3.	20	6.8	10	1.2	52	0.018	5.85	4.3
4.	25	5.2	9	1.4	46	0.011	5.02	4.2
5.	35	5.01	8	1.5	44	0.011	5.02	4.2
6	45	5.01	7	1.7	42	0.010	5.02	4.1
7	0	3.5	0.8	1.0	20–25	0.0045	3.90	–

By study of thermal oxidation destruction of obtained composites it is established that the polymers are more high thermal proof systems in comparison with initial epoxy resin.

The Fig. 11 shows that at increasing of the ladder fragment length of a chain increases mainly relative stability of the composites. Nearly same situation has place at study of stability of the composites with different hardeners as in preliminary case in accordance with the Fig. 11.

All composites G-D based on epoxy resin ED-22 modified by synthetic tetraepoxypolyphenylsilsesquioxanes and hardened by amine type hardeners are divided on two stages. On the first stage composites destruction is followed by high loss of the mass. This process proceeds at temperatures from 180^0 up to 420°C. Probably on this stage the compounds organic groups burn out (Fig. 12).

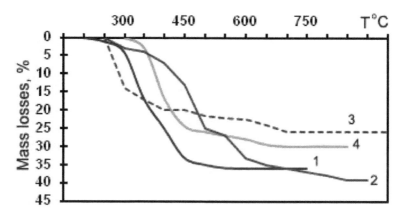

FIGURE 12 Thermogravimetric curves of thermal oxidative destruction for the compound G-2 hardened by hardeners D-1 (1), D-2 (2), D-3 (3) and D-4 (4).

In search of polymers with different properties along with the research of new methods for the synthesis of oligomers and polymers special place choice of optimal curing conditions modified composites. In this regard, particular importance has an elaboration of rational designs temperature curing of composites. We have studied the hardening of composite G-D with different amine hardener D-1, D-2, D-3, D-4, D-5 at different temperatures from 80°C up to 160–180°C.

From Table 14 full curing of the composite G-1 with binder D-2 at a temperature of 150°C is reached within 4–5 hours.

TABLE 14 Yields of gel-fraction of the composite K–1 with hardener D-1 in dependence of hardening temperature.

No.	Hardening time, min	Hardening tempera-ture, °C	Yield of gel-fraction, %
1	20	18	0
2	60	45	0,1
3	120	60	0,2
4	160	80	2,8
5	180	110	10,0
6	240	120	40,0
7	270	130	60,0
8	300	140	80,0
9	360	150	100,0

At obtaining of superposed composites of epoxyorganosilicon oligomers (IX/XV) and epoxy-dian resin ED-22 the chemical interaction between them and decreasing of the functional epoxy groups were not observed. However at hardening their behavior is another. The present of the functional groups with high reaction ability even at present of amine type hardeners leads to chemical interaction and step-by-step disappearing of the functional groups.

The monomers containing organosilicon fragments and epoxy groups in the molecule are cured in the presence of amine curing agents and give a three-dimensional cross-linked polymers, deformation heat resistance which is in the range 250–300°C. Combinations of different types behave differently when cured. During curing of composites by hardeners of cold cure, for example polyethylene polyamine (PEPA), increasing of the hardening time till 3 days is observed.

On the Figs. 13 and 14 the curves of hardening kinetics for composites G-D and for pure ED-22 at using of hardeners PEPA and MTGPA are presented. From these figures it is shown that hardening rate for ED-22 essentially higher than in case of other compounds. At heating of these composites the fullness of hardening increases, however the hardening rate for epoxy resin is higher than for all organosilicon composites. This fact proves once more that the modified composites are characterized with lower reaction ability and more high viability than that for epoxy-resin. Therefore, these composites are more suitable embedding materials.

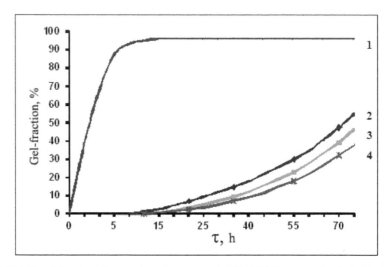

FIGURE 13   Kinetics of hardening of epoxy resin ED-22 (1), composite G-1 (2), G-2 (3) and G-3 (4) hardened by PEPA in the cold.

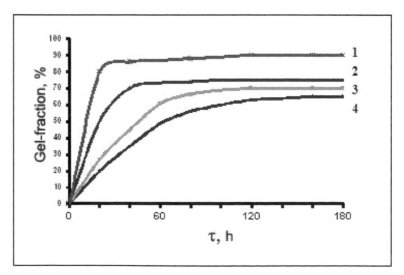

FIGURE 14   Kinetics of hardening of epoxy resin ED-22 (1), composite G-1 (2), G-2 (3) and G-3 (4), hardened by MTGPA at temperature 120°C.

At account of noted above it may be concluded that hardening of the epoxy resin modified by organosilicon oligomers (IX/XV) may be divided on two stages. On the first stages proceeds the process of the slow for-matting of the linear polymers and increasing of polymer grid because of interaction between functional groups and hardener, the consumption on which reaches about 30%. The temperature of reaction mass on the first stage of hardening must be not exceed 60°C. To it ensures the application of MTGPA instead of PEPA.

The second stage of the process takes place by heating the composition up to 120°–150°C during 3 hours. Under such conditions, insoluble and infusible polymers are obtained by the interaction of the functional groups and the formation of new cross-links due to the reaction of the secondary hydroxyl groups.

In the threefold system (ED-22 + epoxyorganosilicon oligomer + hard-ener) interaction of hardener occurs with ED-22, and with epoxyorgano-silicon modifier. And the second reaction is intense during heat treatment. The hardened material presents relatively rare grid of epoxy polymer filled with thermoplastic product epoxyorganosilicon modifier + hardener and therefore has more high than pure epoxy polymer strengthening and ther-mal stability.

Increasing of hardener amount in the threefold system leads to obtain-ing of the polymer with high density of the grid. Such polymer is character-ized by more stress structure, which leads to decreasing of its mechanical strengthening and polymer becomes fragile. For hardening of composites of the threefold system the optimal amount of hardener is 10–15 mass. part (20–25%). For the samples hardened during 5 h at 150°C decreasing of softening temperature and increasing of high elastic deformation are observed. This fact explained by full finishing of the chemical processes of hardening of thermal treated samples.

So, it seems that polymer materials based on epoxy dian resin ED-22 modified by epoxyorganosilicon oligomers possess with the complex of the valuable exploitation properties: high thermal stability, dielectric and physical and mechanical properties. The complex of these properties are fully corresponds to demands presented to them the modern technique and is perspective.

Epoxy including oligomer IX obtained by us also was used for elaboration and investigation of thermal stable glue composites of the cold hardening with high elasticity. For hardening of the binder the ester of boric acid with dihydroxydiphenylsilane was proposed with following structure: $(HO)_nB(OSiPh_2OH)_{3-n}$.

For preparing of glue compositions mixing of components was conducted at 75°C during 5min. (at lower temperature the hardener did not mix with binder). The hardening of composites proceeded at room temperature. However the composites did not hardened. Therefore to the composition was added the aminoethoxysilane as a catalyst. Using noted binder the composite (G-4) with following composition was prepared: binder –3 mass.part, hardener –1 mass.part and aminoethoxysilane (of type ADE–3)– 0,6 mass.part. In this case composite G-4 was prepared at room temperature after 24 h.

However it was impossible to define the strength of agglutination for composite G-4 (under small loading the samples destruct). The glue film is elastic.

After the composite G-5 with hardener diethylenetriaminomethyl phenol (hardener UP–583) of the following composition was investigated: binder –3 mass.part and hardener 1 mass.part. It was defined the strengthening of glue to steel at shear at 20°C and 300°C. Testing results are presented on the Table 15.

TABLE 15   The strength of agglutination of steel by glue of composition G-5.

No.	Agglutinated materials	Testing temperature, C^0	τ shear kg/cm^2	Type of destruction
1	Steel 08ПС+ Steel 08ПС	20	$\underline{50}$ 46–54	
2	Steel 08ПС+ Steel 08ПС	300	$\underline{0,34}$ 0,34–0,34	

On the basis of conducted works it was established that the use of epoxy contained organosilicon compounds as binder is actual and gives principally new possibility of creation of agglutinated cold hardened composites enable of work at high temperatures.

In this chapter, we will discus the synthesis and investigation of some properties of copolymers produced in result of copolymerization of epoxy-silicon oligomers and novolac phenol-formaldehyde oligomers have been carried out.

Epoxy functionalized polyphenylsilsesquioxanes have been synthesised by condensation reaction of industrial products tetraethoxypolyphenylsilsesquioxane [PhES–50 – n≈1, PhES–80 – n≈2] with epichlorohydrin. The reaction has been carried out at 80–100°C temperature range in neutral medium in the presence of catalytic amount of iron (III) chloride and aluminum chloride. The reaction proceeds according to the following Scheme 10 [50]:

SCHEME 10. CONDENSATION OF TETRAETHOXY-POLYPHENYLSILSESQUIOXANE (PHES–50 AND PHES–80) WITH EPICHLOROHYDRIN IN THE PRESENCE OF CATALYST.

where, n=1 (XVI), n=2 [XVII (AlCl$_3$), XVII' (FeCl$_3$)].

It is interesting to note, that iron (III) chloride paints the reactionary mix in dark brown color. Decolorizing of an obtained product neither processing on a centrifuge, nor adsorption of a solution of organosilicon oligomers on the activated coal was not possible. Therefore the further researches were carried out in the presence of catalytic amounts of aluminum chloride.

The structure and composition of synthesized oligomers were determined by means of elementary and functional analyzes by finding of molecular masses, FTIR and ^{1}H NMR spectra data. In the FTIR spectra of compounds XVI–XVII' the absorption bands both for n$_{as}$ asymmetric, and for symmetric valence oscillation of ≡Si-O–Si≡ bonds are kept in the field of absorption at 1060–1010 cm^{-1}, and also there are new absorption bands at 820–840, 917 cm^{-1} characteristic for epoxy rings. In the spectra does not observe the absorption bands, characteristic for ethoxy groups that testifies to full replacement of ethoxy groups by epoxy one. In the ^{1}H NMR spectra of synthesized oligomers one can observe multiplet signals with center of chemical shift δ= 3.3 ppm characteristic for methylene protons in the –CH$_2$O– group; multiplet signal characteristic for methine group of oxirane cycle with center of chemical shift δ=3.01 ppm and also multiplet signal characteristic for methylene group of oxirane cycle with center of chemical shift δ=2.5 ppm. Yields, amount of epoxy groups and molecular masses $\overline{M}_n$, $\overline{M}_\omega$, $\overline{M}_z$ and polydispersity of synthesized epoxy group containing siliconorganic oligomers XVI–XVII' are presented in Table 16.

TABLE 16　Yields, amount of epoxy groups and molecular masses of synthesized tetraethoxypolyphenylsilsesquioxanes (XVI–XVII').

No.	Yield, %	Epoxy* group, %	M*	$\overline{M}v$	$\overline{M}\omega$	$\overline{M}\zeta$	$\overline{M}\omega/\overline{M}v$
XVI	94	4.74	3626	–	–	–	–
		4.80	3672				
XVII	93	13.52	1272	1260	2460	6250	1.95
		13.58	1277				
XVII'	95	13.52	1272	1200	2080	4100	1.83
		13.48	1268				

***In numerator** the calculated values are presented and in denominator – experimental values; Molecular masses were calculated from the values of epoxy groups.

For oligomers XVI and XVII' gel permeation chromatographic investigations were carried out. On the Fig. 15, gel permeation chromatographic curves of synthesized oligomers are presented.

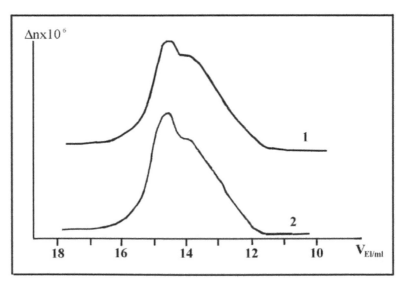

FIGURE 15　Gel permeation chromatographic curves of tetraepoxypolyphenylsilsesquioxanes. Where curve 1 corresponds to oligomer XVI, curve 2 – XVII'.

The form of curves, proximity of correlation of $\overline{M}_\omega / \overline{M}_n$, and also designed values of average molecular masses, provided that the microstructure of oligomeric chain does not undergo to regroupings and reaction of condensation does not proceed inter-molecularly, that it would be possible to expect meaning activity of functional groups entering PhES–80. Most likely reactions of condensation PhES–80 with epichlorohydrin at the presence of $AlCl_3$ and $FeCl_3$ basically proceed with formation of corresponding tetra epoxy-derivatives.

The synthesized tetraepoxypolyphenylsilsesquioxanes do not change their characteristic properties during three months that specifies their long viability (Table 17).

TABLE 17   Change of specific viscosity, amount of epoxy groups for oligomers XVI and XVII depending on duration of their storage

No.	η_{sp} 50% benzene solution at 20°C					Amount of epoxy groups, %				
	After reaction	After 10 days	After 1 month	After 2 months	After 3 months	After reaction	After 10 days	After 1 month	After 2 months	After 3 months
XVI	2.6	2.8	2.8	2.8	2.9	24.30	24.29	24.28	24.28	24.28
XVII	5.2	5.2	5.3	5.5	5.5	7.19	7.18	7.17	7.17	7.17

Epoxy group functionalized polyphenylsilsesquioxane oligomers were used for modification of phenol-formaldehyde oligomer (novolac resin – SF-0112) for increasing their flexibility, their hydrophobic surface properties, thermal stability and flame-retardant properties. Novolac resin – SF-0112 contains a little amount of free phenols (0.9%) compared with other phenol-formaldehyde oligomers.

Polymerization reaction of phenol-formaldehyde oligomer (novolac resin – SF-0112) with tetraepoxypolyphenylsilsesquioxanes was carried out in melt condition at 125±5°C temperature at various ratios of initial compounds, without catalyst. The samples were tested on contain of epoxy groups during the reaction. It was shown that during synthesis of block-copolymers the reaction rate decreases, which may be explained by presence of free phenols in used phenol-formaldehyde oligomer. In accordance with [52–57], the presence of free phenols in phenol-formaldehyde oligomer evokes an acceleration of interaction reaction between epoxy cycles and phenol hydroxyl groups. In this case blocking of epoxy groups which hinder the formation of branched structures will be avoided. The reaction proceeds according to the following Scheme 11 [58–60]:

SCHEME 11. MODIFICATION REACTION OF NOVOLAC PHENOL-FORMALDEHYDE OLIGOMER BY EPOXY GROUP CONTAINING POLYPHENYLSILSESQUIOXANE OLIGOMER. N=1–2, WHERE, N=1 (XVIII); 2 (XIX).

Synthesized block-copolymers are yellow-orange like vitreous products well soluble in ordinary organic solvents.

Besides of these reactions it may be occurred some co-reactions in view of polymerization of epoxy groups with secondary hydroxyl groups and partial condensation reactions of secondary hydroxyl and phenol hydroxyl groups at high temperature conditions.

All the above noted reactions evoke increasing of block-copolymer molecular masses with consequence branching of the macromolecular structure and formation of net like cross-linking structures.

On the Fig. 16, the dependence of amount of epoxy groups on time in the reaction mixture with different ratio of phenol-formaldehyde oligomer to tetraepoxypolyphenylsilsesquioxane is presented. In accordance with Fig. 16 one can see that with an increase of amount of phenol-formaldehyde oligomer in reaction mixture the concentration of epoxy groups decreases rapidly. The main reaction is the interaction of epoxy groups with phenol hydroxyl groups of phenol-formaldehyde oligomer. In the next stage, when the amount of secondary hydroxyl groups increases in the reaction blend the probability of co-reactions increases. The reaction was carried out up to saving of fluidity and solubility of block-copolymer in the solvents (acetone, toluene and acetone–toluene mixture).

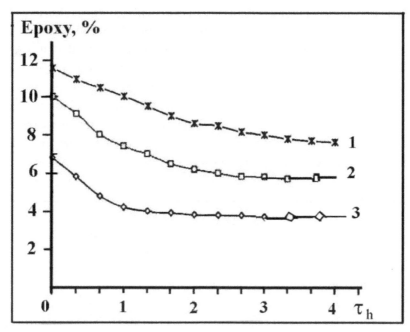

FIGURE 16 The dependence of amount of epoxy groups on time in reaction mixture with different ratio of phenol-formaldehyde oligomer to tetraepoxypolyphenylsilsesquioxane: 1 – 60:40; 2 – 40:60; 3 –30:70.

FTIR spectra of the reaction products contain the absorption band at 920 cm^{-1}, which corresponds to deformation oscillations of epoxy groups. The absorption bands at 760 and 830 cm^{-1} corresponds to ortho- and para-substituted phenyl groups and absorption band at 1365 cm^{-1} is characteristic for phenol hydroxyl groups. About presence of hydroxyl groups in block-copolymers proved the absorption band in the region of 3250–3450 cm^{-1}. About proceeding of copolymerization reaction proved decrease of intensity of absorption bands characteristic of epoxy groups.

The hardening processes of obtained block-copolymers were carried out both in the presence of accelerator and without it. Amine type accelerator hexamethylenetetramine (HMTA) and isomer mixture of methyltetrahydrophthalic anhydride *(iso*-MTHA) was used as hardened activators.

The hardening of epoxy-novolac block-copolymer at the presence of amine type curing agent does not proceed at room temperature during 24 h. This fact is confirmed by good solubility of reaction products in ordinary organic solvents and by definition of epoxy group's amount. This phenomenon may be explained with very weak interaction between solid polymer and curing agent molecules or this interaction is so low that it doesn't leads to visible change of block-copolymer structure. Only at 100°C and at more high temperatures the hardening reaction proceeds more rapidly.

The hardening reaction was carried out at 160 and 180°C temperatures in the polymers with definition of gel-fraction (Fig. 17). It was established that in block-copolymers the formation of gel-fraction occurs too, which is due to above mentioned co-reactions.

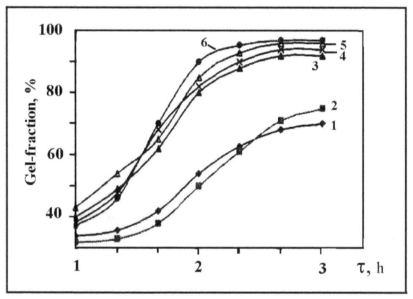

FIGURE 17   Dependence of gel-fraction yields on the time during hardening of block-copolymer at 160°C without accelerator (1), with *iso*-MTHP (3), with HMTA (5) and at 180°C without accelerator (2), with *iso*-MTHP (4), with HMTA (6).

With an increase of temperature from 160 up to 180°C the cross-linking processes rises due to secondary hydroxyl group's intermolecular interaction. Comparison of hardening processes initiated by HMTA and iso-MTHP shows that in first case the gel-fraction degree is more than in second one. This may be explained with acceleration by HMTA both reaction of epoxy groups with phenol hydroxyl and rapid formation of the net like structure, because this accelerator is high effective curing agent of the phenol-formaldehyde oligomer blocks as well as of block-copolymer. It is possible that the part of the functional groups does not enter in reaction.

This fact will be take place in such part of polymer matrix in which the cross linking bonds with large amount will be formed at first hardening stages, which leads to breaking of polymer segments mobility in the polymer and consequently the reactions between residual epoxy groups and hydroxyl group's decreases essentially.

Thermogravimetric investigation of hardened samples with various content of polyphenylsilsesquioxane fragment has been carried out (Fig. 18). As it is seen from Fig. 18 with an increase of amount siliconorganic fragments from 40 up to 70% in composites the thermal oxidative stability increases. 10% mass losses for composite materials one can observe in the 270–330°C temperature range.

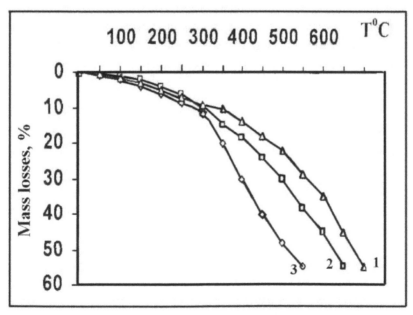

FIGURE 18 Mass losses of the block copolymers at ratio PFO: ESO 30:70 (1); 40:60 (2) and 60:40 (3) (in open area, heating rate 10°C).

The comparison of FTIR spectra of the epoxy-novolac block-copolymer hardened and unhardened samples (with curing agent HMTA and without it) shows that at hardening the intensity of deformation absorption bands of epoxy groups at 920 cm^{-1} and deformation absorption bands of hydroxyl groups at 1367 cm^{-1} decreases. Using of HMTA as a curing agent essentially decreases the above mentioned absorption bands in comparison with hardened polymers obtained without curing agent. This change is connected with acceleration of the intermolecular interaction reactions between epoxy and phenol hydroxyl groups.

The condensation reaction of tetraethoxypolyphenylsilsesquioxane with the excess of epichlorohydrin in the presence of catalysts Fe (III) and Al chloride has been investigated and corresponding tetra epoxy-derivatives has been obtained. By polymerization reaction of tetraepoxypolyphenylsilsesquioxanes with phenol-formaldehyde novolac oligomers epoxy-siliconorganic and phenol-formaldehyde novolac block-copolymers have been obtained.

Novolac-epoxy networks can be modified with reactive epoxy functional polysiloxane oligomers. In accordance with FTIR spectra of the reaction products obtained at various temperatures with an increase of amount of phenol-formaldehyde oligomer in reaction mixture the concentration of epoxy groups decreases rapidly. The composite materials on the basis of organo-inorganic block-copolymers has been obtained and studied.

10.3 COMPOSITES BASED ON EPOXY RESINS FILLED WITH MODIFIED BENTONITE

In the recent time, mineral fillers attract generally the attention as active filling agents with the purpose of using them in polymer composites [58, 59]. Thanks to these fillers many properties of the composites are improved – increases the durability and rigidity, decrease the shrinkage during hardening process and water absorption, improves thermal stability, fire proof and dielectric properties and finally the price of composites became cheaper [61, 62]. At the same time it must be noted that the mineral fillers especially at their high concentrations in composites lead to some impair of different physical properties of composites. Therefore the attention of the scientists is attracted to substances, which would be remove mentioned leaks. It is known that silicon organic substances (both low and high molecular) reveal hydrophobic properties, high elasticity and durability in wide range of temperature [63, 64].

The purpose of presented chapter is the investigation of effect of modify by tetraethoxysilane (TEOS) of the mineral-bentonite as filler on some physical properties of composites based on epoxy resin.

High temperature condensation reaction between bentonite and TEOS was carried out in anhydrous toluene solution (~35%). The masses of TEOS were 3, 5 and 10% from the mass of filler. The reaction system was heated at the solvent boiling temperature (~110°C) during 5–6 hours by stirring. After there was separate from precipitate solvent ethyl alcohol and remain tetraethoxysilane on the rotor-evaporator and work thermal the modified bentonite as at the first stage pure bentonite up. The reaction goes by elimination of ethyl alcohol (the direction of reaction was improved by FTIR spectra analysis) [65].

On the basis of modified bentonite and epoxy resin (of type ED-20) the polymer composites with different content of filler were obtained after careful wet mixing of components in mixer. After the blends with hardening agent (polyethylene– polyamine) were placed to the cylindrical forms (in accordance with standards ISO) for hardening at room temperature during 24 h. The samples hardened later on undergo to temperature treatment at 120°C during 4 h.

It was shown that the modification goes more successful when the masses of TEOS were near 5% from the masses of the filler. The concentration of powder bentonite (average diameter up to 50 micron) was changed in the range 10–60 mass %. Filler was modified with 3, 5 and 7 mass % of TEOS.

Following parameters were defined for obtained composites: ultimate strength, softening temperature and water absorption.

The dependence of ultimate strength on the content of bentonite (modified and unmodified) presented on the Fig. 19 shows that they have an extreme character. However the positions of corresponding curves maximums essentially depend on amount of modified agent TEOS. The character of these curves show that an optimal content of modifier is near 5 mass %. The general view of these dependences is in full conformity with well known dependence of $\sigma - C$ [66]. Investigation of composites softening temperature was carried out by apparatus of Vica method. Fig. 19 shows the temperature dependence of the indentor deepening to the mass of the sample for composites with different concentration of the fillers.

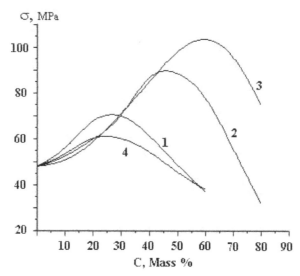

FIGURE 19 Dependence of ultimate strength of the composites based on ED-20 with unmodified (1) and modified by 3% (2), 5% (3) and 7% (4) TEOS bentonite.

Effect of silane modifier on the investigated polymer composites reveals also in the water absorption. In accordance with Fig. 20 this parameter is increased at increasing of filler contain. However if the composites contain the bentonite modified by TEOS this dependence becomes weak (feeble).

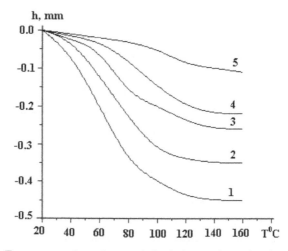

FIGURE 20 Temperature dependence of the indentor deepening in the mass of the sample for composites with different concentration of the fillers.

The obtained experimental results may be explained in terms of composite structure peculiarities.

Silane molecules displaced on the surface of bentonite particles leads to activation of them till formation of chemical reactions between active groups of TEOS (hydroxyl) and homo polymer (epoxy group) (Fig. 21). Silane molecules create the "buffer" zones between filler and the homo polymer.

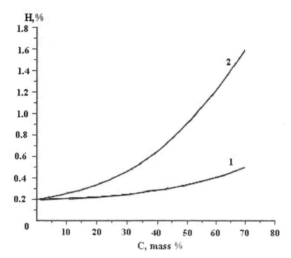

FIGURE 21   Dependence of water absorption (%) of composites based on ED-20 with modified by 5% TEOS and unmodified bentonite on the filler concentration.

This phenomenon may be one of the reasons of increasing of strength of composite in comparison with composites containing unmodified bentonite. The composites with modified bentonite display more high compatibility of the components than in case of same composites with unmodified filler. The modified filler has more strong contact with polymer matrix (thanks to silane modifier) than unmodified bentonite. Therefore mechanical stresses formed in composites by stretching or compressing forces are absorbed effectively by relatively soft silane phases, i.e. the development of micro defects in carbon chain polymer matrix of composite closes in silane part of the material.

The structural peculiarities of composites display also in thermo-mechanical properties of the materials. It is clear that softening of composites with modified by TEOS composites begins at relatively high temperatures.

This phenomenon is in good correlation with corresponding composite mechanical strength. Of course the modified filler has more strong interactions (thanks to modifier) with epoxy polymer molecules, than unmodified filler.

The amplified competition of the filler particles with macromolecules by TEOS well displays also on the characteristics of water absorption. In general loosening of micro-structure because of micro empty areas is due to the increasing of filler content. Formation of such defects in the microstructure of composite promotes the water absorption processes. Water absorption of composites with modified bentonite is lower than that for one with unmodified filler to some extent. The decreasing of water absorption of composites containing silane compound is result of hydrophobic properties of ones.

Comparison of mechanical strengthening, softening temperature and water absorption for polymer composites based on epoxy resin and mineral filler bentonite with ones but containing modified by low amount (less than 7 mass %)of TEOS leads to conclusion that modify agent stipulates the formation of heterogeneous structures with higher compatibility of ingredients and consequently to enhancing of noted technical characteristics. The increasing of mechanical and thermal properties of the composites with relatively high contain of modified by TEOS bentonite filler has practical aspect: the higher is content of mineral fillers the cheaper is the polymer composite.

10.4   POLYFUNCTIONAL METHYLSILOXANE OLIGOMERS WITH GLYCIDOXYPROPYL SIDE GROUPS

Epoxy containing organosiloxane may be obtained by hydride addition of mono– and dihydride-containing linear organosilanes and siloxanes to allyl glycidyl ether [67, 68] in the presence of platinum hydrochloric acid in anhydrous iso-propanole.

During the hydrosilylation of allyl glycidyl ether with 1,4-bis(dimethylsilyl)benzene, or diphenylsilane in isothermal condition by calorimetric method at various concentration of initial compounds and catalyst the hydrosilylation rate constants and reaction order have been calculated [69].

The epoxy containing organosiloxanes, (with Si-C bonds) are obtained by oxidative epoxidation of α,ω-divinylorganosiloxanes the yield of synthesized diepoxyorganosiloxanes in this case is about ~85% in tetrachlorocarbon area [70].

Epoxy organosiloxanes with various length of chain may be obtained by anionic co-oligomerization of organocyclosiloxanes with 1,3-bis(3-glycidoxypropyl)tetramethyldisiloxane [67].

The epoxy ring-opening reaction with organic amines and organic acids [69] showed that it proceed with the formation of β and α–adducts.
The epoxy ring-opening reactions with primary and secondary amines showed that in case of primary amines the reaction proceeds completely and has a macromolecular character [69].

In literature there is some information about synthesis of polyfunctional organosiloxane containing glycidoxypropyl groups in the side dimethylsiloxane chain [71].

The present chapter deals with synthesis and studies of the properties of comb-type organo-siloxanes containing epoxy groups in the side chain.
Organosiloxanes and with epoxy group in the side chain have been synthesized by the reaction of hydride addition of α,ω-bis(trimethylsiloxy)methylhydridesiloxane (n≈30.53) to allyl glycidyl ether at 1:30 or 1:53 ratio of initial compounds, in the presence of platinum hydrochloric acid or Pt/C as a catalyst has been investigated.

The reaction of hydrosilylation was investigated without solvent in the presence of catalysts at temperature range 80–90°C. In both cases the hydrosilylation proceeds with an inductive period, after that the reaction proceeds autocatalytic with the formation of three-dimensional systems.

For obtaining of soluble, epoxy group containing comb-type organosiloxanes, hydrosilylation reactions were investigated in dilute solutions of anhydrous toluene. The optimal concentration is about 0.1 mole/l. Lower this concentration; sewing processes does not take place. The hydrosilylation was carried out at various temperatures 80/90°C. The course of the reaction was observed by decrease of amount of active ≡Si-H groups. As it is clear from Fig. 22, with the rise of the temperature, the rate of hydride addition increases. At one and the same temperature the rate of hydrosilylation in the presence of platinum hydrochloric acid is tighter than in case of platinum on carbon. As it is obvious from Fig. 1 the reaction mainly proceed vigorously during the first hours. After 2 hours, the conversion slowly changes.

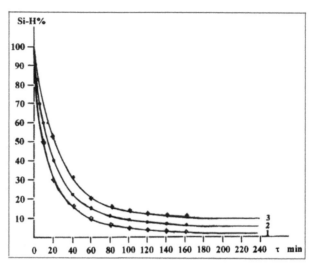

FIGURE 22 Dependence of changes of active Si-H% groups on time upon hydride addition of allyl glycidyl ether to α,ω-bis(trimethylsiloxy)methylhydridsiloxane, where curve 1 is at 90°C, curve 2 is at 85°C and curve 3 is at 80°C.

From Fig. 22, we can see that conversion of active ≡Si-H groups is not complete and decreases from 98% (t=90°C) to 92% (t=80°C).

The synthesized oligomers with epoxy groups in the side chain, after removal of solvents are liquid colorless (or light yellow) transparent products well soluble in ordinary organic solvents, with $\eta_{sp} \approx 0.04/0.08$. Some physical-chemical properties of synthesized oligomers are presented in Table 18.

TABLE 18 Some physical-chemical properties methylhydridesiloxanes with organic branching side groups

No.	Yield %	η_{sp}*	d_1, Å	$\overline{M}_\omega$**	Elementary composition, %*				
					C	H	Si	N	Br/S
XX	92	0.05	–	5384	48.13	8.19	16.77	–	–
				5200	47.74	8.01	16.30		
XXI	94	0.04	–	–	48.13	8.19	16.77	–	–
					47.98	8.08	16.45		

TABLE 18 *(Continued)*

No.	Yield %	η_{sp}*	d_1, Å	$\overline{M}_\omega$**	Elementary composition, %*				
					C	H	Si	N	Br/S
XXI'	92	0.04	–	–	48.13	8.19	16.77	–	–
					47.72	7.93	16.56		
XXII	97	0.05	–	–	48.13	8.19	16.77	–	–
					48.01	8.21	16.14		
XXIII	93	0.06	–	9384	48.13	8.19	16.77	–	–
				9000	48.81	7.87	16.60		
XXIV	97	0.06	–	7576	53.21	10.21	11.85	5.54	–
				7500	53.30	10.13	11.60	5.12	
XXV	97	0.07	–	8596	59.46	8.29	10.44	4.88	–
				8600	59.13	7.87	10.13	4.41	
XXVI	98	0.06	–	9016	60.68	8.58	9.96	4.66	–
				8900	60.31	8.30	9.63	4.20	
XXVII	–	–	–	9134	52.02	7.09	9.80	4.59	10.51
				8700	51.76	6.70	9.31	4.70	9.78
XXVIII	97	–	–	8654	54.91	7.48	10.35	4.85	–
				8300	54.28	6.97	9.86	4.41	
XXIX	98	0.07	6.94	7576	53.21	10.21	11.85	5.54	–
				7600	52.78	9.89	11.56	5.08	
XXX	97	0.07	6.94	8596	59.46	8.29	10.44	4.88	–
				8500	59.22	8.45	10.20	4.61	
XXXI	98	0.06	6.90	9258	59.10	10.96	9.70	4.53	–
				9100	58.76	10.57	9.34	4.23	
XXXII	98	0.08	–	11026	46.36	6.69	8.12	3.80	21.76
				10900	46.12	6.45	8.33	3.54	20.89
XXXIII	98	0.08	7.08	10006	40.29	7.97	8.95	4.19	23.98
				9900	40.02	7.55	8.76	3.80	23.40
XXXIV	97	0.08	7.02	9436	50.35	9.40	9.49	4.45	–
				9400	50.11	9.16	9.13	4.09	

*In 1% toluene solution (XXVIII and XXIX in water) at 25°C.
**In numerator there are calculated values, in denominator experimental values.

Average molecular masses were determined by gel permeation chromatography.

The FTIR spectra of oligomers revealed that the absorption bands characteristic for ≡Si-H bonds in the range 2160–2170 cm^{-1} do not vanish completely. In the FTIR spectra one can see the absorption band at 850, 920, and 4545–4550 cm^{-1} characteristic for epoxy groups, the absorption bands at 1020–1100 and 1150 cm^{-1} corresponds to asymmetric valence oscillation for the linear ≡Si-O-Si≡ and C–O–C simple ether bonds, as well as absorption bands in the range 2900–2950 cm^{-1}, typical for valence oscillations of the CH bonds in the CH$_2$ groups in the ethylene bridges and absorption bands at 1270 cm^{-1} characteristic for ≡Si-Me bonds.

Hydrosilylation reaction proceeds according to the following Scheme 12 [72, 73].

SCHEME 12. HYDROSILYLATION REACTION OF METHYLHYDRIDESILOXANE TO ALLYL GLYCIDYL ETHER, IN THE PRESENCE OF CATALYSTS.

where, [(a)+(b)](x)=n; n≈30,53. n=30, XX (80°C); XXI (85°C); XXII (90°C) – [H$_2$PtCl$_6$]. n=53, XXIII (85°C) – H$_2$PtCl$_6$. n=30, XXIII'(85°C) – Pt/C.

In the ^{1}H NMR spectrum of oligomer XXII Fig. 23 one can observe the singlet signals for methyl protons in trimethylsiloxy group with chemical shift δ=0.05 ppm, for methyl group at δ=0.01 ppm, 0.47 ppm (s– C^{6}H$_2$), 1.50 ppm (s– C^{5}H$_2$), 3.20; 3.30; 3.45; 3.50 ppm (m– C^{3}H$_2$, C^{4}H$_2$) ; 3.0 ppm (s– C^{2}H) and 2.41 and 2.65 ppm (d-C^{1}H$_2$).

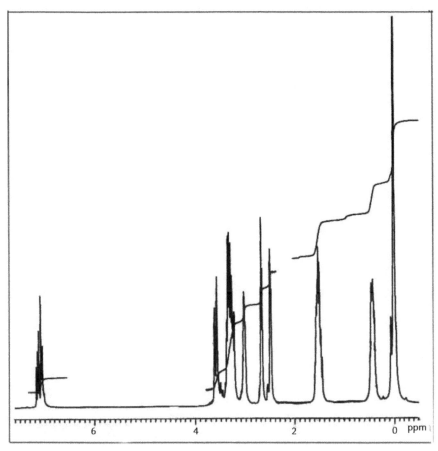

FIGURE 23   ¹H NMR spectrum of oligomer XXII.

In the ¹³C NMR spectrum of oligomer XXII Fig. 24 one can observe the signals 1.85 ppm (C⁷); 13.9 ppm (C⁶); 23.56 ppm (C⁵); 71.86 ppm (C⁴); 74.43 ppm (C³); 51.29 ppm (C²) and 44.70 ppm (C¹).

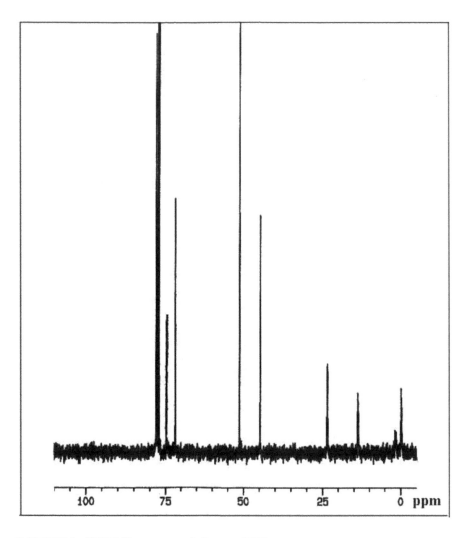

FIGURE 24 ^{13}C NMR spectrum of oligomer XXII.

Figure 25 shows the time dependence of the reverse concentration, from which it is evident, that at the beginning stages the hydrosilylation reaction is of a second order. This result is in opposition with literature data [65], where the authors proved that depending on the origin concentration of reacting compounds the hydrosilylation reaction may be at first either at low value of concentration, or zero-at higher value of concentration.

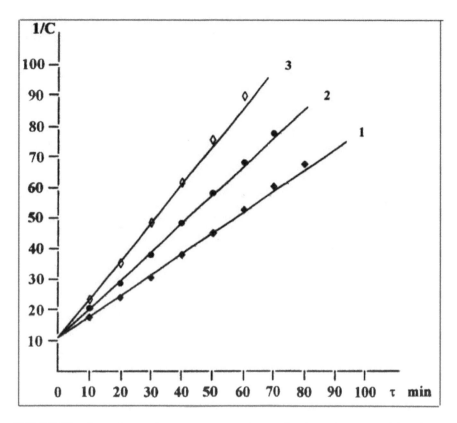

FIGURE 25 Dependence of reverse concentration on the time, where curve 1 is at 80°C, curve 2 at 85°C and curve 3 is at 90°C.

From the dependence of hydrosilylation rate constants on the time at various temperatures were calculated: $k_{85°C} \approx 0.6878$, $k_{90°C} \approx 0.9325$ and $k_{95°C} \approx 1.2571$ l/mol×min. There was found that for each increase in temperature of 10°C the reaction rate constants increase approximately 1.35 times.

From the dependence of logarithm of hydrosilylation rate constants on the reverse temperature, from which the activation energy was calculated $E_{activ} \approx$ 30.6 kJ/mol.

Comb-type organosiloxanes with epoxy groups in the side chain-reacted with primary amines. During the epoxy ring-opening reaction in melting condition the reaction proceeds with the formation of three dimensional systems, this fact may be explained after the ring-opening reaction with the present of reactionable secondary (–NH–) groups, which enter the ring-opening intra– and intermolecular condensation. In order to receive soluble comb-type organosiloxanes containing along hydrophobic methylsiloxane chain, hydrophilic hydroxyl and amino groups the reaction of epoxyorganosiloxanes with primary amines were carried out in 10% solution of anhydrous toluene. The reaction products were heated up to the boiling point of toluene for 12 hours. The completeness of the reaction was confirmed by ^{1}H NMR spectra by disappearance the resonance signals characteristic for epoxy groups. As it's seen from Table 18 aminohydroxysiloxanes were obtained in good yields. The reaction proceeds according to the following Scheme 13.

SCHEME 13. REACTIONS OF EPOXYORGANOSILOXANES WITH PRIMARY AMINES.

where, $[(a)+(b)](x) \approx 30$. $R=C_4H_9$ (XXIV), $C_6H_5CH_2$ (XXV), $2,3\text{-}Me_2C_6H_3$ (XXVI), $1,4\text{-}HOC_6H_4$ (XXVII), $o\text{-}HSC_6H_4$ (XXVIII).

Synthesized secondary aminohydroxysiloxanes (XXIV–XXVI and XXVIII) are colorless or slight yellow compounds soluble in ordinary organic solvents. Unfortunately all experiments of receiving the soluble product XXVII were without a success. It's possible that during the reaction of p-aminophenol with epoxymethyhydridelsiloxanes the secondary intra- and intermolecular reactions takes place. The composition and structure of synthesized compounds were determined by elementary analysis, by finding the molecular masses, by FTIR and NMR spectra data.

In the FTIR spectra of all aminohydroxymethylsiloxanes one can see the absence of absorption bands characteristic for epoxy bonds at 910 cm^{-1} and absorption bands characteristic for –NH– and –OH bonds at 3400–3600 cm^{-1} (those absorption bands are put on each other). Hence it follows that all epoxy groups take place in ring-opening reaction. There the absorption bands for trimethylsiloxy are observed, $\equiv$Si-Me, $\equiv$Si-O–Si$\equiv$ and COC bonds at 840, 1275, 1000–1130 cm^{-1} accordingly.

In the ^{1}H NMR spectra of all secondary aminohydroxymethylhydridesiloxanes (Figs. 26–29) one can observe the singlet signals with chemical shifts 0.1–0.15 ppm for methyl protons at silicon (7 and 8 position), $\delta=0.5$ ppm (C^6H_2); $\delta=1.5$ ppm (C^5H_2); $\delta=1.9 \div 2.1$ ppm (NH); $\delta=2.9 \div 3.0$ ppm (C^4H_2); $\delta=3.2$ ppm (C^3H_2); $\delta=3.3$ ppm(C^2H); $\delta=3.6$ ppm (C^1H_2); $\delta=3.7–3.8$ ppm (OH) (see Fig. 26 for compound XXIII).

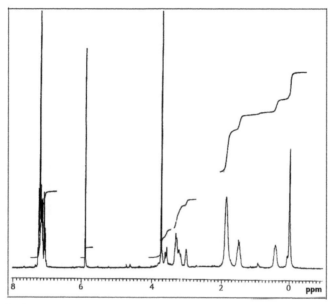

FIGURE 26 ¹H NMR spectrum of oligomer XXV.

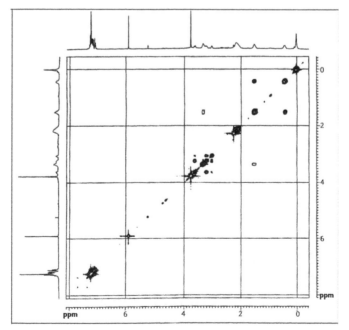

FIGURE 27 COSY NMR spectrum of oligomer XXV.

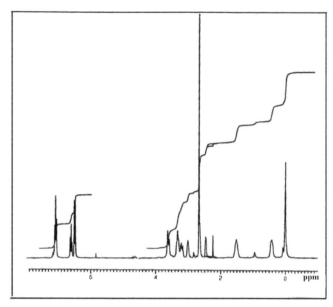

FIGURE 28   ¹H NMR spectrum of oligomer XXIV.

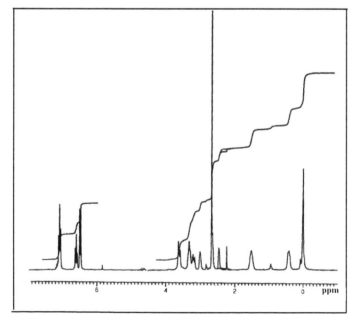

FIGURE. 29   ¹H NMR spectrum of oligomer XXV.

In the ^{1}H NMR spectrum of compound XXV one can see broadened singlet signal, which may correspond to labile protons linked to electronegative nitrogen and sulfur groups. By integral intensity this signal corresponds approximately to two-proton absorption intensity, but this does not exclude the existence of –SH groups, because there is possible proton exchange reaction, which may be seen from broadened singlet signals. That is why in compound XXVIII there is possible existence of –NH$_2$ or =NH and –SH groups.

Full identification of aminohydroxymethylsiloxanes was received for example from benzylaminohydroxylsiloxane XXV by using COSY NMR spectrum. On Fig. 27, one can see the resonance signals on surface. Where the resonance signals at $\delta{\approx}0.1$ corresponds to methyl groups at 7 and 8 positions. The resonance signals with chemical shift at $\delta{\approx}0.5$; 1.5 and 2.2 correspond to methylene groups in 7,8; 6 and 5 positions and resonance signal with chemical shift in the range $\delta{\approx}$ 7.0÷7.4 ppm corresponds to phenyl protons.

Comb-type organosiloxanes with epoxy groups in the side chain-reacted with secondary amines. The reaction was carried out in 10% solution of anhydrous toluene. The reaction products were heated up to the boiling point of toluene for 12 hours. The completion of the reaction was confirmed by ^{1}H NMR spectra by disappearance of the resonance signals characteristic for epoxy groups. As it's seen from Table 1 the yields of comb-type aminohydroxysiloxanes higher because this reaction proceeds without any complication, according to the following Scheme 14.

SCHEME 14. REACTION OF EPOXYORGANOSILOXANES WITH SECONDARY AMINES.

where, $[(a)+(b)](x){\approx}30$. R=R'=C$_2$H$_5$ (XXIX), R=Me, R'=C$_6$H$_5$ (XXX), R=R'=C$_4$H$_9$ (XXXI).

The composition and structure of tertiary aminohydroxysiloxanes was established by means of elementary analysis, by finding the molecular masses by FTIR and NMR spectra data. In ^{1}H NMR spectra of compounds XXIX÷XXXI one can observe resonance signals with chemical shifts 0.1– 0.15 ppm for methyl protons at silicon (in 7 and 8 position), δ=0.5÷0.6

ppm (C^6H_2); additionally complicated signals with chemical shift $\delta=$ 0.9÷1.1 ppm for methyl protons in alkaline fragment (compounds XXIX and XXIX); $\delta=$1.5÷1.6 ppm (C^5H_2); $\delta=$2.3÷2.7 ppm (NC^1H_2 and NCH_2); $\delta=$2.9÷3.0 ppm (C^4H_2); $\delta=$3.2 ppm (C^3H_2); $\delta=$3.3 ppm (C^2H); $\delta=$3.6 ppm (C^1H_2); $\delta=$3.7 ppm (OH) and $\delta=$4.2 ppm ($\equiv$Si-H) (see Figs. 30 and 31 for compound XXIX). Full identification of tertiary aminohydroxyme-thylsiloxanes was received for example of benzylaminohydroxysiloxane XXIX by using COSY NMR spectra. In the ^{13}C NMR spectra of compound XXIX, one can see the signals for carbon attached with hydroxyl groups at $\delta=$67.13 ppm and signals for carbon attached with nitrogen at δ 47.57 and δ 56.50 ppm. On Fig. 31 one can see the resonance signals on surface.

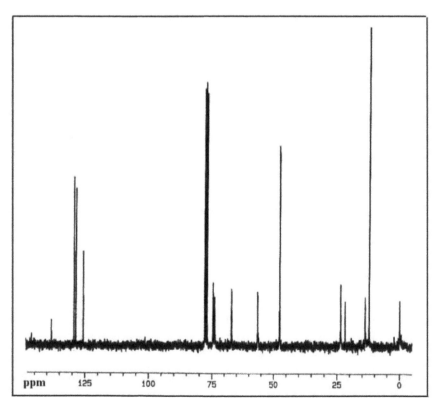

FIGURE 30 ^{13}C NMR spectrum of oligomer XXIX.

where, the resonance signal at ~0.1 ppm, corresponds to methyl groups at 7 and 8 positions. The resonance signal at $\delta \approx 0.5$; 1.5 and 3.4 ppm corresponds to methylene groups in 6, 5 and 4 positions.

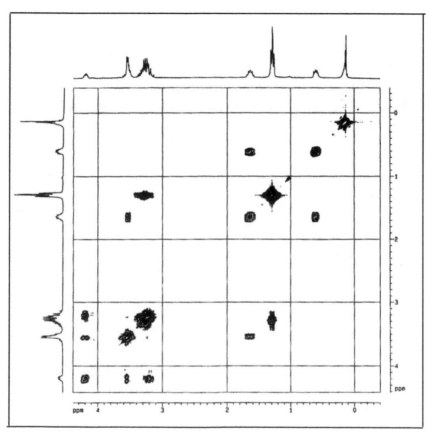

FIGURE 31   COSY NMR spectrum of oligomer XXIX.

The triplet signal in the range $\delta = 0.9/1.1$ ppm corresponds to methyl groups in ethyl fragment. In the FTIR spectra of tertiary aminohydroxysiloxanes XXIX/XXXI one can see the narrow absorption bands characteristic for secondary hydroxyl group at 3600 cm^{-1}, from which it is evident that the reaction of amines with epoxyorganosiloxanes mainly proceeds with the formation of α–adduct. The tertiary amines did not react with epoxy containing comb-type organosiloxanes because the tertiary amines did not have the active hydrogen groups, which may enter the ring-opening reaction.

For the purpose of synthesis of siliconorganic quaternary ammonium salts the reaction of hydrogen bromide or acetic acid with aminohydroxy organosiloxanes has been investigated. It was very interesting to test the stability of siloxane chain and to determine that such strong acid as hydrogen bromide (or acetic acid) can to break linear siloxane chain. The reaction was carried out at 0°C temperature to prevent the break of $\equiv$Si-O-Si$\equiv$ backbone. The hydrogen bromide in equimolar amount was added and stirred for 4 hours it proceeds according to the following Scheme 15.

SCHEME 15. REACTION OF AMINOHYDROXY ORGANOSILOXANES WITH ACIDS.

where, $[(a)+(b)](x) \approx 30$; X=Br: R=Me, R'=C$_6$H$_5$ (XXXII); R=R'=C$_2$H$_5$ (XXXIII). X=CH$_3$COO, R= R'=C$_2$H$_5$ (XXXIV).

The synthesized salts are vitreous transparent colorless products the compounds XXXIII and XXXIV are soluble in water and the compound XXXII in organic solvents such as CH$_2$Cl$_2$, chloroform and so on. By GPC analysis it was shown that during the reaction of aminohydroxysiloxanes with the above-mentioned acids break down reactions of $\equiv$Si-O-Si$\equiv$ backbone do not take place (see Table 18). The compounds XXXII–XXXIV were characterized by ^{1}H, ^{13}C NMR and FTIR spectra data. As it's seen from NMR spectra of protonated compounds XXXII–XXXIV Figs. 32–35, one can see resonance singlet signals for methyl protons at silicon atom $\delta \approx 0.1$–0.13 ppm (for hydrogen protons at 8 and 7 position accordingly), for protons in methylene groups with chemical shifts at $\delta \approx 0.5$; 1.5 ppm; in the range 2.9÷3.6 ppm (6,5, 4 and 3 position).

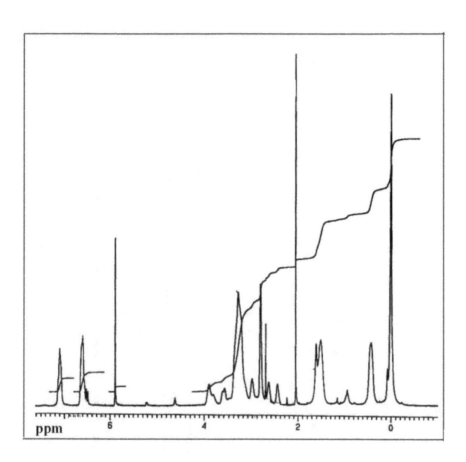

FIGURE 32 ¹H NMR spectrum of oligomer XXXII.

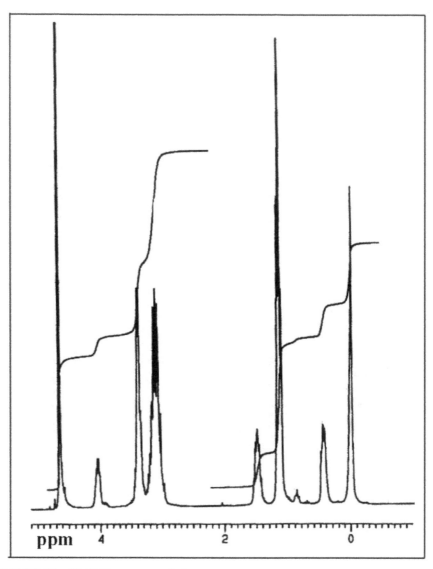

FIGURE 33 ^{1}H NMR spectrum of oligomer XXXIII.

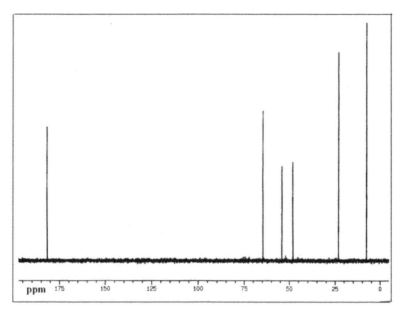

FIGURE 34 ¹H NMR spectrum of oligomer XXXIV.

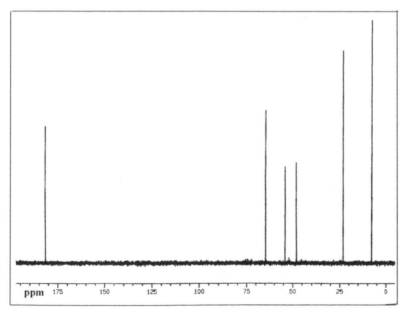

FIGURE 35 ¹³C NMR spectrum of oligomer XXXIV.

The comparison of chemical shifts for $N-CH_2-$ groups in tertiary and quaternary aminohydroxysiloxanes, shows a significant shift in low field for quaternary compounds.

Full identification of protonated aminohydroxyl methylhydridesiloxanes COSY NMR spectra was received for compound XXXIII (Fig. 36). On surface COSY NMR spectra (Fig. 33) the resonance

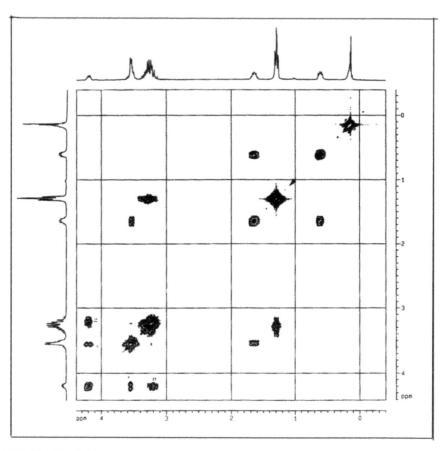

FIGURE 36   COSY NMR spectrum of oligomer XXXIII.

signals at $\delta \approx 0.1$ and $\delta \approx 0.13$ ppm correspond to methyl protons at 7 and 8 positions. The resonance broadened signals with chemical shift and with center at $\delta \approx 0.5$; 1.3; 1.7; 3.2 and 3.5 ppm correspond to protons of methylene groups in 6÷1 positions and in ethyl fragment.

The large absorption band in the region 3200–3600 cm⁻¹ corresponds to the associated hydroxyl and cationic nitrogen groups. The absorption bands in the region 1590–1650 cm⁻¹ corresponds to NH_2^+ group.

For some synthesized aminohydroxyl methylsiloxanes the roentgeno-graphic investigations have been carried out. As it is seen from Fig. 37 the synthesized methylhydridesiloxane with amino and hydroxyl groups in the side chain as well as protonated corresponding compounds are repre-sented as amorphous systems. On the diffractograms one can observe two maximums $2\theta^0 = 12.5^0$ and $2\theta = 18.0^0 \div 22.0^0$. The main maximum $2\theta^0 = 12.5^0$ which corresponds to interchain distance, changes from $d_1 = 6.94$ Å (for compounds XXIX and XXX) up to $d_1 = 7.08$ Å, these increasing values of interchain distances correspond to protonated aminohydroxyl methylh-ydridesiloxane XXVI.

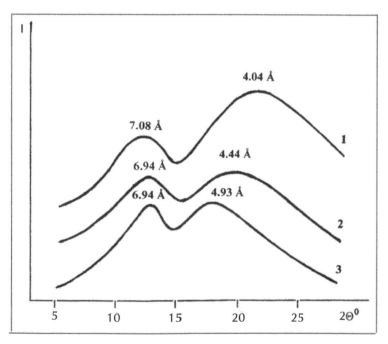

FIGURE 37   Diffractograms of comb-type organosiloxanes with aminohydroxyl groups in the side chain. Where curve 1 corresponds to oligomer XXXI, curve 2–XXX and curve 3 – XXIX.

Thermogravimetric investigation of some synthesized compounds in isothermal condition (200°C) was carried out. As it is seen from Fig. 38 (curves 1 and 2), for aminohydroxysiloxanes XXIX and XXXI mass losses for 10 h do not exceed 21–24%, while for protonated compound XXXI for first 3 h mass losses consist about 74% (Fig. 38, curve 3), then the destruction process proceeds slightly.

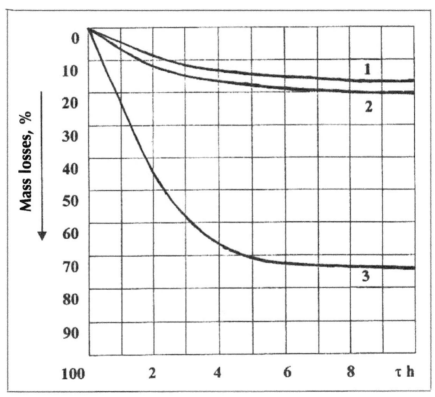

FIGURE 38   Thermogravimetric curves in isothermal (200°C) conditions (where curve 1 corresponds to oligomer XXX, curve 2 – XXIX and curve 3 – XXXI).

So, by us for the first time was synthesized methylhydridesiloxane with hydroxyl and amino functional group in the side chain, and by protonation of aminohydroxysiloxanes protonated aminohydroxysiloxanes were obtained.

10.5 CONCLUSIONS

The synthesis of organosiloxane monomers and polymers of linear and cyclic structure with epoxy groups via catalytic oxidation reactions of compounds with vinyl group and condensation reactions of compounds with hydroxyl or ethoxyl groups with epichlorohydrin have been summarized. The hydrosilylation reaction of linear polyhydromethylsiloxane with allyl glycidyl ether is also considered. Some ring opening reactions of epoxy groups have been studied.

The composite materials on the basis of epoxy containing organosiloxane compounds and their physical-chemical properties have been considered.

KEYWORDS

- **diglycidoxypolymethylsiloxane (IV)**
- **epoxides**
- **hexamethylenetetramine**
- **organosiloxanes**
- **phenol-formaldehyde oligomer**
- **siliconorganic oligomers**
- **tetrahydroxyoligoorganosilsesquioxanes**

REFERENCES

1. Prilezhaev, E. N. *Prilezhaev Reaction*. Electrophilic oxidation, Moscow, "Nauka", **1974**, (In Russian).
2. Huirong, Y.; Richardson, D. E. Epoxidation of Alkenes with Bicarbonate-Activated Hydrogen Peroxide. *J. Am. Chem. Soc.*, **2000**, *122*, 3220.
3. United States Patent: *Selective catalysts for the synthesis of epoxysilicone monomers and polymers*, **2000**, 6124418.
4. Valetski, P. M. *Epoxide polymers*. Polymer Chemistry and Technology, **1966**, *2(64)* (In Russian).

5. Mukudan, A. L.; Balasubramanian, K.; Srinivasan, K. S. V. *Synthesis of epoxypoly-organosiloxane Polym.Commun.* **1988**, *29(10)*, 310.
6. Iskakova, M. K.; Markarashvili, E. G.; Mindiashvili, G. S.; Shvangi-Radze, G. M.; Gvirgvliani, D. A.; Mukbaniani, O. V. Synthesis of Epoxyorganosilicon Oligomers on the Basis of, α,ω-Dihydroxydiorganosiloxanes Abstract of Communications of X All Russian Conference «Organosilicon Compounds: Synthesis, Properties and Application», Moscow, Russian Federation, 25-30 May, **2005**, 3C12.
7. Metelitsa, D. I. Reaction mechanisms of the direct epoxidation of alkenes in the liquid phase. *Uspekhi Khimii*, **1972**, *41*, 1737.
8. House, H. O. *Modern synthetic reactions*, N.-Y., **1972**.
9. Swern, E. D. *Organic Peroxides*, Wiley, D.; Intersci, N. -Y., **1970**, *1*, 313.
10. Kucher, R.V.; Tikhomin, V. I.; Shevchuk, I. P.; Vasiutin, I. A. M. *Gass-liquid oxidation of unsaturated compounds in olefine oxide.* Kiev, Ukraina, USSR «Naukova Dumka», 1986 (In Russian).
11. Emmons, W.D. Peroxytrifluoroacetic Acid. I. The oxidation of nitrosamines to nitramines, *J.Amer. Chem. Soc.*, **1954**, *76*, 3468.
12. Plueddeman, E. P.; Fanger, G. Epoxyorganosiloxanes. *J. Amer. Chem. Soc.*, **1959**, *81(11)*, 2632.
13. Peterson, P. E.; Nelson, D. J.; Risener, R. Preparation of vinylsilanes and vinyl halides containing alkene or peroxide functional groups. *J. Org. Chem.*, **1986**, *51(12)*, 2381.
14. Sheng, M. N.; Zajacek, J. G. Hydroperoxide oxidations catalyzed by metals. III. Epoxidation of dienes and olefins with functional groups. *J. Org. Chem.*, **1970**, *35*, 1839.
15. Emanuil, N. M.; Zaikov, G. E.; Maizus, Z. K. Environment role in radical-chain reactions of oxidation of organic compounds. Moscow, "Nauka", **1973** (in Russian).
16. Nikolaev, A. F.; Rodchenko, M. V.; Karkozov, V. G. Curing epoxy resin by liquid anhy-drides of dicarboxylic acids. *Zhurn. App. Chem.*, Leningrad (USSSR), depositing Scientific work, **1989**, *27* (in Russian).
17. Bazant, V. Epoxidation of silicon unsaturated compounds. *Chem. Communs*, **1959**, *24*, 3758.
18. Kucher, R. V.; Tikhomin, V. I.; Shevchuk, I. P.; Vasiutin, I. A. M. *Liquid phase oxidation of unsaturated compounds in olefin oxides.* Kiev, (USSR) "Naukova Dumka", **1986** (in Russian).
19. Odian, G. *Principle of polymerization.* NewYork: John Wiley and Sons; **1991**.
20. Hshieh, F.; Beeson, H. D. Flammability testing of flame-retarded epoxy composites and phenolic composites. *Fire Mater.*, **1997**, *21(41)*.
21. Kunz, H. E. Flame retarding materials for advanced composites. *Macromol. Chem. Macromol. Symp.*, **1993**, *74*, 155.
22. Arends, C. B. *Polymer Toughening.* NewYork: Marcel Dekker; **1996**.
23. Ho, T. H.; Wang, C. S. Modification of epoxy resin with siloxane containing phenol aralkyl epoxy resin for electronic encapsulation application. *Eur.Polym J.*, **2001**, *37*, 267.
24. Feng, S.; Cui, M. Study of polysiloxanes containing epoxy groups: I. Synthesis and charac-terization of polysiloxanes containing 3–(2,3-epoxypropoxy)propyl groups. *React. Funct. Polym.*, **2000**, *45*, 79.
25. Garin, S.; Lecampf, L.; Yousse, B.; Bunel, C. New polysiloxanes bearing heterocyclic groups-synthesis and curability. *Eur Polym J.*, **1999**, *35*, 473.

26. Rutnakornpituk, M. Modification of epoxy-novolac resins with polysiloxane containing nitrile functional groups: synthesis and characterization. *Eur. Polym. J.*, **2005**, *41*, 1043.

27. Mukbaniani, O.; Markarashvili, E.; Iskakova, M.; Mindiashvili, G. Synthesis and Investigation of Properties of Diepoxydiorganosiloxane and Composite Materials on their Basis. *Oxidation Communications*, **2008**, *31(1)*, 116.

28. Ivanov, S.; Boeva, R.; Tanielyan, S. Catalytic epoxidation of propylene with tert-butyl peroxide in presence of molybdenum complexes on polymer carriers. *React. Kinet. Catal. Letter*, **1976**, *5(3)*, 297.

29. Markarashvili, E. G.; Iskakova, M. K.; Mindiashvili, G. S.; Mukbaniani, O. V.; Compounds on the basis of epoxy pitch eD-22, modificated by epoxysiliconorganic oligomers. Abstracts of Communications of III All Russian Scientific Conference «*Physical-Chemistry Polymer Reprocessing Processes*», Ivanovo, **2006**, 10–12 October, 4–25 (in Russian).

30. USA Patent, # 4354013, 12. 10. 82. *Process for preparing epoxy-modified silicone resins*. Kimura; Hiroshi (Oota, JP).

31. Iskakova, M.; Markarashvili, E.; Mukbaniani, O.; Mindiashvili, G.; Zaikov, G. Epoxyorganosilicon Oligomers on the Base of α,ω-Dihydroxydiorganosiloxanes. *Abstracts of Co-mmunications of III International Conference on Times of Polymers (TOP) & Composites*, Italy, Ischia, **2006**, 18–24 June, 179.

32. Odian, G. *Principle of polymerization*. NewYork: JohnWiley and Sons; **2004**.

33. Hshieh, F.; Beeson, H. D. Flammability Testing of Flame-retarded Epoxy Composites and Phenolic Composites. *Fire Mater*, **1997**, *21*, 41.

34. Kunz, H. E. Flame retarding materials for advanced composites. *MACROMOL. Chem. Macromol. Symp.,* **1993**, *74*, 155.

35. Mark, H. F.; Bikales, N. M.; Overberger, C. G.; Menges, G.; Kroshwitz, J. I. *Encyclopedia of Polymer Science and Engineering*, NewYork: JohnWiley Sons; **1984**, *11*, 45.

36. Borowitz, J.; Kosfeld, R. About the influence of hexamethylene tetramine content on the dynamic-mechanical properties of the polymeric system: Phenolic novolac resin (PF) and nitrile rubber (NBR-29). *Rheol Acta*, **1980**, *19*, 770.

37. Tiberg, C. S.; Bergeron, K.; Sankarapandian, M.; Shih, P.; Loos, A. C.; Dillard, D. A.; Mcgrath, J. E.; Riffle, J. S.; sorathia, U. Structure-property relationships of void-free phenolic-epoxy matrix materials. *Polymer*, **2000**, *41*, 5053.

38. Nikolaev, A. F.; Trizno, M. S.; Voronova, N.; Petrova, L. A.; Topornina, V. M. Properties of hardening and un-hardening epoxy-novolac composition. *Plastich. Massi (Plastics, Russian J.)*, **1965**, *4*, 76.

39. Nikolaev, A. F.; Koctselainen, I. V.; Borisova, V. V.; Korkozov, V. G. Curing of epoxide resins via anhydride of dicarbon acids at medium temperatures. *J. Appl. Chem.*, Academy of Sciences USSR, **1989**, *25*.

40. Abramov, N. K.; Kuzmina, L. V.; Kovalchuk, G. A.; Vinogradov, M. B. Hardening of epoxy oligomers at low temperatures. *Techology of Vysokomolek, Soed.*, **1983**, *2*, 35.

41. Riffle, J. S.; Yilgor, I.; Banthia, A. K.; Wilkes, G. L.; Mcgrath, J. E. Elastomeric polysiloxane modifiers for epoxy networks. *Epoxy Resins Chemistry, ACS Symp.*, **1983**, *221*, 21.

42. Yilgor, I.; Steckle, W. P. JR; Yilgor, E.; Freelin, R. G.; Riffle, J. S. Novel triblock siloxane copolymers: Synthesis, characterization, and their use as surface modifying additives. *J. Polym Sci., A: Polym. Chem.*, **1989**, *27(11)*, 3673.

43. Garin, S.; Lecamp, L.; Youssef, B.; Bunel, C. New polysiloxanes bearing heterocyclic groups-synthesis and curability. *Eur. Polym. J.*, **1999**, *35(3)*, 473.

44. Hou, S. S.; Chung, Y. P.; Chan, C. K.; Kup, C. P. Function and performance of silicone copolymer. Part IV. Curing behavior and characterization of epoxy-siloxane copolymers blended with diglycidyl ether of bisphenol-A. *Polymer*, **2000**, *41(9)*, 3263.

45. Arends, C. B. *Polymer Toughening*. New-York, Marcel Dekker; **1996**.

46. Kinloch, A. J.; Rubber-toughened plastics. In: *Riew*, C. K.; editor. Advances in Chemistry Wa-shington DC, **1989**, *68*, 222.

47. Mukbaniani, O.; Markarashvili, E.; Iskakova, M.; Aneli, J.; Gventsadze, L. D. Synthesis investigation properties of diepoxydiorganosiloxane and composite materials on their basis. *Oxidation Communications*, **2008**, *31(2)*, 300.

48. Aneli, J.; Markarashvili, E.; Titvinidze, G.; Katsitadze, M.; Meladze, S.; Mukbaniani, O. Effect of filler modification by the siloxane compounds on the properties of composites based on epoxy resin. *Abstracts of Communications, of 1st International Caucasian Symposium on Polymers and Advanced Materials.* **2007**, 11–14 September, Tbilisi, Georgia, 62–63.

49. Tvedokhlebova, I. I.; Mamaeva, I. I.; Pavlova, S. A. Conformation molecules of polyphenylsilsesquioxanes. *Vysokomol. Soed.*, **1984**, *26(9)*, 1971

50. Volchek, B. Z.; Purkina, A. B.; Nikitin, V. N. Investigation of polyphenylsilsesquioxane structure via IR spectroscopty. *Vysokomol. Soed.*, **1976**, *28(6)*, 1203.

51. Katz, H. S.; Milewski, J. V. *Handbook of Fillers for Plastics*, RAPRA, (1987).

52. Mareri, P.; Bastrole, S.; Broda, N.; Crespy, A. Mechanical behaviour of polypro-pylene composites containing fine mineral filler: Effect of filler surface treatment. *Comp. Sci. and Techn.,* **1998**, *58(5)*, 747.

53. Tolonen, H.; SjÖlind, S. G. Effect of mineral fillers on properties of composite matrix material. *Mech. of comp. mater.*, **1996**, *31(4)*, 317.

54. Rothon, R. *Particulate filled polymer composites*, RAPRA, N-Y **2003**.

55. Lou, J.; Harinath, V. Effects of mineral fillers on polystyrene melt processing. *J. of Mater. Processing Tech*, **2004**, *152(2)*, 185.

56. Motsarev, G. V.; Sobolevski, M. V.; Rozenberg, V. R. *Organosilanes and organosiloxa-nes.* M. Khimia, **1990** (In Russian).

57. Khananashvili, L. M.; Mukbaniani, O. V.; Zaikov, G. E. The book, New Concepts in Polymer Science, «Elementorganic Monomers: Technology, Properties, Applications». Printed in Netherlands, VSP, Utrecht, **2006**.

58. Mukbaniani, O.; Aneli, J.; Markarashvili, E.; Titvinidze, G.; Katsitadze, M.; Gogesashvili, N. Effect of Modification of Bentonite by Tetraethoxysilane on the Properties of Composites Based on Epoxy Resin. *Oxidation Communications*, **2010**, *33(3)*, 555.

59. Aneli, J.; Khananashvili, L.; Zaikov, G.: *Structuring and conductivity of polymer composites*. Nova Sci. Publ., N-Y, **1998**.

60. Plueddemann, E. P.; Fanger, G. Epoxyorganosiloxanes. *J. Amer. Chem. Soc.*, **1959**, *81*, 2632.

61. Greber, G.; Metzinger, L. Über oligomere siliciumverbindungen mit funktionellen gruppen. 2. Mitt. Über die darstellung von polysiloxanhydriden und deren addition an ungesättigte verbindungen. *Makrom. Chem.*, **1960**, *39*, 189.

62. Riffle, J. S.; Yilgor, I.; Tran, C.; Wilkes, G. L.; Mcgrath, J. E.; Bantia, A. K. Elastomeric Polysiloxane Modifiers for Epoxy Networks. *Synthesis of Functional Oligomers and Network Formation Studies «Epoxy Resin Chemistry II»*, ACS Symp. Ser., **1983**, *221*, 21.

63. Gagnebien, D.; Madec, P. J.; Marechal, E. Synthesis of poly(sulphone-b-siloxane)s-II. Synthesis of poly(sulphone-b-siloxane) by reaction of α,ω–diphenol oligosulphones with α,ω–die-poxy oligosiloxanes. *Eur. Polym. J.,* **1985**, *21*, 289.

64. Girgvliani, D. A.; Khananashvili, L. M.; Davitian, S. P.; Khachatrian, A. R.; Tsomaia, N. G.; Tonoyan, A. O. The kinetics and interaction mechanism of 1,4-bis(di-methylsilyl)benzene with allyl glycidyl ester in the presence of Speier catalyst. *J. Polym. Mater.*, **1994**, *25*, 65.

65. Khananashvili, L. M.; Bochorishvili, I. Z.; Markarashvili, E. G.; TsO-Maia, N. I. Epoxidation reaction of vinyl containing methylsiloxanes. Proceedings of the *Academy of Sci. of Georgia, chem.* Ser. 20-21, **1995**, *35* (in Russian).

66. Khananashvili, L. M.; Tsomaia, N. I.; Murachashvili, D. U.; Girgvliani, D. A.; Karchkhadze, M. G.; Tkeshelashvili, R. Sh. Anionic co-oligomerization of organic cyclosiloxanes with 1,3-bis(3-glycydoxypropyl) tetramethyldisiloxanes and investigation of co-oligomerization products. *J. Polym. Materials*, **1997**, *37*, 65.

67. Rao, B. S.; Madec, P. J.; Marechal, E. Synthesis and characterization of a vinyl ester resin with an oligo(dimethyl siloxane) backbone . *Makromol.Chem. Rapid Commun.*, **1986**, *7*, 703.

68. Murachashvili, D. U.; Khananashvili, L. M.; Kopilov, V. M.; Tsomaia, N. I.; Shkolnik, M. I. The reaction of silicon contain epoxides with acids, amines and oligomers on their basis. *Proc. Academy of Sci. Georgia, Chem. Ser.*, **1993**, *19*, 201. (in Russian).

69. Khananashvili, L.; Gurgenidze, G.; Karchkhadze, M.; Mukbaniani, O. XVI D. *Mendeleev Symposium in General and Applied Chemistry*, Saint-Petersburg, **1998** (in Russian).

70. Murachashvili, D.; Khananashvili, L.; Tsomaia, N.; Kopilov, V.; Shkoln-Ik, M.; Khubulava, E. Investigation of hydrosilylation reaction of allyl glycidyl ether with polyfunctional hydride methylsilane and siloxane. Proceedings of the *Academy of Sci. of Georgia, Chem. Series.*, **1996**, *22*, 36 (in Russian).

71. Mukbaniani, O. V.; Scherf, U.; Gurgenidze, G. N.; Karchkhadze, M. G.; Meladze, S. M.; Khananashvili, L. M. Comb-type Organosilicon Compounds with Epoxy Groups in the Side Chain. *Intern. J. Polym. Maters*, **2001**, *48(3)*, 267.

72. Mukbaniani, O. V.; Tatrishvili, T. N.; Zaikov, G. E. *Modification reactions of oligomethylhydridesiloxanes*. Nova Science Publisher, Inc. Huntington, New York, **2007**.

CHAPTER 11

THE REACTIVITY OF TERPENEPHENOLS AND THEIR PHENOXYL RADICALS IN REACTIONS OF OXIDATION

L. I. MAZALETSKAYA, N. I. SHELUDCHENKO, L. B. DUDNIK and L. N. SHISHKINA

CONTENTS

11.1   INTRODUCTION

The stoichiometric factors of inhibition and the rate constants of the terpenephenols (TP) with isobornyl and isocamphyl substituents were determined by the reaction with peroxy radicals of ethylbenzene. The reactivity was found to decrease for o-alkoxy compared with *o*-alkyl substituent caused by the intramolecular hydrogen bond formation that is conformed by FTIR-spectroscopy. The inhibitory activity for mixtures of terpenephenols with 2,6-di-*tert*-butyl phenols in the initiated oxidation of ethylbenzene was also studied. In spite of the similar antiradical activities of terpenephenols with isobornyl and isocamphyl sunstituents, the reactivity of phenoxyl radicals formed from them are substantially different that is resulted from the kinetic data for mixtures of terpenephenols with sterically hindered phenols.

Natural and synthetic antioxidants (AO) are known to be active regulators of the lipid peroxidation in biomembranes and of interest as the biologically active substances with a wide spectrum of the action [1]. Substituted phenols synthesized on the natural raw material base are considerable as the worthwhile biological active substances due to their comparatively low toxicity [2–4]. Earlier the high reactivity was found for terpenephenols (TP) which differ in the structure of terpene substituents and their arrangement in the aromatic ring with respect to the OH-group [5, 6]. However, values of the rate constant of TP with free radicals (k_7) were detected to depend substantially on the nature of the peroxy radicals, the rate initiation of the oxidative process and the oxidation substrate. Analysis of the literature data is allowed us to consume differences in the reactivity of both TP and their phenoxyl radicals in the dependence on the stereometric arrangement of their substituents.

The aim of this chapter was to study the effect of the stereometry on the kinetic parameters of terpenephenols and their phenoxyl radicals with the isobornyl (IBP) and isocamphyl (ICP) substituents in the initiated oxidation of ethylbenzene.

11.2   EXPERIMENTAL PART

The AIBN-initiated oxidation of ethylbenzene, a model hydrocarbon, with air oxygen was studied by measuring the oxygen uptake on a sensitive volumetric Warburg setup at 333 K and the initiation rate of $W_i = 5 \times 10^{-8}$ mol/(l s). The induction period (τ) as described in [7] and the initial rate of oxygen uptake were determined from the kinetic curves of oxygen uptake in the presence of AO. The substrate of oxidation together with a dissolved initiator was preliminarily held at a constant temperature, and then an antioxidant was added. Ethylbenzene was purified according to the standard procedure.

The combined inhibitory action for mixtures of TP and the sterically hindered 2,6-di-*tert*-butylphenol was studied at a constant total concentration of the mixture components varying the ratio between the components. The effectiveness of the inhibitory action (a) of different components and their mixtures was estimated as the time when a certain amount of oxygen was absorbed.

The studied TP were synthesized by the research group led by A.V. Kutchin, a corresponding member of the Russian Academy of Sciences (Institute of Chemistry of the Ural Branch of the Russian Academy of Sciences) and were kindly provided for the present investigation. The synthesis of IBP was carried out according to the procedures described in [8, 9]. The synthesis of ICP was carried out according to the procedures described in [6]. The studied TP were used without the additional purification. The structures of the compounds under study are given in Table 1.

TABLE 1   Structure, name, abbreviation used terpenephenols, their stoichiometric factor (f) and rate constants of TP with ethylbenzene peroxy radical (k_7).

Structure	Name, abbreviation	$fk_7 \times 10^{-4}$, l/(mol s)	f	$k_7 \times 10^{-4}$, l/(mol s)
	2-Isobornyloxyphenol, **TP-1**	0.8±0.05	–	0.4±0.03*

TABLE 1   *(Continued)*

Structure	Name, abbreviation	$fk_7 \times 10^{-4}$, l/(mol s)	f	$k_7 \times 10^{-4}$, l/(mol s)
	2-Isobornylphenol, **TP-6**	5.2±0.07	2.1	2.5±0.04
	2-Isobornyl-4-methyl-phenol, **TP-5**	11.6±0.8	1.9	6.3±0.4
	2-Isobornyl-6-methyl-phenol, **TP-8**	10.0±0.9	2.0	5.0±0.5
	2,6-Diisobornyl-4-methylphenol, **TP-7**	20.7±0.3	1.8	11.5±0.2
	3-Isobornyl-1,2-hydroxybenzene, **TP-2**	54.9±2.6	2.0	27.5±1.3
	2-isocamphyloxyphenol, **TP-28**	1.06 ± 0.05	-	0.53±0.03*

TABLE 1 *(Continued)*

Structure	Name, abbreviation	$fk_7 \times 10^{-4}$, l/(mol s)	f	$k_7 \times 10^{-4}$, l/(mol s)
	2-isocamphylphenol, **TP-26**	4.1 ± 0.04	1.9	2.05±0.03
	2-isocamphyl-4-methylphenol, **TP-14**	9.7 ± 0.8	1.1	8.8±0.8
	3-isocamphyloxyphenol, **TP-29**	2.6 ± 0.08	1.8	1.4±0.06
	4-isocamphylphenol, **TP-27**	3.9 ± 0.1	2.0	1.95±0.8

*according to the assumption that f = 2.

IR-spectra of the hexane solutions of the individual components at the concentration of 10^{-2} mol/l were registered on a Spectrum 100 IR-Fourier-spectrophotometer.

11.3 RESULTS AND DISCUSSION

The typical kinetic curves of the oxygen uptake in the initiated ethylbenzene oxidation in the absence and in the presence of TP-5 in dependence on its concentrations are shown in Fig. 1. To calculate factor of inhibition (f), the dependence of τ on the initial AO concentration was plotted. It is seen that the f values calculated from the slopes of the straight lines (Fig. 2) are close for all TP under study ($f\sim2$) (Table 1), which agrees well with the mechanism of the action of phenolic AO, according to which two free radicals decay on a phenol molecule. It is necessary to note an unexpectedly low $f=1.1$ value for TP-14 (Fig. 2). So, $f=1.9$ for a similar stereoisomer, TP-26, which does not contain the CH_3 group in the o-position and the f value is also 1.9 for TP-5 which has the isobornyl group in the o-position distinct from TP-14 (Table 1).

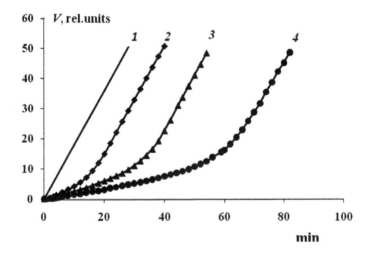

FIGURE 1   Kinetics of oxygen uptake in the oxidation of ethylbenzene without additives (*1*) and in the presence of $[\text{TP-5}]_0 \times 10^4$ mol/l 1: *2* – 0.3; *3* – 0.6; *4* – 1.0. Temperature 333 K; $W_i = 5 \times 10^{-8}$ mol/(l s).

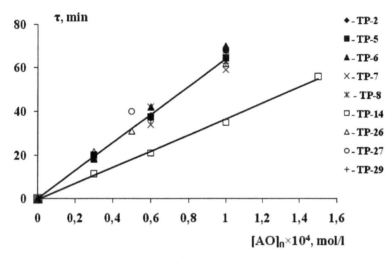

FIGURE 2 Induction time as a function of the initial AO concentration. Ethylbenzene, 333 K, $W_i = 5 \times 10^{-8}$ mol/(l s).

To calculate the rate constants of the interaction of AO with peroxy radicals (k_7), the values of the initial oxidation rates were presented in the coordinates of Eq. (1):

$$W_0/W - W/W_0 = f k_7 [AO]_0 / k_6^{0.5} W_i^{0.5}, \tag{1}$$

where, W_0 and W are the oxidation rates in the absence and in the presence of AO, respectively, and $k_6 = 1.9 \times 10^7$ 1/(mol s) is the rate constant of the bimolecular recombination of the peroxy radicals [10].

The parameter $f k_7$ was calculated from the slope of the straight lines (Figs. 3, and 4). The calculated values are listed in Table 1. The phenols under study can be arranged in a row according to the increase in their antiradical activity: TP-1 < TP-6 < TP-8 < TP-5 < TP-7 < TP-2 (for iso-bornylphenols) and ТФ-28 < ТФ-29 < ТФ-27 < ТФ-26 < ТФ-14 for (for isocamylphenols).

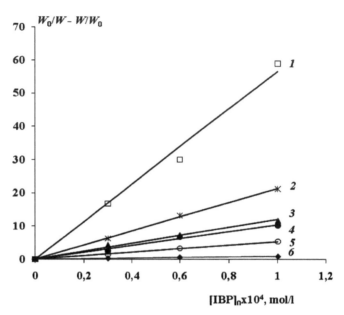

FIGURE 3   Initial rate of oxidation as a function of the initial IBP concentration in the coordinates of Eq. (I): *1* – TP-2, *2* – TP-7, *3* – TP-5, *4* – TP-8, *5* – TP-6, and *6* – TP-1. Temperature 333 K, $W_i = 5 \times 10^{-8}$ mol/(l s).

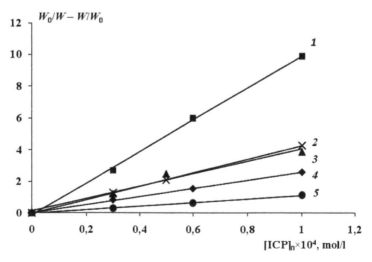

FIGURE 4   Initial rate of oxidation as a function of the initial ICP concentration in the coordinates of Eq. (I): *1* – TP-14, *2* – TP-26, *3* – TP-27, *4* – TP-29, and *5* – TP-28. Temperature 333 K, $W_i = 5 \times 10^{-8}$ mol/(l s).

TP-1 and TP-28 possess the lowest reactivity in respect to ethylbenzene peroxy radicals. A distinctive feature of these compounds in comparison with the other terpenephenols is the presence of 2-alkoxysubstituent. According to published data, alkoxy (-OR) substituents are characterized by the higher electron-donating properties in comparison with alkyl (-R) substituents: the Hammett constants for $-OCH_3$ and $-CH_3$ substituents are $\sigma = -0.778$ and -0.170 [11], respectively. The enhancement of the electron-donating properties of the substituent in the o-position to the hydroxyl group results in an increase in the rate constant k_7. For example, the values of k_7 (333 K) = 2×10^5 and 1.3×10^4 1/(mol s) for 4-methoxy- and 4-methylphenol, respectively [12].

The opposite effect of the substituents $-OR$ and $-R$ on the antiradical activity takes place when the substituents are introduced in the o-position. This follows from the data in Table 1. The replacement of the alkyl substituent (TP-6, TP-26) by the alkoxy substituent (TP-1, TP-28) results in a decrease in the value of fk_7 by a factor of 6.5 for IBP and about 3.9 for ICP. The decrease in the antiradical activity of TP with the alkoxy in comparison with TP with the alkyl substituent in the o-position is probably caused by the intramolecular hydrogen bond between the H atom of the hydroxyl group and the oxygen atom of the o-substituent. The shifts of the stretching mode from 3618.33 cm^{-1} (TP-6) and 3622.6 (TP-26) to 3561.51 cm^{-1} (TP-1) and 3560.3 (TP-28) respectively confirm the fact of hydrogen bond formation (Table 2). It is well documented that intramolecular hydrogen bonding with participation of the H atom of the phenolic OH group results in a drastic decrease in their reactivity [11]. This is confirmed by low values of k_7 for oxyanthraquinones [13] and ketophenol [14]. The decrease in the reactivity of 2-alkoxy-substituted phenols in comparison with their 4-substituted analogs in the reaction with the peroxy radicals agrees with the difference in the energy of the OH bond breaking for these AO.

TABLE 2 Stretching vibration band (v) in substituted phenols.

TP	TP-1	TP-5	TP-6	TP-7	TP-26	TP-28
v, cm^{-1}	3561.6	3620.8	3618.8	3612.5	3622.6	3560.3

The values of bond breaking are 345.8 and 352.1 kJ/mol for 4-methoxy- and 2-methoxypenols, respectively [12]. As was found in [15], the linear correlation between the energy of the OH bond breaking and the log k_7 is observed. Thus, the formation of the intramolecular hydrogen bond between the hydrogen atom of the hydroxyl group and the oxygen atom of the o-substituent accounts for the low antiradical activity of TP-1 and TP-28. A decrease in antiradical activity caused by H-bond formation is very substantial; indeed, the fk_7 value for TP-29 with an alkoxy substituent in the m-position is higher by 2.45 times compared with TP-28.

The antiradical activity for monophenols with the alkyl substituents increases with an increase in the amount of donor substituents in the benzene ring. TP-6, TP-26 and TP-27 with an only substituent in the o- or p-position to the OH group is the less active of the alkyl-substituted phenols. According to the table 1, the antiradical activities of TP-26 and TP-27, which only contain one isocamphyl substituent in the o- or p-position, respectively, are similar. The fk_7 values for these TP are not for certain differ (table 1) and close to fk_7 for TP-6, the molecule of which also contains only one (isobornyl) substituent in the o-position The addition of the CH_3 group at the o- (TP-8) or at the p-position (TP-5 and TP-14) results in an increase in fk_7, with the values of fk_7 being close to $(9.7-11.6) \times 10^4$ $1/(mol~s)$.

The presence of two o,o'-isobornyl and p-CH_3 (TP-7) groups results in the following increase in fk_7 (Table 1).

Compound TP-2 containing two OH groups has the highest antiradical activity, which agrees with the literature data that polyphenols of the pyrocatechol series are characterized by the high values of the rate constant k_7 [16].

Thus, the reactivity of terpenephenols with respect to peroxy radicals of ethylbenzene increases in amount of alkyl substituents and OH-groups in aromatic ring, decreases when o-alkyl substituents are replaced by o-alkoxyl groups owing to the of intramolecular H-bond formation and almost do not change in the dependence on differences in the structure of the terpene substituents and their arrangement with respect to the OH group.

Special attention must be given to TP-7, the molecule of which contains two bulky substituents at the o,o'-positions. We assume that TP-7 belongs to sterically hindered phenol. 2,6-di-*tert*-butylpenols are a classical example of these. The decrease in the rate constant k_7 in comparison to 2-*tert*-butyl-substituted analogs is characteristic. The antiradical activity of 2,6-di-*tert*-butylphenol is approximately fourfold lower than that for 2-*tert*-butylphenol $(k_7 = 0.96 \times 10^4$ and 4×10^4 $1/(mol~s)$, respectively) [17]. If TP-7 with isobornyl substituents has the properties of the sterically hindered phenols, we might expect a decrease in the value of k_7 in comparison with TP-5. However, the experimentally obtained value of k_7 for TP-7 is not only almost twofold higher than that for TP-5, but significantly exceeds the antiradical activity of known sterically hindered phenols (2,6-di-*tert*-butyl-4-methylphenol $k_7 = 2.04 \times 10^4$ l/(mol s) under similar conditions [15]) On the contrary, the antiradical activity of TP-7 is comparable with sterically unhindered (unshielded) phenols with alkyl substituents at 2,6-and 2,4,6-positions. For example, the values of k_7 are 8.6×10^4 and 1.9×10^5 l/(mol s) for 2,6-dicyclohexyl- and 2,4,6-trimethylphenol, respectively [17].

From the comparative data on the rate constants k_7, TP-7 can be classified as sterically unhindered phenol. According to [18], the synergetic action of the mixtures of TP-7 with sterically hindered phenols may be expected. It is known that unhindered and sterically hindered phenol form synergetic mixtures [18 — 20], with the efficiency changing in the dependence on the substrate and conditions of the oxidation and the antiradical activity of the mixture components. In particular, the efficiency of the inhibiting action of a mixture should not differ from the efficiency of the most active component (unhindered phenol), when the oxidation is carried out under condition of unbranched chain reaction and constant overall concentration of added antioxidants. This result is caused by regeneration of the unhindered phenol as a result of interaction of its phenoxyl radical with the sterically hindered phenol [18]. One can see in Fig. 5 (curves *2, 3, 4*) that the inhibiting action of the equimolar mixture of TP-7 with sterically hindered 2,6-di-*tert*-butylphenol (DTBP) is additive, because the values a of the mixture are close to $a_{mix, add}$ For example, if we take the time during which the oxygen uptake is $V = 4$ relative units as the efficiency of inhibition, then the experimental $a_{mix} = 11,5$ min does not differ from $a_{mix, add} = (a_{DTBP} + a_{TP-7})/2 = (3.5 + 19.5)/2 = 11.5$ min. Hence, TP-7 being more active in the reaction with peroxy radicals does not regenerate in the presence of DTBP.

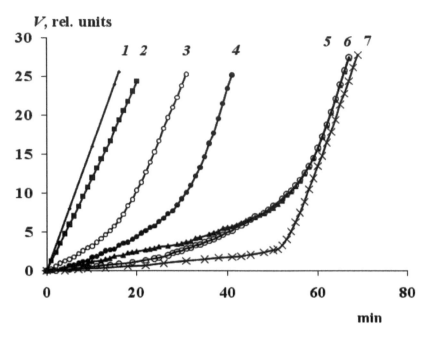

FIGURE 5 Kinetics of oxygen uptake in ethylbenzene oxidation without additives (*1*) and in the presence of AO, mol/l: *2* – 6×10⁻⁵ DTBP; *3* – 3×10⁻⁵ DTBP + 3×10⁻⁵ TP-7; *4* – 6×10⁻⁵ TP-7; *5* – 1×10⁻⁴ TP-7; *6* – 5×10⁻⁵ TPh + 5×10⁻⁵ TP-7; and *7* – 1×10⁻⁴ TPh; 333 K, W_i = 5×10⁻⁸ mol/(l s).

Probably, the two bulky isobornyl substituents at *o,o'*-positions shield the OH groups, and TP-7 plays the role of the sterically hindered phenol forming synergetic mixtures with unhindered phenols. Since TP-7 itself has a high value of k_7 in comparison with *o,o'*-di-*tert*-butyl-substituted phenols, the synergetic mixtures on its basis should use AO with still higher reactivity of the OH group and capable of forming synergetic mixtures with sterically hindered phenols. *α*-Tocopherol (TPh) is one of these phenols with k_7 = 1×10⁶ 1/(mol s) [21]. One can see in Fig. 5 that TPh is more efficient (curve *7*) in comparison with TP-7 (curve *5*), and the inhibiting action of their equimolar mixture (*6*) is close to additive: at the start, TPh inhibits the reaction and oxygen absorption occurs at a rate typical of the process inhibited by TPh. When TPh is completely consumed in accordance to the stoichiometric factor, the oxygen uptake rate curve of the mixture (curve *6*) hardly differs from that in the presence of TP-7 (curve *5*), i.e., the inhibiting effects of the antioxidants are independent of one another.

Thus, these data, the additivity of the inhibiting action of TP-7 mixtures with the unhindered phenol and the value of k_7 indicate that TP-7 can be classified as unhindered phenol.

Our results allow for the assumption that the absence of the expected synergistic effect of the mixtures of unhindered TP-7 with sterically hindered DTBP (Fig. 5) is due to the specifics of the isobornyl substituents, which hinder hydrogen transfer from the OH group of DTBP to the phenoxyl radical formed from TP-7.

To verify this assumption, the joint action of DTBP with the other unhindered phenol TP-5 with an only o-isobornyl substituent was studied. The efficiency of the inhibition of the separate components and an equimolar mixture was estimated at a concentration of 1×10^{-4} mol/1. The time of oxygen uptake of $V = 4$ relative units was used as a criterion for the inhibition period. The values of a are 5 and 25 min for DTBP and TP-5, respectively; $a_{mix} = 15$ min coincides with $a_{mix,add}$. Hence, the action of the mixture of TP-5 with the sterically hindered DTBP is additive similarly to TP-7.

Thus, the phenolic antioxidants containing one (TP-5) or two (TP-7) isobornyl substituents in the o-position, unlike TPh and unhindered phenols (with $tert$-butyl substituent at o-position, with methyl or cyclohexyl substituents at o,o'-positions), are not regenerated in the mixture with the sterically hindered phenols: the inhibiting effect of their mixtures is additive.

Howerver, on the contrary, phenoxyl radicals of ICP can abstract a hydrogen atom from sterically hindered phenols. We studied the joint action of sterically hindered DTBP and o-isocamphylphenol (TP-26) and p-isocamphylphenol (TP-27) to reveal the influence of the terpene substituent structure and its spatial arrangement on the reactivity of the phenoxyl radials. As stated above, mixtures of two phenols, sterically hindered and unhindered, can exhibit a synergetic effect in the inhibition of the oxidation processes. The mechanism of synergism is caused by the regeneration of the more active mixture component (unhindered phenol) from its radical form because of the hydrogen abstraction from the sterically hindered phenol. Accordingly, we selected DTBP as a sterically hindered phenol, the antiradical activity of which (Fig. 6, curve 2) was lower than those of TP-26 (Fig. 6, curve 5) and TP-27. We see that the inhibiting action of a mixture is higher than additive and increases as the content of TP-26 (%) grows (Fig. 6, curves 3, 4; Fig. 7, curve 1). The kinetics of oxygen absorption by an equimolecular mixture of *o*-isocamphylphenol and DTBP (Fig. 6) and the effectiveness of mixture inhibition (Fig.7) are insignificantly different from those of the more active component, TP-26. This result is evidence of the interaction of the phenoxyl radical from TP-26 with DTBP resulting in the formation of initial TP-26 molecules. However, it is need to note that the regeneration of phenol TP-26 is incomplete and decreases as the fraction of DTBP in the mixture grows (Fig. 7). This is explained by insignificantly different antiradical activities of mixture components. Indeed, the k_7 constant of DTBP is 0.95×10^4 l/(mol s) [17]. Therefore, at a DTBP : TP-26 = 4 : 1 ratio, we have $fk_7[DTBP]_0 > fk_7[TP-26]_0$, that is, DTBP successfully competes with TP-26 and is partially consumed in the reaction with the peroxy radicals. For an equimolecular mixture, $fk_7[TP-26]_0 > fk_7[DTBP]_0$. Therefore the peroxy radicals predominantly interact with TP-26 molecules. Phenoxyl radicals formed as a result of interaction react with DTBP with the regeneration of initial more active TP-26 molecules. The consumption of DTBP than mainly occurs in interaction with the phenoxyl radical from TP-26. As a result, the effectiveness of the equimolecular mixture action is close to the effectiveness of TP-26.

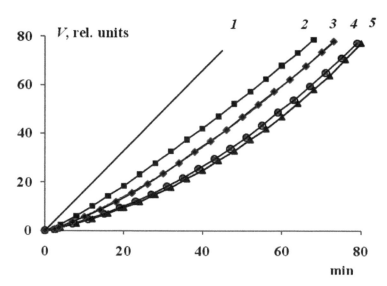

FIGURE 6   Kinetic curves of oxygen uptake in ethylbenzene oxidation without additives (*1*) and in the presence of (mol/1): *2* – 1× 10⁻⁴ DTBP, *3* – a mixture of 0.8×10⁻⁴ DTBP and 0.2×10⁻⁴ TP-26, *4* – a mixture 0.5×10⁻⁴ DTBP and 0.5×10⁻⁴ TP-26 and 5 – 1×10⁻⁴ TP-26. Temperature 333 K, W_i = 5×10⁻⁸ mol/(l s).

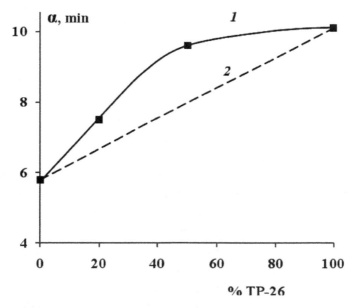

FIGURE 7 Dependences of the effectiveness of inhibitory action (*a*) of a mixture of TP-26 and DTBP on the concentration of TP-26 (in %): *1* − experimental and *2* − calculated data according to the assumption of additive action of components. Concentration of mixture is constant.

For TP-26 phenol with an *o*-isocamphyl substituent, as distinct from TP-5, we observe regeneration in a mixture with the sterically hindered phenol. The effectiveness of the action of a mixture of TP-26 and DTBP (Fig. 7, curve *1*) is higher than additive (Fig. 7, curve *2*), that is, mixtures exhibit synergetic effects.

A similar result was obtained for TP-27, in the molecule of which the isocamphyl substituent is situated in the *p*-position with respect to the OH group and, for this reason, does not prevent the abstraction of hydrogen from the sterically hindered phenol molecule.

11.4 CONCLUSON

The rate constants of the reaction of the series of substituted isobornyl and isocamphyl phenols with the ethylbenzene peroxy radicals were measured.

The phenols under study are arranged in a row according to the increase in their antiradical activity: TP-1 < TP-6 < TP-8 < TP-5 <TP-7 < TP-2 for isobomylphenols and TΦ-28 < TΦ-29 < TΦ-27 <TΦ-26 < TΦ-14 for isocamylphenols.

The presence of the oxygen atom in oposition with respect to the OH group of terpenephenols was found to decrease the antiradical activity of *o*-alkoxy compared *o*-alkyl substituted compounds by a factor of 6.5 for isobomylphenols and about 3.9 for isocamphylphenols owing to the intramolecular H-bond formation.

The differences in the structure of terpene substituents and their arrangement with respect to the OH group almost do not influence the antiradical activity of the investigated phenols, but change the reactivity of the phenoxyl radicals formed from them. When a phenoxyl radical interacts with a sterically hindered phenol molecule, its reactivity increases if the terpene substituent has the isocamphyl structure. Thanks to the interaction of the phenoxyl radicals formed from isocamphyl phenols with the sterically hindered phenols, mixtures of DTBP with TP-26 and TP-27 exhibit synergetic effects in the inhibition of the ethylbenzene oxidation [19].

KEYWORDS

- alkoxy substituent
- ethylbenzene
- isocamphyl substituent
- ketophenol
- oxyanthraquinones
- phenoxyl radicals
- terpenephenols

REFERENCES

1. Burlakova, E. B. Bioantioxidants: yesterday, today, tomorrow/ In: *Chemical and Biological Kinetics. New Horizons.* Biological Kinetics. Eds. Burlakova E.B., Varfolomeev S.D. Leden. Boston: VSP, **2005**, *2*, 1–33.
2. Plotnikov, M. B.; Smol'yakova, V. I.; Ivanov, I. S.; Kutchin, A. V.; Chukicheva, I.Yu.; Krasnov, E. A. Antithrombogenic and antithrombocyte activity of the derivative of the orto-isobornylphenol. *Bull. Exp. Biol. Med.* **2008**, *145(3)*, 296–299 (in Russian).
3. Plotnikov, M. B.; Chernysheva, G. A.; Smolyakova, V. I.; Ivanov, I. S.; Kutchin, A. V.; Chukicheva, I. Yu.; Krasnov, E. A. Neuroprotective effects of dibornol and mechanism of action in the case of cerebral ischemia. *Herald of Russ. AMS.* **2009**, *11*, 12–17 (in Russian).
4. Chukicheva, I. Yu.; Fedorova, I. V.; Buravlev, E. V.; Lumpov, A. E.; Vikharev, Yu. V.; Anikina, L. V.; Grishko, V. V.; Kutchin, A. V. Anti-immflamatory activity of isobornylphenol derivatives. 2010. *Chemistry of Natural Compounds.* **2010**, *3*. 402–403 (in Russian).
5. Mazaletskaya, L. I.; Sheludchenko, N. I.; Shishkina, L. N.; Kutchin ,A. V.; Fedorova, I. V.; Chukicheva, I. Yu. Kinetic parameters of the reaction of isobornylphenols with peroxy radicals. *Petroleum Chemistry.* **2011**, *51(5)*, 348–353.
6. Mazaletskaya, L. I.; Sheludchenko, N. I.; Shishkina, L. N.; Kutchin, A. V.; Fedorova, I. V.; and Chukicheva, I. Yu. Inhibiting activity of isocamphyl substituted phenols and their mixtures with 2,6-di-*tert*-butylphenol in the initiated oxidation of ethylbenzene. *Russian Journal of Physical Chemistry A.* **2012**, *86(6)*, 929–926.
7. Emanuel', N. M.; Gladyshev, G. P.; Denisov, E. T.; Tsepalov, V.F.; Kharitonov, V. V.; Piotrovskii, K. B. *Testing Procedure for Chemical Compounds as Stabilizers of Polymer Materials. Preprint.* Chernogolovka, **1976**, 11 (in Russian).
8. Chukicheva, I. Yu.; Kutchin, A. V. Natural and synthetic terpenophenols. *Ross. Khim. Zh.* **2004**, *48(3)*, 21–37 (in Russian).
9. Chukicheva, I. Yu.; Timusheva, I. V.; Shirikhin, L. V.; Kutchin, A. V. Alkylahion of pyrocatechol and resorcinol by camphene. *Chem. Nat. Compounds,* **2007**, *43(3)*, 245–249 (in Russian).
10. Tsepalov, V.F.; Shlyapintokh, V.Ya. The rate constants of the elementary reactions of the ethylbenzene oxidation by molecular oxygen. *Kinet. Katal.* **1962**, *3(6)*, 870–876 (in Russian).
11. Roginskii, V.A. *Phenolic Antioxidants: Reactivity and Performance.* Moscow: Nauka, **1988**, 247 (in Russian).
12. Denisov, E. T.; Denisova, T. G. *Handbook of Antioxidants; Bond Dissociation Energies, Rate Constants, Activation Energies and Enthalpies of Reaction,* 2nd Ed. CRS Press LLC. Boca Ration L. N.-York Washington, DC, **2000**.
13. Boguslavskaya, L.V.; Khrapova, N. G.; Maksimov, O. B. Polyhydroxyphenols — a new class of natural antioxidants. *Izv. Akad. Nauk SSSR, Ser. Khim.*, **1985**, *7*, 1471–1476 (in Russian).
14. Tsepalov, V. F.; Kharitonova, A. A.; Gladyshev, G. P. Determination of the rate constants and the inhibition coefficients of phenols — antioxidants by means the model chain reaction. *Kinet. Katal.* **1977**, *18(5)*, 1261–1267 (in Russian).

15. Belyakov, V.A.; Shanina, E. L.; Roginskii, V.A.; and Miller, V. B. The binding energies of O-H and inhibitory ability of sterically hindered phenols *Izv. Akad. Nauk SSSR, Ser. Khim.* **1975**, *12*, 2685–2691 (in Russian).

16. Azatyan, N. A.; Karpukhina, G.V.; Belostotskaya, I. S.; and Komissarova, N. L. The mechanism of inhibition of the oxidation processes of diatomic phenols. *Neftekhimiya.* **1973**, *13(3)*, 435–440 (in Russian).

17. Karpukhina, G. V.; Maizus, Z. K.; and Matienko, L. I. *Interaction of inhibitors — phenols and aromatic amines in the oxidation reactions of hydrocarbons.* Neftekhimiya. **1966**, *6(4)*, 603–607 (in Russian).

18. Karpukhina, G. V.; Emanuel, N. M. Classification of synergistic mixtures of antioxidants and the mechanisms of synergy. *Dokl. Akad. Nauk SSSR,* **1984**, *276(5)*, 1163–1167 (in Russian).

19. Makconey, L. R.; Rooge, M. A. Inhibition of free-radical reactions. IV. The synergistic effect of 2,6-di-t-butylphenols on hydrocarcon oxidation retarded by 4-methoxyphenol. *J. Am. Chem. Soc.* **1967**, *89(22)*, 5619–5629.

20. Azatyan, N. A.; Zolotova, T. V.; Karpukhina, G.V.; Maizus, Z. K. *Mechanism of the inhibition of the oxidation reactions by phenol mixtures.* Neftekhimiya, **1971**, *11(4)*, 568–573 (in Russian).

21. Burlakova, E. B.; Krashakov, S. A.; Khrapova, N. G. Kinetic features of tocopherols as antioxidants. *Khim. Fiz.,* **1995**, *14(10)*, 151–182 (in Russian).

CHAPTER 12

REACTIONS OF OZONE WITH HYDROCARBONS – KINETICS AND MECHANISM

S. RAKOVSKY, M. ANACHKOV, V. GEORGIEV, M. BELITSKI, and G. ZAIKOV

CONTENTS

12.1   INTRODUCTION

The ozonolysis of hydrocarbons is the rare reactions that going on at low and ambient temperatures and can transform these non-active organics into new valuable oxygen-containing products [1–12] suitable for synthesis and for special usage.

This reaction has been studied by a number of authors [13–15]. But the conclusions in most cases differ from one another, perhaps because of differences and complicated contents of formed products. Special interest is pointed out on the action of ozone as an initiator in the oxidation processes [16–18], modifying polymer agent [19–21], oxidizer at the preparation of alcohols, ketones, etc. [22–24]. The kinetics and mechanism of this reaction have been intensively discussed [25–46]. The main goal of our present paper is dedicated to the fundamental problems of the kinetic and mechanism of these reactions [47–64].

12.2   REACTION MECHANISMS

The first studies on the reaction of ozone with paraffins have been carried out with methane, ethane, propane and butane in gas phase [2–8]. It has been established that the absorption of 1 mol ozone yields 1 mol products. The analysis of the experimental data reveals two mechanisms of ozone action: 1) ozone is decomposed to atomic oxygen, which initiates the oxidation process [2] mechanism – (M.1.1) and 2) ozone interacts directly with the alkane [3–5] mechanism – (M.1.2.):

$$O_3 \rightarrow O_2 + O$$

$$RH + O \rightarrow R^{\bullet} + {}^{\bullet}OH \quad (M.1.1)$$

$$RH + O_3 \rightarrow RO^{\bullet} + {}^{\bullet}O_2H \quad (M.1.2)$$

At the analysis of M.1.1 becomes clear that this mechanism can be realized only in gas phase at the temperature range of >50–60°C, because in it the value of the rate constant of thermal decomposition of ozone is $<\sim 10^{-4}$ M^{-1} s^{-1} [28, 29] is enough low to ensure the occurrence of the oxidation process. Thus, this reaction is of no importance in liquid phase whereby the reactions are usually carried out in the range of ≤ 50–0°C.

Mechanism M.1.2 supposes the formation of H_2O_2 in quantities proportional to the absorbed ozone which is however not observed experimentally since only traces of H_2O_2 has been found in the liquid phase. Really only a few amounts of hydroperoxide is formed [1]. It allowed us to suggest that the role of M.1.2 in the ozonataion of alkanes can be negligible.

The ozonolysis of alkanes in gas phase leads to the formation of excited species, which determines the peculiar character of these reactions [65].

The first studies in liquid phase have been carried out by Azinger & co., on ozonation of octadecane [16]. On the basis of the composition of the carboxylic acids and their yield they predict that ozone attacks directly the CH_2 groups. Upon ozonolysis of decalin and adamantine containing the *tert*-CH-bonds the formation of *tert* – alcohols is observed. In cyclohexane ozone reacts with the CH_2-groups and giving and increasing amounts of cyclohexanol and cyclohexanone [16].

The kinetics of intermediate and end products formation has been studied and the mechanism of ozone reaction with tetradecane is suggested [66] as:

$$RH + O_3 \rightarrow RO_2^{\bullet} + {}^{\bullet}OH \quad (M.1.3)$$

Although the basic reaction in M.1.3 is exothermal its occurrence in two steps appears to be more entropy favorable than in one as assumed by this mechanism.

Hamilton & co., [6, 67] consider as the most probable mechanism the ozone attack on the CH-bonds via a 1,3-transition state (TS):

$$RH + O_3 \Leftrightarrow TS \rightarrow [R^{\bullet} + HO_3^{\bullet} \Leftrightarrow R^{\bullet} + HO_3^-] \rightarrow ROOOH (1) \quad (M.1.4)$$

Cage

↓↓

$ROH + O_2$ (s) (2); $R^{\bullet}\uparrow + \uparrow{}^{\bullet}OH + O_2$ (t) (3)

↓

$RO_2^{\bullet} + O_2$,

which is followed by H-atom abstraction and hydrotrioxide (1), alcohol, singlet oxygen (2), peroxy radical and triplet oxygen (3) outlet from the kinetic cage into the bulk liquid. They postulate that the products in the cage have a radical character, while according to Myurei they are more likely ionic pairs [68].

Nangiya and Benson [28] assume mechanism M.1.4 as more probable, but the interaction between ozone and CH-bonds is not synchronic one.

Denissov and co. [26] considered the reaction of ozone with C-H-bonds in the frames of parabolic model of the potential surface and also adheres to mechanism M.1.4.

Another concept of the mechanism of alkane ozonolysis is that R, HO -radicals are kinetically responsible for the proceeding of the reaction and ROH, R•, HO•, and O_2 leave the kinetic cage and pass into the solution volume [51]:

$$RH + O_3 \rightarrow [R^\bullet + HO^\bullet + O_2], \quad (M.1.5)$$

However, one cannot exclude the possibility for the $RO_2{}^\bullet$-radicals formation from ROOOH as a precursor via the mechanism of ozone incorporation into the CH-bonds.

$$RH + O_3 \rightarrow ROOOH \quad (M.1.6)$$

ROOOH could act as a precursor in all mechanisms discussed in the literature [9, 67]. This reaction probably have to be entropic disadvantageous, because it goes through a cyclic TS and it is necessary to use the assumption five coordinated C-atom.

12.2.1 THERMOCHEMICAL ANALYSIS

We have determined the heats of ozone reaction with paraffins, according to the different mechanisms on the example of methane. The latter has been chosen since the energy of its CH-bond is the highest and thus the results from the calculations for it can be extended for other paraffins. For this purpose we have used the experimentally and calculated, marked by (*), values of the bond energies and heats of formation [69, 70]:

CH_3-H = 104; OO-O = 24; CH_3O-H = 102, CH_3OO-H = 77?, CH_3OOO-H = 70?, CH_3OO-OH = 51? , CH_3O-OOH = 47?, CH_3-OH = 91, CH_3-OOH = 69 (estimated), CH_3-OOOH = 61?, H-O• = 102; H-OO• = 77; H-OOO• = 62; C-O• = 78; C-OO• = 59; C-OOO• = 48 kcal/mol,

After substituting with the corresponding thermochemical value the heats of the reactions (Q) are:

Mechanism:	M.1.1	M.1.2	M.1.3	M.1.4	M.1.5	M.1.6
Q, kcal/mol:	26	−27	−33	66	26	−5

The M.1, M.1.3, M.1.6 are exothermal while those included in the other mechanisms are endothermal, and therefore the former reactions are thermodynamically more favorable.

For alkanes with secondary and tertiary C-H-bonds with lower bond energies, Q for each one reaction will be higher by 10–15 kcal.

12.3   STRUCTURAL-KINETIC INVESTIGATIONS

As we have shown above there exist two approaches concerning the interaction of ozone with paraffins: ozone introduction along the CH-bonds and H-atom abstraction. We should note, however, that the effect of paraffins composition in this reaction has not been systematically studied which limits the possibilities for analysis and elaborations of the mechanism of these reactions [25, 59].

The variety of the proposed mechanisms and the insufficient kinetic data impose the performance of a systematic investigation of the reaction applying a wide spectrum of conditions and theoretical and experimental methods of study.

12.3.1   LINEAR AND ISOPARAFFINS

We have studied the kinetics of ozone reaction with a series of paraffins of different structure [47, 59]. The rates of the reactions (W) have been determined in a close and open system. By following the change of $[O_3]$ in time (t) in static conditions we have determined the rate constant of the pseudomonomolecular reaction $k' = k [RH]$ and the bimolecular constant k.

$$W = d[O_3]/dt = k [RH]_0.[O_3] = k'[O_3] \qquad (1)$$

The constant k' has been determined during the half-decomposition time:

$$k' = \ln2/\tau_{1/2} \qquad (2)$$

or from the slope of the curve in coordinates $ln([O_3]_o/[O_3]/t$.

The reaction in open system has been carried out at bubbling of ozone through the paraffin and following spectrophotometrically its concentration in the UV-region (254–300 nm). The course of the concentration change has the following pattern: first it drops off abruptly from $[O_3]_o$ – the concentration at the reactor inlet to $[O_3]_g$ – the concentration at the reactor outlet after which for a long time it remains parallel to the x axis, at a distance of $\Delta[O_3] = \{[O_3]_o - [O_3]_g\}$ The analysis of the kinetic curves has been done on the basis of its stationary part when the rate of the chemical reaction becomes equal to the rate of ozone consumption $W = W_{O3}$ or:

$$k\,[RH].[O_3]_1 = \omega.\Delta[O_3] \qquad (3),$$

From which:

$$k = \omega\,\Delta[O_3] / ([RH].[O_3]_1), \qquad (4)$$

where $\omega = v/V$ is the relative rate; v – the rate of the gas flow; V – the volume of the liquid phase; in one of the used models $[O_3]_1 = \alpha.[O_3]_g$ – the ozone concentration in the liquid; α – the Henry's coefficient [31]. This model is valid in all cases when the rate of ozone absorption is bigger than the rate of chemical reactions.

Several criteria could by applied if the Henry's low is performed during the course of absorption and proceeding of irreversible first order chemical reaction (k_1):

$$k_1.\tau < 1,$$

where, $\tau – 1/\omega$
    Or

$$D_{O3}.k_1/k_L^2 < 1$$

where, D_{O3} – diffusion coefficient of ozone in solution; $k_L = D_{O3}/\delta$ – coefficient of mass transfer in liquid phase and δ – thickness of the bound layer in hydrodynamic model – renovation surface or $k_L = (D_{O3}.s)^{1/2}$, where s – time of renovation; at bubbling method with small bubbles with diameters to 2.5–3 mm – $k_L = 0.31 \cdot (gv)^{1/3} \cdot (D_{O_3}/v)^{2/3}$, where $v = \eta/\rho$ – kinematics viscosity of the solvent, η – viscosity of the solvent, ρ – solvent density, g – earth acceleration. Usually k_L value is 0.1–0.05 cm/s.

In all other cases $[O_3]_l$ is a complicated function of $[O_3]_g$ è $[O_3]_0$ and it requires special investigations [54].

Conventionally, $v = 0.1 l/min$, $V = 10$ ml and respectively $\omega = 0.167$ s^{-1}, $[O_3]_0 \approx 10^{-6} - 10^{-3}$ M, solvent – CCl$_4$, for which depending on the temperature $\alpha \approx 1.8$–3, $[RH] \approx 10^{-3} - 10^1$ M and for the paraffins $\alpha \approx 1.8$ –2.5. The bubbling reactors are glassy with grating G2, G3 or G4 built in its lower end, with cylindrical form with inner diameter $\phi = 1.7$–3.7 cm and height $h = 7$–15 cm. The accuracy of temperature keeping was ±0.1°C.

By applying the kinetic methods pointed above we have studied the influence of concentrations, composition, solvent nature and temperature on the kinetics of a definite reaction.

The rates of the reaction studied were linearly dependent on $[O_3]$ and with some exceptions – linearly on [RH] (for example adamantane in the first case and 3-methylpentane for the second one (Fig. 1).

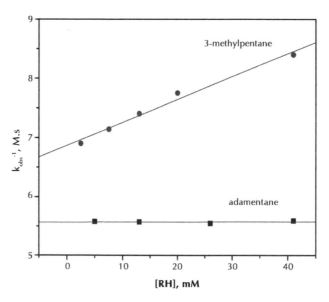

FIGURE 1   The dependence of $1/k_{obs}$ v [RH].

The observed kinetic constant for 3-methylpentane linearly dependence on [RH] and the constant for adamantane does not depend on [RH]. The dependence is true only at [RH] < 0.3 M and ratio of [RH]/ [O_3] < 1.10^4.

Williamson and Cvetanovich [17] have also registered an increase of W at low [RH] and ratios [RH]/ [O_3] < 1.10^4 at ozonolysis of n-octane, n-pentane, 3-methylpentane and *iso*-pentane. They explain this fact rather by the ozone interaction with microimpurities and by the initiation of the chain reactions at low paraffins concentration.

We have tried to explain the dependence of $1/k_{obs}$ from [RH] assuming that the reaction takes place via a complex multistep mechanism.

For this purpose we calculated and analyzed several kinetic schemes including steps of complex formation between the paraffin and ozone. Eq. (1) provides a good agreement between the applied kinetic and chemical criteria and the experimental data:

SCHEME 1

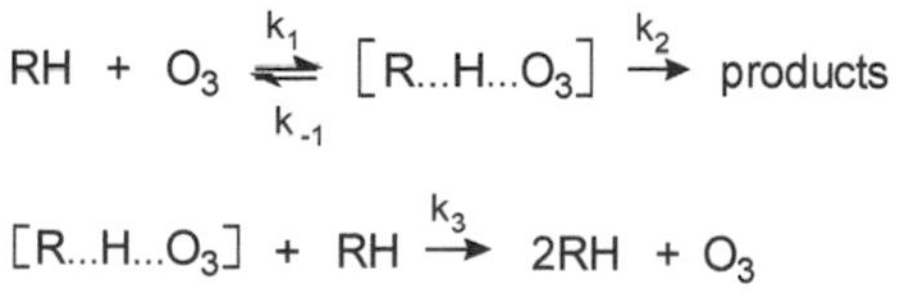

$$RH + O_3 \underset{k_{-1}}{\overset{k_1}{\rightleftharpoons}} [R...H...O_3] \overset{k_2}{\rightarrow} products$$

$$[R...H...O_3] + RH \overset{k_3}{\rightarrow} 2RH + O_3$$

which includes the formation of complex – [$R...H...O_3$].

We have studied the case when the equilibrium concentration of the complex is not affected by the occurrence of Eqs. (2) and (3).

The material balance relative to the complex is Eq. (4):

$$[R...H...O_3] = [O_3]_0 - [O_3] \, ,$$

At quasi-stationary approximation in relation to [$R...H...O_3$] and taking into consideration the material balance, we obtain for k_{obs}:

$$k_{obs} = k_1 k_2 / \{k_{-1} + k_2 + (k_1 + k_3) \, [RH]\}, \tag{5}$$

And in reciprocal form:

$$1/k_{obs} = (1/k_1 + k_{-1}/k_1 k_2) + \{(k_1 + k_3)/k_1 k_2\}.[RH] \tag{6}$$

It follows from Eq. (6) that $1/k_{obs}$ will increase linearly with the increase of [RH]. From the ordinate intercept one can determine $(1/k_1 + K_d/k_2)$, where $K_d = k_{-1}/k_1$ – the dissociation constant of the complex and from the slope angle the ratio $(k_1 + k_3)/k_1 k_2$. At our calculations we have assumed that $k_2 < k_{-1}$ according to Eq. (1). Hence the intercept gives K_d/k_2 and at $k_1 > k_3$ $1/k_2$ can be found from the slope. The last assumption transforms Eq. (1) into a mechanism of Michaelis-Menten type.

SCHEME 2

$$RH + O_3 \underset{k_{-1}}{\overset{k_1}{\rightleftharpoons}} [R...H...O_3] \overset{k_2}{\rightarrow} \text{ products}$$

Then Eq. (5) is transformed in Eq. (7):

$$k_{obs} = k_1 k_2/(k_{-1} + k_2 + k_1[RH]), \tag{7}$$

And Eq. (6) in:

$$1/k_{obs} = (1/k_1 + k_{-1}/k_1.k_2) + (1/k_2)[RH] \tag{8}$$

In the case when the equilibrium constant of $[R...H...O_3]$ varies as a result of the proceeding of Eqs. 2 and 3, $1/k_{obs}$ will be equal to:

$$1/k_{obs} = (1/k_1 + k_{-1}/k_1 k_2) + k_3/k_1 k_2.[RH] \tag{6'}$$

And $1/k_1$ will be determined from the intercept since $K_d < k_2$, and $k_3/k_1 k_2$ from the slope, respectively.

In order to assess the validity of the kinetic schemes we have experimentally determined the rate constants of ozone interaction with 24 paraffins of various structures. The rate constants depending on [RH] are given in Table. 1 whereby in column 3 are shown the constants calculated from the ordinate' intercepts and in column 4 are those obtained from the slope according to Eq..(6)'/Eq. (8).

TABLE 1   Rate constants of ozone reactions with paraffins in CCl_4 at 20°C [47].

No	Hydrocarbon	k ±10% [$M^{-1}. s^{-1}$]	$(k_3/k_2)/k_2$ [M^{-1}]	No	Hydrocarbon	k ±10% [$M^{-1}. s^{-1}$]	$(k_3/k_2)/k_2$ [M^{-1}]
1.	n-Pentane	0.015	0.001/15	13.	2-Me-hexane	0.13	5.2/0.025
2.	n-Hexane	0.019	0.0013/15	14.	3-Me-hexane	0.20	7.4/0.027
3.	n-Heptane	0.021	0.0012/17	15.	2,2-di-Me-hexane	0.015	<0.001/<15
4.	n-Octane	0.023	–	16.	2,4-di-Me-hexane	0.13	–
5.	n-Nonane	0.026	–	17.	2,2,5-tri-Me-hexane	0.19	9.0/0.021
6.	n-Decane	0.029	<0.001<29	18.	3-Me-heptane	0.20	9.0/0.022
7.	n-Tetradecane	0.036	–	19.	Hexa-Me-ethane	0.0002	–
8.	n-Octadecane	0.048	<0.001/<48	20.	Ethyl-n-propane	0.05	–
9.	3-Me-pentane	0.15	0.75/0.2	21.	Cyclohexane	0.01	<0.001/<14
10.	2,3-di-Me-pentane	0.29	11/0.026	22.	Norbornane	0.014	<0.001/<14
11.	2,4-di-Me-pentane	0.08	–	23.	Adamantane	0.22	<0.1/<2.2
12.	2,3,4-tri-Me-pentane	0.59	40/0/015	24.	Isooctane	0.14	–

The analysis of the data in Table .1 shows that the kinetic models, most likely, do not confirm the experimental results although they formally describe some of the experimental observations in Fig. 1 and Table 1. For some of the representatives among n-, iso- and cyclo-paraffins no relationship of $1/k_{obs}$ on [RH] has been established: Nos 4, 5, 7, 11, 16, 19, 20, and 24 or this dependence is within the limits of the graphic presentation: Nos 1, 2, 3, 6, 8, 15, 21, 22, and 23. A good correlation has been found for Nos 9, 10, 12, 13, 14, 17, and 18, since all these are iso-parrafins with 1, 2 or 3 CH_3- groups in the side chains. These isomers contain always hardly separable microimpurities, which are more reactive than the parent paraffins and can affect the processes of di-, trimerization of the hydrocarbon molecules.

It is seen that the values of k_1 depend on the structure and type of the CH-bonds in the alkane molecule as for the primary ones (I) it is 0.0002 No 19, for the secondary ones (II) – 0.048 No 8 and for the tertiary (III) – 0.20 No 18. These bonds differ basically in their energies which are in the ratio: 1 : 0.98 : 0.96 [69, 70].

The strong effect of the bond energies supposes that the CH-bond in the transition state is highly distorted and the heat of the reaction may be zero or negative [75]. The enthalpy of the process experimentally determined is within the limits of 0–10kcal. The Hammond law states that the TS would be more similar to the reaction products than to the initial compounds, i.e. the reaction coordinate is relatively high. This justifies to a great extent that the dependence of $1/k_{obs}$ on [RH] is not so much related to the chemistry of the reaction but to the experiments conditions because it is observed only in separate cases and is concentration limited. In addition W has a constant value in solvents of various properties (CCl_4, $CHCl_3$, CH_3CN, decane). The experimentally determined activation energies for the ozonolysis of paraffins are in the range of 14–15 kcal/mol when the energies of complex formation are $E_a < 10$ kcal/mol. The UV-spectrum of ozone in n-hexane solution is similar to that in the gas phase. The insignificant bathochromic shift of 6 nm does not lead to broadening of the absorption band. The hypochromic effect leads to change in the molecular extinction (ε) by 10% from 1820 to 2002 $M^{-1}.cm^{-1}$. The blue color of ozone is preserved in the n-hexane solution. The IR-spectra of the gaseous ozone and of its solution in n-hexane are practically identical and new bands are not identified. In the NMR-spectra of the n-hexane solutions of ozone at temperature –80°C and after a long stay no changes in the spectrum of n-hexane are observed. Only upon ozonolysis of olefins at –100°C in the IR-cell and the analysis of the IR spectra is presumed the formation of π-complexes [76, 77].

Thus the reaction of ozone with paraffins is more likely a bimolecular process, excluding preliminary complex formation, which leads directly to the formation of the reaction products:

RH + O$_3$ → products (k)

And then the data extrapolated to $[RH]_{\to 0}$ in column 3 of Table 1 will correspond to the bimolecular constant k.

We have assumed that the reaction of ozone with the homologies series of n-paraffins is isoentropic and the difference in their reactivity is determined by the energies of C-H-bonds. In order to eliminate random factors we have applied the relative rate constants on example of n-hexane Eq. (9):

$$k^n/k^6 = \exp(-0.3\Delta D/RT) \tag{9}$$

where $\Delta D = D^n - D^6$ is the difference between the energies of C-H bonds in kcal/mol; n – index for the corresponding paraffin; 0.3 – empirical coefficient.

At 293K, R = 1.99 cal/(mol.K) and k_6 = 0.019 M^{-1}.s^{-1}:

$$k^n = 0.019.\exp(-0.514.\Delta D) \tag{10}$$

The energies of C-H-bonds attached to primary (I), secondary (II) and tertiary (III) C-atom can be calculated by the empirical equations as follow [70] Eqs. (11)–(13):

$$D_I = 93.6 + 8\Sigma 0.4 \text{ } ^{f, g, I, j,...} \tag{11}$$

$$D_{II} = 85.6 + 8\Sigma 0.4 \text{ } ^{f, g, I, j,} \tag{12}$$

$$D_{III} = 77.6 + 8\Sigma 0.4 \text{ } ^{f, g, I, j,...} \tag{13}$$

where, f, g, I, j,... are the number of C-atoms in each linear hydrocarbon chain bonded to C-atom, whereby the energy of the C-H-bonds is calculated.

The estimated by Eq. (10) k values for paraffins comprising only primary C-H-bonds (for example methane, neopentane) are within the limits of 0.00078/0.000013; with secondary C-H bonds (n-paraffins)–0.48/0.00078 and with tertiary CH-bonds (sec-butane, alkylsubstituted n-paraffins) – 2.93÷0.048. The found ration of reaction ability is: 1/60/3692.

The calculated values of k and D of the C-H-bonds are presented in Table 1 and a comparison with the experimental data have been made Table 2.

TABLE 2   Calculated and experimental kinetic and thermodynamic parameters of ozone reactions with some paraffins.

No.	Paraffin	k_{cal}/k_{exp} $M^{-1}.s^{-1}$	D_{C-H} kcal	No.	Paraffin	k_{cal}/k_{exp} $M^{-1}.s^{-1}$	D_{C-H} kcal
1.	Methane	$1.2.10^{-5}/6.6\square10^{-4}$	101.6	18.	2,3-di-Me-pentane	0.08/0.29	84.64
2.	Ethane	$1.52.10^{-4}/6.3\square10^{-3}$	96.8	19.	2,4-di-Me-pentane	0.06/0.08	85.02
3.	n-Propane	$1.8.10^{-3}/3.4\square10^{-3}$	92	20.	2,3,4-tri-Me-pentane	0.04/0.59	85.92
4.	n-Butane	$5.10^{-2}/5.8\square10^{-3}$	90	21	2-Me-hexane	0.10/0.13	84.20
5.	n-Pentane	0.013/0.015	88.2	22.	3-Me-hexane	0.23/0.20	82.59
6.	n-Hexane	0.019/0.019	87.4	23.	2,2-di-Me-hexane	0.01/0.015	88.42
7.	n-Heptane	0.029/0/021	86.6	24.	2,4-di-Me-hexane	0.17/0.13	83.10
8.	n-Octane	0.033/0.023	86.3	25.	2,2,5-tri-Me-hexane	0.08/0.19	84.61
9.	n-Nonane	0.039/0.026	86	26.	3-Me-heptane	0.26/0.20	82.28
10.	n-Decane	0.041/0.029	85.9	27.	Ethylcyclopropane	0.15/0.05	88.6?
11.	n-Undecane	0.043	85.8	28.	Neopentane	$1.12\square10^{-4}/3.8\square10^{-5}$	97.4
12.	n-Dodecane	0.046	85.7	29.	Hexamethylethane	0.00016/0.00002	96.67
13.	n-Tridecane	0.046	85.66	30.	Cyclohexane	0.01/0.01	88.6?
14.	n-Tetra-decane	0.047/0.036	85.66	31.	Norbornane	0.014/0.014	99?
15.	n-Octa-decane	0.048/0.048	85.607	32.	Adamantane	0.22/0.22	82.6?
16.	Polyethylene	0.048/	85.6	33.	2,2,4-tri-Me-pentane	0.05/0.14	85.54
17.	3-Me-pentane	0.15/0.15	83.36				

It is seen that the calculated values of k and D are in a very good fit with the experimental results of Table 1 and are in a complete conformity with the literature data [2–15] too. A slight difference is noticed for hydrocarbons containing primary C-H-bonds. Because of their lower reactivity insignificant amounts of microimpurities might increase the experimental values of k. Thus the difference between the calculated and experimental values of k could be used as a measure for checking the purity of paraffins.

The energies of C-H bonds, given in column 4 for linear and branched paraffins are calculated according to Eqs. (11)–(13) and for cycloparaffins Nos. 27, and 30–32 from the rate constants. The latter can be regarded as a very good method for determination of thermodynamic parameters on the basis of kinetic data.

The good correlation between the calculated and experimental values of k testifies the H-atom abstraction mechanism is more adequate for describing the interaction of ozone with alkanes.

12.3.2 CYCLOPARAFFINES

We have studied the kinetics and mechanism of ozone reaction with cycloparaffins ranging from cyclopentane to cyclododecane. From thermodynamically point of view cycloparaffins contain only equivalent C-H-bonds and their energies calculated according to Eqs. (11)–(13), using a model of infinite chain, are 88.9 – 88.5 kcal/mol and for cyclopropane and cyclobutane are 91 and 89.5 kcal/mol. The literature values for the former are within the limits of 94±3 kcal/mol and for the latter are 100.4±2 and 95±3 kcal/mol [28]. The deviations in the calculated and literature values could be attributed to the approximations of the model chosen which considers both the small and big cycles as infinite large. In both cases, however it could be stated that after cyclobutane, D does not actually depend on the cycle size.

From stereochemical point of view, however, cycloparaffins differ considerably with respect to number, stability and symmetry of the conformers in which they exist, to the number of equatorial, axial, pseudoequatorial and pseudoaxial, exo- and endo H-atoms, to the deviations from the equilibrium values of the bond lengths, valent and torsion angles and steric hindrance in the cycles. Upon ozonolysis the energy of the system would either decrease and the reaction will be accelerated or increase and the reaction will be retarded and the reactivity of cycloparaffins would be different irrespective of the similar energies of the C-H bonds in them. The rate will be also determined by the product of the numbers of symmetry:

$$\sigma = \prod_{i=1}^{j} C_v$$

where, C_v – the axis of symmetry; v – the order of identity per one motion; j – the number of axes of symmetry (for example, cyclopropane has one axis of symmetry of third order or three-fold C_{3v} and then $\sigma = 3$, cyclobutane – three axis of symmetry of second order – C_{2v} and $\sigma = 8$, cyclohexane – one axis of third order C_{3v} and one of second C_{2v} and $\sigma = 6$) of the initial compounds and the activated complex (AC), which upon other other conditions the increase of symmetry will increase W.

Using *PCMOD4* program based on molecular mechanics methods [78, 79] we have calculated the steric energies (SE) of cycloparaffins and their radical forms (c-C_nH_{2n} and c- $C^{\bullet}_nH_{2n-1}$). First of all, we have carried out calculations on the various stereoisomers and select the conformer with the lowest energy (for example envelope for C_5 and bath for C_6) and SE_{RH}. Further we simulate the H-atom abstraction at consecutive move of the reaction center at each nonequivalent C-H-bond in this conformer and thus we select the model with minimum SE_R. Five calculations have been made for cyclo-C_5, 2 for cyclo- C_6, 7 – for cyclo- C_7, etc.

Thus by the difference $\Delta SE = SE_{RH} - SE_R$ we have estimated whether the transition $sp^3 \rightarrow sp^2$ is favorable or not. The results from the calculations are given in Table .3.

TABLE 3   Conformations, numbers of symmetry (σ) calculated SE_{RH} and SE_R, ΔSE and SE_{RH} and the experimental SE_R of cycloparaffins.

No.	cyclo-C_n	Conformation	σ	SE_{RH}, kcal/mol	SE_{Rx}, kcal/mol	ΔSE, kcal/mol	SE_{RH}, (exp) kcal/mol
1.	Propane	△	3	26.2	39.9	−13.7	27.6

TABLE 3   *(Continued)*

No.	cyclo-C_n	Conformation	σ	SE_{RH}, kcal/mol	SE_{Rx}, kcal/mol	ΔSE, kcal/mol	SE_{RH}, (exp) kcal/mol
2.	Butane		4	24.0	21.7	2.3	26.4
3.	Pentane		1	5.5	3.4	2.1	6.5
4.	Hexane		6	0.0	0.0	0.0	0
5.	Heptane		2	7.1	4.1	3.0	6.3
6.	Octane		8	12.5	8.0	4.5	9.6
7.	Nonane		1	15.6	9.8	5.8	12.6
8.	Decane		10	15.3	9.3	6.0	12.0
9.	Undecane		1	14.6	9.9	4.7	–
10.	Dodecane		16	10.2	6.2	4.0	3.6

Note: the designations of the conformations of cycloparaffins from top to bottom are: triangle, plane square, envelope, chair, twist chair, twist crown, twist crown, twist crown, twist crown and square with 4 butane segments in chair conformation. For the cycles over C_7, the conformations shown are among those differing not more than by 1 kcal/mol but have the highest symmetry. In real conditions the intrinsic (real) number of symmetry is between 1 and its maximum magnitude and will be determined by the partial participation of each conformer: we have looked for a coincidence between the theoretical and the experimental values of SE [80]; we have calculated SE_{RH}.(exp) from the thermochemical values of the heats of combustion of cycloparaffins in gaseous phase according to the expression $(Q_n/n–Q_6/6)$, where n is the number of methylene groups in the cycle, and Q_n and Q_6 are the heats of combustion of cycloparaffin and cyclohexane, respectively.

The reaction with cyclopropane and cyclohexane appears to be the slowest due to the increase in energy cycle upon $sp^3 \rightarrow sp^2$ transition. In the other cases the energy decreases and this leads to raise of the reaction rates.

In Table 4 are depicted the parameters of cycloalkane ozonolysis [48, 51] experimentally obtained.

TABLE 4   Kinetic parameters of ozone reaction with cycloparaffins (20°C, CCl_4).

Parameters	Pentane	Hexane	Heptane	Octane	Nonane	Decane	Dodecane
$k.10^3 \pm 10\%, M^{-1}.s^{-1}$	24.9	10.0	229.0	480.0	890.0	408.0	62.0
$A.10^{-7}, M^{-1}.s^{-1}$	2.7	38.2	2.2	1.7	24.3	170.0	168.0
$E_a \pm 0.5$, kcal/mol	12.2	14.3	10.8	10.2	11.4	13.0	14.1

For elimination of some factors general for the reactivity of cycloparaffins we have used the values of the relative constants:

$$k^n/k^6 = (A_n/A_6).\exp(-\Delta E/RT) \tag{14}$$

where, n is the number of carbon atoms in the cycle; $\Delta E = E_n - E_6$ is the difference between the activation energies of n-th cycloparaffin and that of cyclohexane.

Taking into consideration the experimental data of Table 1 the magnitudes of the parameters in the right term of Eq. (14) will be:

	C5	C6	C7	C8	C9	C10	C12
ΔE, kcal/mol	−2.1	0	−3.5	−4.1	−2.9	−1.3	−0.2
$\exp(-\Delta E/RT)$	35.9	1	383.7	1064.2	138.4	9.1	1.4
A_n/A_6	0.071	1	0.058	0.045	0.636	4.45	4.34

The decrease of ΔE to the left and right from cyclohexane indicates the $sp^3 \rightarrow sp^2$ transition in cyclohexane as energetically most unfavorable. The variations in A_n/A_6 and ΔE show the occurrence of compensation effect

(CEF), which in this particular case could be ascribed to the different SE of the cycles and their symmetry.

On the base of the expression for the rate constant in the theory of the activation complex including the entropy of the activation complex [81, 82]:

$$k = e^2(RT/N_a h).exp(\Delta S/R).exp(-\Delta E/RT) \tag{15}$$

where, e -the base of natural logarithm, N_a – Avogadro's number, h – Planck's constant.

The values of ΔS have been calculated:

	C5	C6	C7	C8	C9	C10	C12
ΔS, cal/(mol.K)	−19.46	−14.46	−19.87	−20.39	−15.06	−11.17	−11.20

The values of entropy increases for small – C_5 and medium – $C_7 - C_9$ – cycles, and are lower for the large C_{10} and C_{12} but even at the latter the rate constants are higher than that of cyclohexane.

The combined solution of Eq. (14) and Eq. (15) leads to Eq. (16):

$$k^n/k^6 = exp(\Delta\Delta S/R).exp(-\Delta E/RT) \tag{16}$$

where, $\Delta\Delta S = \Delta S_n - \Delta S_6$

The change of $\Delta\Delta S$ and ΔE is an additional evidence for the occurrence of CEF.

We have found out an empiric relationship between the reactivity of cycloparaffins and their steric energy (Eq.(17)) [48]:

$$lg(k^n/k^6) = \xi.(SE)^{1/2} \tag{17}$$

where, ξ is an empiric coefficient; SE – steric energy

Thus by applying Eq. (17) (the data for the steric energies are available) it is possible to evaluate k for new compounds.

The experimental dependence of the reactivity of cycloalkanes on their steric energy is shown in Fig. 2.

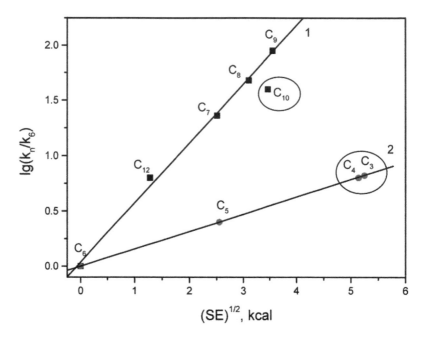

FIGURE 2 Dependences of the cycloalkanes reactivity on their steric energy. 1 – Cycles > C_6; 2 – Cycles < C_6.

It follows from Fig. 2 that a=0.54 and 0.16 for curve 1 and curve 2, respectively. The experimental point for C_{10}, which deviates from the straight line by 16% (the experimental value of lg (k_n/k_6) is 1.6 and the extrapolated one is 1.9 (designated by a circle on curve 1). The same designation is made on curve 2 for the extrapolated points for lg (k_n/k_6) of C_4 and C_3 with values of 0.8 and 0.83 respectively. The calculated rate constants for the ozone reaction with these two cycloalkanes will amount to 63 and 67 $M^{-1}.s^{-1}$, respectively. The difference in the slopes of the two curves is most likely associated with the more hindered $sp^3 \rightarrow sp^2$ transition in smaller cycles. This fact is also supported by the data for the values of SE in Table .3.

The theoretical and kinetic studies reveal that ozone reacts with cycloalkanes via hydrogen atom abstraction. The reactivity of cycloalkanes depends on the direction of the steric energy change during the $sp^3 \rightarrow sp^2$ transition in AC.

In Fig. 3, is presented the dependence of the rate constant of cyclooctane ozonolysis on its concentration.

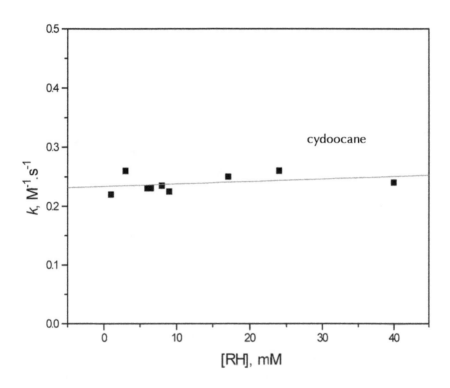

FIGURE 3   Dependence of the rate constant of ozone reaction with cyclooctane at 10°C.

The variation of [RH] and [O] up to three orders does not affect k and the rate of the reaction is described by the following equation:

$$W = k \, [RH]^1 . [O_3]^1 \tag{18}$$

The semilogarithmic anamorphosis of the kinetic curves of ozone consumption at [RH]>[O$_3$] upon ozonolysis of H$_{12}$ and D$_{12}$ cyclo-C$_6$ is shown in Fig. 4.

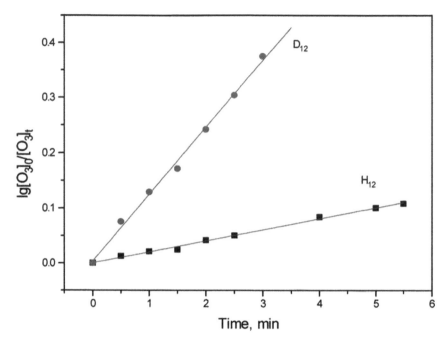

FIGURE 4   Semilogarithmic anamorphosis of the kinetic curves of ozone consumption upon ozonolysis of H_{12} and D_{12} cyclo-C_6 ozone consumption at 28°C.

The reaction of ozone decomposition follows first order kinetics and the rate of the reaction with D_{12} is found to be higher. The ratio $k_H/k_D = 5.4$ is indicative of the occurrence of primary isotope effect which testifies that the H-atom abstraction is the rate limiting step of the reaction.

12.3.3   METHOD FOR ESTIMATION OF REACTION MECHANISMS (ERM)

The method for estimation of reaction mechanism (ERM), developed by us, is based on the comparison of calculated by means of activated complex theory and experimental values of preexponents. According to the concepts developed in the AC theory [81–83] the equation for the rate constant k of a bimolecular reaction is:

$$k = \gamma \cdot \chi \cdot [RT/(N_a h)] \cdot [F^{\#}/(F^a F^b)] \cdot F_{fr} \exp[-E_a/(RT)] \qquad (19)$$

where, $F^{\#}$, F^a and F^b are the corresponding total statistical sums of the activation complex and the initial compounds A and B.

The total statistical sums are a product of the separate statistical sums of the state:

$$F = F_t \cdot F_r \cdot F_v \cdot F_e \cdot F_n \tag{20}$$

where, the indexes t, r, v, e and n denote the corresponding translational, rotational, vibration, electronic and nuclear sums of the state.

The translational sum of the state at 3 orders of freedom is described by the following equation:

$$F_t = h^{-3}(2\pi mkT)^{3/2} \tag{21}$$

where, m is the mass of the molecule; k – Boltzmann constant; T – absolute temperature.

The values of F_t are within the limits of 10^{24}–10^{25} cm^3 per molecule.

The rotational sum of the state for nonlinear molecule is:

$$F_r = 8\pi^2(8\pi^3 I_A I_B I_C)^{1/2}(kT)^{3/2}h^{-3} \tag{22}$$

where, $I_A I_B I_C$ – are the principal moments of inertia of the molecule when the origin of the coordinate lies on the mass center of the molecule.

The values of these sums do not exceed 10^2 –10^3 cm^3/molecule.

The equation for estimation of the statistical sum of the vibration movement is:

$$F_v = (1 - e^{-h\nu/kT})^{-1} \tag{23}$$

where, ν – vibrational frequency.

The magnitude of F_v for one degree of freedom is of the order of 1–10 cm^3 per molecule, as the vibrations in the molecules are within the limit of 100–3000 cm^{-1}, i.e., ν is in the interval of 3.10^{12}–9.10^{13} c^{-1} that is close to the value of $kT/h = 6.1.10^{12}$ c^{-1} at 100 K.

The statistical sum of the free rotator is:

$$F_{fr} = (8\pi^3 I'kT)^{1/2}h^{-1}\sigma^{-1} \tag{24}$$

where, I' is the reduced moment of inertia towards the axis of rotation.

The electronic sum of the state for adiabathic processes, including also ozone reactions, is demonstrated by:

$$F_e = 1 \qquad (25)$$

The nuclear sum of the state for Frank-Condon processes, such as are most of the chemical transformations, is given by:

$$F_n = 1 \qquad (26)$$

The transmitional coefficient (γ) takes account of the AC probability to be converted into reaction products. In case of adiabathic reactions it is equal to 1, and for multiplied nonadiabathic reactions it may be much smaller than 1.

The coefficient λ takes account of the tunnel effects associated with the reaction pathway via quantum way through percolation "below barrier" of the potential surface.

The probability for tunneling depends on: (1) the mass of the particle m and the smaller is it, the greater is the probability for transition; (2) the barrier width δ – its decrease leads to increase of the tunneling probability; (3) its height E_a – with its decrease the probability for tunneling increases; and (4) the steepness of the potential surface at the depression point – (d^2U/dx^2) – its rise results in increasing the tunneling probability.

According to Gamov [83] the probability for tunneling via triangle barrier is:

$$\chi = [2m \, (E_a - U)]^{1/2} . \exp[-8\pi\delta/(3h)] \qquad (27)$$

where, U is the height of tunneling occurrence.

In case of reactions carrying out with redistribution of atoms at room and higher temperatures the tunnel effect is extremely small and λ is equal to 1. This is can be applied to the reactions studied by us.

For calculation of the rate constant according to Eq. (19) it is necessary to estimate the potential surface of the reaction and the statistical sums of AC, RH and O_3.

The calculation of the potential surfaces is a difficult and labor-consuming problem and it has been solved only for small particles. This is a typical quantum-chemical task, which has been the subject of intensive studies of various researchers, and we will not focus our attention on the difficulties and successes in this respect [84–88]. Easier methods for calculation of potential surfaces are semi-empirical ones, like *mopac6* [87].

The estimation of the statistical sums is simple problem but it provides important information for the mechanism of chemical reactions. For performing the calculations it is necessary to know the geometry of the starting molecules and AC and the vibration frequency of the bond on the reaction coordinate. The geometry of the starting molecules can be usually determined with high degree of confidence, because of the great number of experimental and theoretical data concerning the interatomic interactions, the conformational states, bond lengths, the angles between atoms and their masses [70]. The question about the determination of the geometrical parameters of the activated complex is more complicated since the potential surface of the reaction is unknown. It is difficult, usually only by indirect data, to determine the reaction coordinate. In order to find an acceptable solution of this problem we have applied the Franck-Condon principle [90] and the Hammond empirical rule [75], which connects the thermal effect of the reaction with the reaction coordinate. These two principles allow the determination of the geometrical parameters of the activated complex with accuracy up to 10%.

The preexponential in the Arrhenius equation in the framework of the AC theory can be described in the following way:

$$A = \gamma.\chi.[RT/(N_a h)].[F^{\#}/(F^a F^b)].F_{fr} \qquad (28)$$

These considerations were used in developing a computer program for calculating A and the statistical sums of internal rotation [89]. In addition the program includes also calculation of A and the steric factor – p, according to the collision theory. It also takes into account the number of equivalent H-atoms and carrying out the reaction in liquid phase.

For calculation of A it is necessary to define the geometrical parameters of AC. Principally there exist two forms of the activated complex – linear and cyclic. In the first one an inner rotation is possible while in the second it is absent. The values of A calculated for the linear AC in each case will be higher than those calculated with the cyclic form.

The calculated values of A were compared with the experimental ones and on the bases of the coincidence degree one or another mechanism could be preferred. By this manner the method ERM gives additional possibilities to clarify the investigated reaction mechanism.

12.3.4 APPLICATION OF ERM

Using ERM we have made the analysis of mechanisms M.1.1–1.6. The reactions according to them take place via two forms of the activated complex – linear (LC) and cyclic (CC):

SCHEME 3

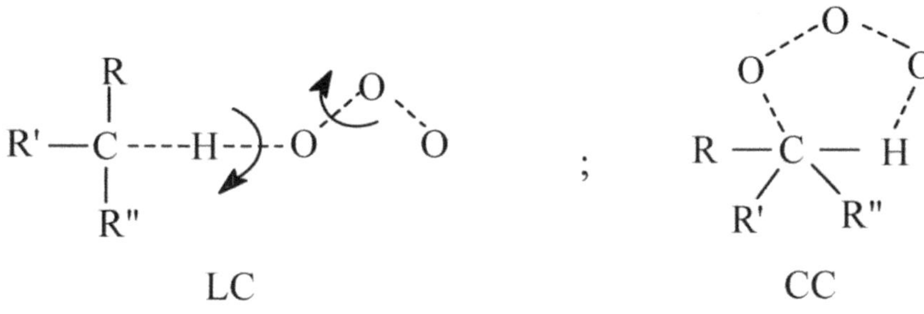

LC   CC

LC is associated with H-atom cleavage mechanism and CC with ozone incorporation along the C-H bonds. LC is characterized by the ozone interaction with the C-H bond via one of its terminal atom and thus there is a possibility for free rotation around H-O and O-O while in the CC it forms five member cycle with participation of five coordinated carbon. In Table 5, are summarized the alculated values of A for the two AC forms.

TABLE 5. Preexponentials (A) of the ozone reaction with cycloalkanes at 300 K calculated by the method of the activated complex.

Cycle	σ	$J.10^{-7}$, $M^{-1}.s^{-1}$	$I_A I_B I_C.10^{-6}$, au.^{2}	$\sigma.F_{fr}$	$^\#I_A I_B I_C.10^{-8}$ au.^{2}	$^\#F_r.10^{-6}$	$F_{fr}.10^{-3}$	$A_0.10^{-7}$	$A.10^{-7}$, $M^{-1}.s^{-1}$	$A_1.10^{-7}$, $M^{-1}.s^{-1}$
C_5	1	2.54	0.8	0.71	0.44	0.52	1.24	6.84	0.59	1.65
C_6	6	2.29	2.5	1.23	0.59	0.60	1.28	26.6	2.29	6.40
C_7	2	2.12	8.3	2.25	2.02	1.10	1.32	8.56	0.74	2.06
C_8	6	1.99	13.5	2.88	2.54	1.24	1.34	21.5	1.85	5.17
	1	1.99	13.5	–	2.54	1.24	1.34	3.6	0.31	0.86
C_9	1	1.88	34	4.53	5.40	1.81	1.37	3.2	0.28	0.77
C_{10}	2	1.81	97	7.68	12.2	2.72	1.39	5.6	0.48	1.34
C_{12}	16	1.69	278	13	32	4.40	1.41	40.3	3.55	9.95

where, $J = (kT/h) (F_t^{\#}/F_t^{RHF}F_t^{O3})$.

Further, we will focus our attention on estimation of the value of A calculated for each separate reaction [48, 52].

The ratio between the statistical sums of electronic and nuclear movement of the activated complex and the initial compounds is 1 since the change of the energy is smooth and the electron-nucleic states are not changed; γ and λ are equal to 1 under the experimental conditions.

The value of A can be presented by two multipliers [63]: A_0 is a geometric factor, which is determined by the AC geometry and A_1 – the contribution by the internal degrees of freedom. In this case we can write the following expressions Eqs. (29)–(30):

$$A_0 = (kT/h) . (F_t^{\#}/F_t^{RHF}F_t^{O3}).(F_r^{\#}/F_r^{RHF}F_r^{O3}).F_{fr}^{\#} \tag{29}$$

$$A_1 = (F_v^{\#}/F_v^{C-H}).(F_v^{R\#}/F_v^{R}).(F_v^{O3\#}/F_v^{O3}).\varphi \tag{30}$$

where, F_v are the statistical sums of the vibrations of the reaction center, i.e., C-H-bond in the initial alkane;

$\varphi = \varphi_0 . \exp(-E_h/RT)$ – coefficient considering the highest energy that the rotator must overcome at its twist to 360°.

The sums for 300 K were calculated by using the following equations Eqs. (31)–(34):

$$F_t = 3.86.10^9[(m_{O3} + m_{RH})/m_{O3}.m_{RH}] \tag{31}$$

$$\sigma.F_r = 78 \, (I_A I_B I_C)^{1/2} \tag{32}$$

$$\sigma.F_{fr} = 6.24 \, I'^{1/2} \tag{33}$$

$$F_v = (1 - e^{-h\nu/kT})^{-1} \tag{34}$$

It should be noted that the values of A_o thus calcultated are relatively reliable. For ozone are used literature values [29, 64] for the length of O-O bond 1.278±0.003 A and for the OOO angle – 116° 49' ±30'; the main inertia moments $I_A I_B I_C$ are $7.875.10^{-40}$, $62.844. \, 10^{-40}$ and $70.888. \, 10^{-40}$, respectively; $\sigma = 2$, F_e and $F_n = 1$. For cycloalkane the values of the length of the C-C and C-H bonds are 1.54 A and 1.09 A respectively and 109.5 for the CCC=CCH=HCH angles have been used. In Table 5, the values of $I_A I_B I_C$ are calculated according to well-known methods [1, 90]. We have selected the most stable conformers of the cycles known [80] confirmed also by the calculations according to Alinjer method (Table 3). Thus the preferred conformation for cyclopentane is envelope (C_s), for cyclohexane-chair, for cycloheptane-twisted chair (C_{2v}), for cyclooctane-twisted crown with 3 axes C_{2v}, regular crown with 2 axes C_{2v} or knock crown without any elements of symmetry. The product $I_A I_B I_C$ is nearly the same for all conformations. However, bearing in mind that the different degrees of symmetry would affect the calculation of A we have used two conformations with a number of symmetry of 6 and 1.

The conformation state of cyclononane is a crown with a symmetry number of 1 but it is insufficiently studied and for this reason we have combined the calculations by the Alinjer method with the X-ray data for cyclononylamine bromohydrate [80]. The conformation of cyclododecylamine-1,6 dihydrohydrate ,which has only one axis of symmetry of second order, is taken as a base for cyclodecane. According to the X-ray data cyclododecane has a square shape conformation with butane segments as its sides. This conformation has one axis of fourth and two of second order and its number of symmetry is 16. Some of the most stable conformations of cycloalkanes are presented in Fig. 5.

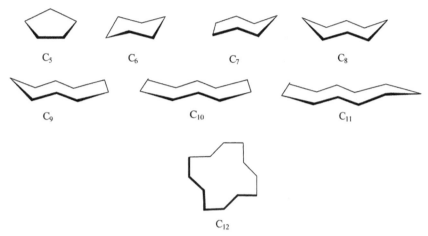

FIGURE 5   Most stable cyclopalkanes conformations.

The structure of LC was elaborated on the basis of the following geometrical parameters: the H...O length was estimated according to Pauling formulae [91]; $d_{H...O} = d_{H-O} - 0.6.\ln(r)$, where, r is the bond order, $d_{H...O}$ the bond length in the complex and d_{H-O} – the length of ordinary H-O – bond equal to 0.95 . Considering that the coordinate of the maximum on the potential surface can be approximately determined the bond order was defined in a wide range from 0.2 to 0.8 with $d_{H...O}$ varying from 1.92 to 1.08 . The actual value of $d_{H...O}$ lies most likely in this interval and thus the value of 1.35 at r=0.513 was used for the calculations. The lengths of C-H and O-O bonds in LC are increased by 10% as compared to their normal values, the H...O-O angle is accepted as 100° (in water HOH is 104.5° and in HOOH the HOO angle is 96.6°).

The calculation of F_{fr} was carried out for two asymmetry rotators: one with rotation about axis x coinciding with the H...O bond and the second with rotation about y axis coinciding with the O-O bond:

The reduced moments of inertia about y axis are found to be equal for all cycles and have the value of 21 au. ^{2}. However those about the x-axis increase with the rise of the cycloparaffins masses and their values are:

C5	C6	C7	C8	C9	C10	C12
50	53	56	58	60	62	64

In Table 5, the presented values of A_0 depend on the LC geometry only and to a lesser extent on the potential surface. The values of A_1 could be calculated with less accuracy for they depend to a greater extent on the form of the potential surface.

For the accurate calculation of F_{fr} it is necessary to know the values of the rotation barrier E_{fr}. The latter were calculated by us using *MOPAC 6* and for the rotation around x and y axes they are 1.8 and 7.5 kcal, respectively. These values correlate well with the energy of NO rotation in the nitromethane molecule [91]. The correction coefficient φ was also determined using the Pitzer tables [92]. Although the latter are designed for symmetric rotators as Pitzer has shown, they can be successfully applied for the approximate estimation of the thermodynamical functions of asymmetric rotators.

The ratio of the vibration statistical sums $(F^{\#RH}_v.F^{\#O3}_v)/(F^{RH}_v.F^{O3}_v)$ which stands for the variations of the vibration frequencies is unit. At the same time for the ozone molecule which is characterized by two valent vibrational frequencies at 1106 and 1036 cm^{-1} and one deformational at 703 cm^{-1} one can expect a decrease of the frequency at 703 cm^{-1}, but not lower than the value within the range of 150–200 cm^{-1} to became kinetic active at 300K.

The contribution of the vibration sums in the thermodynamical functions is small and is accepted unit for calculating the ratio $F^{\#rc}_v / F^{C-H}_v$ at such moderate temperatures like 300K, where $F^{\#rc}_v$ is the vibration statistical sum of the reaction center. For higher temperatures, however, this ratio may be much higher [93, 94].

The accurate determination of the vibration frequency of the reaction center requires knowledge of the form of the potential surface near the maximum point. For that reason this value could be only approximately evaluated.

The frequency of the deformational vibrations of the reaction center C...H...C lies in the range of 100–400 cm^{-1} [94]. For the reaction center C-H...O this center is probably shifted to the higher frequency kinetically nonactive range. This is associated with the fact that v_{O-H} lies in the range of 1000–1600 cm^{-1} and that of C-H in 400–1000 cm^{-1}. Hence the contribution of the vibrational sums should not differ from 1. However one cannot rule out the probability of the vibration sum of the reaction center for bigger cycles to reach values by 5–10% higher than those for the small cycles. Even in this case the trend in the calculations remains the same. Upon increasing the inertia moment of cycloalkanes as shown above, the vibration frequency should also increase since it is inverse to its square root. For example, if we accept that $v = 200$ cm^{-1} for C_5, then for C_{12} it will be 120 cm^{-1}. In this case the vibration term will be increased 3.5-fold.

It should be noted, however, that the calculations have been made for reactions in the gas phase, while for those in the liquid medium the values of A would be different. According to the collision [95] and activated complex [81–83, 96] theories the number of collisions in the liquid phase is by 2–3 times higher than those in the gas phase. The both theories predict that at $E_l = E_g$ the value of A will be 3 times higher than that in the gas one. In [91] is supposed that the increase of k for a bimolecular reaction proceeding in liquid phase will be about 5 times.

The same tendency is observed at comparing the experimental results from the ozonolysis of acyclic hydrocarbons in gas phase [2–10] and in an inert solvent – CCl$_4$ [47]. The ratio of the preexponential factors in them is in the range of 2–3.

The aforementioned considerations give us ground to introduce a coefficient for the ratio A_l/A_g equal to 2.8 (Table 5, column 11).

At comparing the experimental results with the theoretical data the experimental value of A is usually related to the number of the active centers. It is well known [6] that the hydrogen atoms in cycloparafins exist in various conformations in respect to the cycle planes they can be oriented in various positions from equatorial to axial and from endo- to exo-. In cyclohexane, for example, the three equatorial H atoms are 7 times more reactive that the three axial H atoms [74]. The H-atoms in the bigger cycles are even more nonequivalent since besides the axial and equatorial atoms in them there are also hydrogen atoms lying within the cycles for which a transanular interaction is characteristic. This can be clearly demonstrated by the models of Dreiding and the various stereochemical programs. The conformation analysis performed points out that the hydrogen atoms distribution in the cycles studied is as follows:

	C5	C6	C7	C8	C9	C10	C12
ax	10	6	8	8	10	10	16
eq	–	6	4	4	6	6	–
exo	–	–	2	4	2	4	8

In order to compare the experimental with the calculated values of A we relate the experimental values of A per the number of the equatorial H atoms (Table 6).

TABLE 6 The experimental and calculated values of A

A, $M^{-1}.s^{-1}$	C5	C6	C7	C8	C9	C10	C12
$A_{exp}.10^{-7}$	0.32	3.34	0.36	0.29	2.48	22.8	12.9
$A_{cal}.10^{-7}$	1.65	6.40	2.06	0.86	0.77	1.34	9.90

The experimental values of A vary from $0.3.10^7$ to 23.10^7 $M^{-1}.s^{-1}$. It has been found that the values of A calculated for the activated complex with LC lie within this range $(0.8–10).10^7$, while those for the CC are in the range of $10^5–10^6$ $M^{-1}.s^{-1}$. The lower values in the latter are related to the lack of free rotation (about 130 times) and to the more packed structure

(2–4 times) of this complex. However in this case the eventual increase of A, on account of the occurrence of low frequency vibrations of the reaction center which according to us can bring about not more than 2–3 times increase, should be taken into account. Thus the calculated with CC values of A should be lower than the experimental ones by about 100 times. The coincidence of A calculated with LC with the experimental values of A becomes even better at their comparison for each cycle alone. In fact the differences in the values of A for all cycles do not exceed 3 which fact we will discuss further.

The results obtained show that the probability of the reaction to occur via a cyclic transition state is negligible. Consequently, the probability of the ozone reaction with the C-H bonds to take place through a mechanism of incorporation or formation of oxy-radicals in one step is also small. It follows from these two mechanisms that the reaction should take place minimum in two steps – the first is H-atom abstraction giving rise to $(R^• +$ $^•OOOH)$ or to$(R^• + HO^• + O_2)$. Since the endothermity of the latter is lower by 16 kcal/mol it is thermodynamically more favorable.

In our opinion, one of the reasons for the existence of such a majority of schemes describing reaction $RH + O_3$ is due to the lack of entropy factors consideration. As a result of their taking into account it can be seen that the abstraction of H-atom is the limiting step in the ozonolysis of paraffins. By its own possibilities the ERM is method universal, because it can be applied to every typs of chemical reactions.

12.4 CYCLOHEXANE

The study of the relationship between the rate of ozone consumption and [ozone] and [cyclohexane] [52] in a wide range reveals a first order kinetics in regard to each reagent.

The typical kinetic curves of reaction products formation are given in Fig. 6.

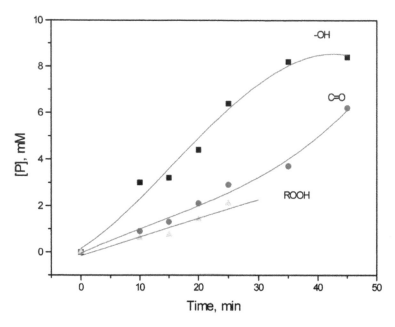

FIGURE 6   Kinetics of basic products formation during cyclohexane ozonolysis at [RH] = 9.29 M, [O$_3$] = 0.7 mM, 20°C

The profile of the curves of products formation shows that cyclohexanone, peroxide and acids including adipic acid (the latter is not shown in the figure because of its extremely low concentration) are obtained in parallel reactions. The initial rate of alcohol formation is by about 3 times higher than that of the ketone. The same ratio for this reaction has been also reported by other authors [9, 67, 97].

The ratio between the rates of peroxides formation is $W_{al}:W_{ket}:W_{ROOH} = 4:1:1$. The nature of the peroxides formed in the course of the reaction has been clarified by studying the kinetics of iodine liberation during the oxidation reaction with HI solution in CHCl$_3$:CH$_3$COOH. The rate of iodine evolution was measured at $\lambda = 470$ nm. In order to ignore the formation of peroxide compounds arising from the further oxidation of the reaction products (alcohol and ketone), we have used reaction mixtures of very small conversion degrees, ca. 0.07%. The results obtained (Fig. 7) justify the probable formation of H$_2$O$_2$, cyclohexylhydroperoxide (ROOH) and dicyclohexylperoxide (ROOR) in the reaction mixture, which react with different rates with HI (KI) [76].

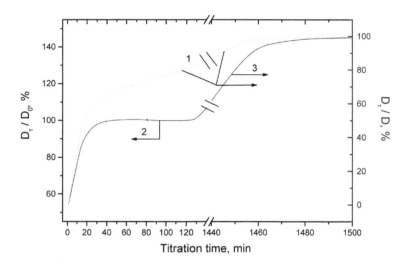

FIGURE 7 Dependence of the relative optical intensities D_τ/D (%) on the titration duration : D_τ, D – determined at time τ and after 24h at 470 nm; 1 – 0.07% conversion; 2, 3 – reference mixture (solution) – *tert*-butyl hydroperoxide (2) /di-*tert*-butylperoxide (3) = 1:1

The formation of the peroxides abovementioned was confirmed by the iodometric procedure modified by us. 0.5 ml saturated water solution of KI and 0.5 ml CH_3COOH were added to 0.5 ml oxidate. The latter was kept at dark and room temperature for 2 h. The quantity of the iodine liberated after 30 min corresponds to H_2O_2 and ROOH, and this after 24 h to ROOR. Further, the separate determination of H_2O_2 and ROOH has been carried out by treatment of the oxidate sample with aqueous solution of catalasa (a specific enzyme decomposing H_2O_2) [98].

The duration of the quantitative iodine liberation was defined using a model system comprising H_2O_2, *tert*-butyl hydroperoxide and di-*tert* butylperoxide and it was found to be 30 min for H_2O_2 and *tert*-butyl-OOH and 24 h for di-*tert*-butylperoxide. These time durations at dark treatment correspond to well known literature data for their rates of interaction with HI (KI) [99–102]. The possibilities of the proposed procedure for the separate identification of H_2O_2, alkylhydroperoxides and dialkylperoxides are illustrated by the data in Table 7.

TABLE 7   Concentrations of H_2O_2, t-BuOOH and t-BuOOBu-t $\times$ 10^2 M.

Compound	0 min	30 min	24 h
H_2O_2	1	1	–
H_2O_2 *	1	0	–
t.-BuOOH	1	1.13	–
t.-BuOOH*	1	1.2	–
t.-BuOOBu-t.	1	0.07	1.0
t.-BuROOBu-t.*	1	0.08	1.0
H_2O_2 : t.–BuOOH=1:1	1=0.5+0.5	0.49	1.08

The yield of peroxides was found to be 1–15%. The authors of Ref. [78] reported a 45% yield but they, in fact, do not discuss and have no explanation of the reasons for this high yield. In our opinion, most likely because of the higher experimental temperature the chain radical oxidation might generate higher quantities of peroxides.

The ratio of the initial rates of peroxides formation as shown by the kinetics of H_2O_2, ROOH and ROOR accumulation is 1:1:4. The relationships of the rates of products formation on the ozone content depicted in Fig. 8 have a linear pattern.

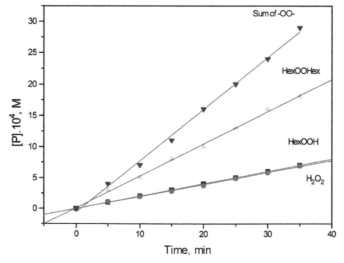

FIGURE 8   Kinetics of peroxides formation.

$[RH] = 9.26m, W_{03} = 1.10^{-5} M.s^{-1}, 20°C.$

The acid numbers of the oxidates determined by titration with aqueous-alcohol solution of NaOH and indicator phenolphthalein is of an order higher than the content of adipic acid as determined by GC. This observation demonstrates the complex composition of the acids. Ref. [97] is considered, however, that the acid number is completely defined by the presence of adipic acid. But this statement obviously contradicts the chromatographic data obtained by us. The latter shows that adipic acid comprises a very small percent of the total acid content.

In a number of experiments we have found out that the content of water is about 50% of the total yield. The same water yield was obtained at tetradecane ozonolysis [1]. The most plausible mechanism of reaction products formation, in our opinion is consistent with the following scheme:

SCHEME 4

$$\rightarrow (\varphi) R^\bullet + O_2 + {}^\bullet OH$$
$$RH + O_3 = [R^\bullet + O_2 + {}^\bullet OH] | \qquad (1)$$
$$cage \rightarrow (1 - \varphi) ROH + O_2$$
$$RH + {}^\bullet OH = R^\bullet + HOH \qquad (2)$$
$$R^\bullet + O_2 = RO_2^\bullet \qquad (3)$$
$$\rightarrow (\alpha) ROH + R'C{=}O + O_2$$
$$RO_2^\bullet + RO_2^\bullet | \qquad (6)$$
$$\rightarrow (1 - \alpha) ROOR + O_2$$

where, [] is the kinetic "cage"; φ – coefficient accounting to the number of radicals in the bulk solution; $(1-\varphi)$ – the part of radical, that recombine in the cage with alcohol formation

The scheme are not included the following reactions:
$$RO_2^\bullet + RH = ROOH + R^\bullet$$
$$ROOH = RO^\bullet + HO^\bullet,$$

as the concentration of the hydroperoxides in the reaction mixture is about only 2% and the experiment temperature is below 40°C and the chain radical reactions are of minor importance in the oxidation process.

The kinetic analysis of Scheme leads to the following expressions for the rates of formation of alcohol, ketone and ROOR:

$$W_{OH} = (1 - \phi) W_{O3} + \alpha\, k_6\, [RO_2^{\bullet}]^2 = (1 - \phi - 2\alpha\phi)W_{O3} \tag{35}$$

$$W_{C=O} = \alpha\, k_6\, [RO_2^{\bullet}]^2 = 2\alpha\phi\, W_{O3} \tag{36}$$

$$W_{ROOR} = (1 - \alpha)\, k_6\, [RO_2^{\bullet}]^2 = (1 - \alpha)\phi W_{O3}, \tag{37}$$

where, W_{O3} – the rate of ozone absorption

The experimentally obtained dependencies of the accumulation rates of the alcohol, ketone and ROOR are in a good agreement with the equations derived from the scheme (Fig. 9).

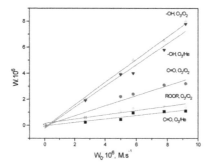 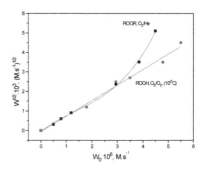

FIGURE 9 Curves of the rates of products formation during cyclohexane ozonolysis by W_{O3} (20°C).

On the basis of Eqs. (35)–(37) the value of α was determined as 0.7 (it stands for this part of alcohol and ketone which has been obtained via the recombination of $RO_2^{\bullet}$-radicals) and ϕ has been found to be 0.3 (the part of the radicals in the solvent bulk) and from the rates of reaction products accumulation was determined the stationary concentration of the peroxide radicals as $1.5.10^{-6}$ M.

It has been supposed in Refs. [11, 18] that ozone reacts with $RO_2^{\bullet}$ – radicals via reaction:

$$RO_2^{\bullet} + O_3 = RO^{\bullet} + 2O_2 \tag{4}$$

and then the alcohol will be obtained through reaction:

$$RO^{\bullet} + RH = ROH + R^{\bullet} \tag{5}$$

However, in this case the dependence of W_{OH} on W_{O3} should be linear in coordinates $W_{OH}/(W_{O3})^{1/2}$, which is nor experimentally observed (Fig. 9). In addition W must be greater than the sum of the rates of all reaction products formation, fact that is also not confirmed be the experimental results. These two facts suggest that Eqs. (4) and (5) could be ignored.

If we accept that similarly to cumene oxidation [14] the square termination step generates RO-radicals:

$$RO_2^{\bullet} + RO_2^{\bullet} = 2RO^{\bullet} + O_2$$

during this reaction, then the additional quantity of alcohol will be provided by Eq. (5). Thus it may be expected that the dilution of cyclohexane with an inert solvent should decrease the rate of Eq. (5), consequently the ratio alcohol:ketone. But the experimental results show that the ratio is not changed at the cyclohexane dilution with tetrachloromethane thus rejecting the generation of oxy radicals from the peroxide radicals recombination.

The yield of H_2O_2 is 2–3%. The formation of H_2O_2 could be explained by its yield via the following reactions [97]:

$$HO^{\bullet} + O_3 \rightarrow HO_2^{\bullet} + O_2$$

$$HO_2^{\bullet} + RH \rightarrow H_2O_2 + R^{\bullet}$$

Adipic acid is obtained at the monomolecular decomposition of $RO^{\bullet}$-radicals:

$$RO^{\bullet} \rightarrow {}^{\bullet}CH_2(CH_2)_4CHO \rightarrow HOOOC(CH_2)_4COOH$$

In addition one should also consider that in the total yield are also included, of course in tenths of percent, some products of the ozone reaction with cyclohexanone, cyclohexanol or cyclohexylperoxide. This shows that the reaction with $RO^\bullet$ – radicals has an insignificant contribution in the total yield.

We have also carried out cyclohexane ozonolysis at reduced concentrations of O_2 by its replacement with Helium. In this case, at equal initial ozone concentration its consumption rate increases two-fold which could be due to the proceeding of the following reaction:

$$R^\bullet + O_3 = RO^\bullet + O_2 \qquad (7)$$

and the increase of the contribution of reaction (5).

Considering Eqs. (7) and (5) W_{OH} and $W_{C=O}$ will depend on W_{O3} in the following manner:

$$W_{O3}{}^{obs} = W_{O3}{}^{RH} + W_{O3}{}^{R\bullet} \qquad (38)$$

$$W_{O3}{}^{R\bullet} = k_7.[R^\bullet][O_3] = 2\varphi \, k_7 [O_3] \, W_{O3}{}^{RH} (k_2 [O_2])^{-1} \qquad (39)$$

$$W_{OH} = \{1-\varphi+2\alpha\varphi+2\varphi k_7[O_3](k_2[O_2])^{-1}\} W_{O3}{}^{RH} \qquad (40)$$

$$W_{C=O} = 2\alpha\varphi W_{O3}{}^{RH} = 2\alpha\varphi k_2 [O_2]\{k_2 [O_2]+2\varphi k_7[O_3]\}^{-1} W_{O3}{}^{RH} \qquad (41)$$

The dependencies of the rates of product formation on W_{O3} using O_3/He and O_3/O_2 demonstrate that at equal rate of Eq. (1) the rates of product formation vary with the replacement of O_2 by Helium (Fig. 9). This fact could be due to the occurrence of Eq. (7), which apparently reduces the efficiency of ozone reaction. The comparison of the equations derived for the processes with and without Eq. (7) provides a reasonable explanation for the relationships given in Table .8.

TABLE 8   The rates of cyclohexanol and cyclohexanone formation at various rates of ozone absorption ($W_{O3} \times 10^6$, $M^{-1}.s^{-1}$)

Rate	O_3/He, $W_{O3} = 7$	O_3/O_2, $W_{O3} = 3.5$	O_3/He, $W_{O3} = 5$	O_3/O_2, $W_{O3} = 2.5$
Cyclohexanol	6.0	3.0	4.8	2.2
Cyclohexanone	0.8	1.2	0.6	0.8
RH	3.6	3.5	2.4	2.5
R$^\bullet$	3.4	–	2.5	–

At low O_2 concentration the rates of alcohol formation will be Eqs. (42)–(43):

$$W_{OH}{}^{He} = (1 - \varphi)\, W_{O3}{}^{RH} + W_{C=O}{}^{He} + k_5[RO^\bullet][RH] \qquad (42)$$

The rate of ozone uptake in its reaction with R -radicals can be estimated solving the following equation:

$$W_{O3}{}^{R\bullet} = W_{OH}{}^{He} - W_{OH}{}^{O2} + W_{C=O}{}^{O2} - W_{C=O}{}^{He} \qquad (43)$$

The values obtained for $W_{O3}{}^{RH}$ and $W_{O3}{}^{R\bullet}$ are in a good agreement with the experimentally observed two-fold rise of W_{O3}.

Under ozonolysis by mixtures O_3/O_2 ROOR are obtained via Eq. (6) (Fig. 9), and with O3/He, because of the higher stationary concentration of $RO_2^\bullet$ – radicals, the rate of the square termination of the alkoxy radicals increases and hence $W_{C=O}$ (Fig. 9). In the chromatograms of the oxidates resulting from the ozonation of cyclohexane solutions in CCl_4 we have identified chlorocyclohexane. This fact confirms the generation of R$^\bullet$ – radicals, which can further abstract Cl – atom from the solvent molecule via Eq. (8):

$$R^\bullet + CCl_4 = RCl + CCl_3{}^\bullet \qquad (8)$$

According to Ref. [14], the rate constant of the reaction at 20°C is 1.59 $M^{-1}.s^{-1}$.

We have estimated the stationary concentration of $R^\bullet$ as 10^{-7} M from the dependence of the rate of cyclohexane formation on the CCl_4 concentration.

The mechanism of ozonolysis includes a stage of radical generation, which has been also assumed by other authors [10]. ESR signals, however, have been registered only at ozonolysis of polyoleffins and cumene [50], which are characterized by relatively low values of k_6 and high $[RO_2^\bullet]_{st}$.

We have confirmed the existence of radical steps in the paraffins ozonolysis by carrying the reaction directly in the ESR resonator. For this purpose we have constructed and worked out a special EPR cuvette (Fig. 10) [50]:

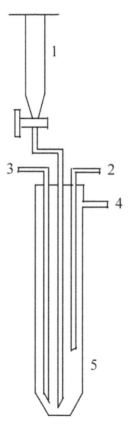

FIGURE 10   Glass unit for carrying out reactions in EPR-spectrometer.

1 – Separating funnel with capillary for liquid reagent feeding; 2 – Capillary connected with the vacuum pump ensuring the necessary level of the liquid in the cuvette; 3 – capillary for the gas inlet; 4 – gas outlet; 5 – EPR-cuvette.

The EPR signals registered are shown in Fig. 11. They have been assigned to by comparison with those obtained in the initiated oxidation of *n*-decane, cumene and polyvinylcyclohexane (PVC) with oxygen. On the basis of the similarities in shape, g-factors and signals width we have assigned them to $RO_2{}^{\bullet}$-radicals.

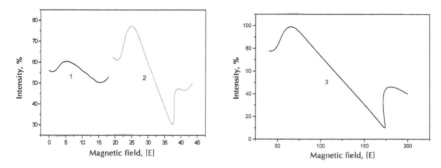

FIGURE 11   EPR spectra of 1 – *n*-decane, 2 – cumene and 3 – PVC at ambient temperature.

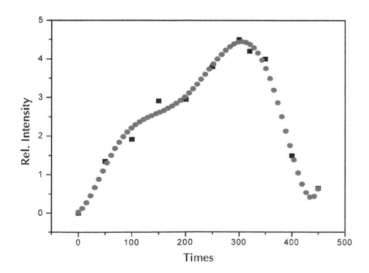

FIGURE 12   Kinetics of $RO_2{}^{\bullet}$ formation at the ozonolysis of powder PVC at 20°C.

The kinetics of ESR signals intensities during the PVC ozonolysis is presented in Fig. 12.

The rise of the intensity of $RO_2^{\bullet}$ – singlet at the beginning has related to the disclosure of new surfaces and ozone interaction with the intermediate products of the reaction. Latter, Ref. [50] has been considered as one of the reasons for the observed course of the CL – intensity in the ozonolysis of 2, 7-dimethyloctane which has been found to be similar.

The drop off of the ESR intensity is due to the consumption of the initial compound (100% conversion after 7 min). When O_3/He mixture was used, as it should be expected, the ESR signal intensity is of an order lower as a result of the active proceeding of reaction 7.

Figure 13 is given as the kinetics of $RO_2^{\bullet}$-radicals formation during cumene ozonolysis. The absolute concentration of $RO_2^{\bullet}$-radicals was determined using Rubin as reference.

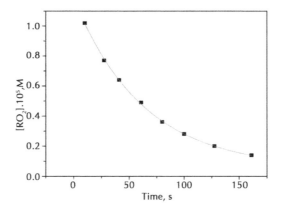

FIGURE 13   Kinetics of $RO_2^{\bullet}$-radicals formation during cumene ozonolysis: $W = 3.5.10^{-4}$ $M.s^{-1}$, 20°C.

The decrease in the intensity of the ESR signal with time is connected to the consumption of cumene because of the high rate of ozone absorption. As the first point in Fig. 13 is registered at 0.6% conversion, the stationary concentration of $RO_2^{\bullet}$-radicals was determined by extrapolation to zero time and amounts to $[RO_2^{\bullet}]_{st} = 1.4.10^{-5}M$. Under the experimental conditions applied the rate of the chain oxidation reaction is very low and thus $RO_2^{\bullet}$-radicals would be consumed exclusively via square termination reaction. The rate of the later as estimated according to $[RO^{\bullet}_2]_{st}$ and $k_6 =$

$1.26.10^4$ $M^{-1}.s^{-1}$ [88] is $5.4.10^{-6}$ $M.s^{-1}$. This value is 64 times lower than that calculated at the assumption that $W_{O3} = W_i$ (initiation rate). This fact confirms that ozone is apparently not an initiator in the studied reaction.

12.5   CUMENE

It has been found that the O_2-oxidation of cumene takes place through a chain-radical route yielding cumylhydroperoxide (CHP), dicumylhydroperoxide (DCHP), acetophenone (AcP), dimethylphenylcarbinol (DMPC), α-methylstyrene, acetone, methanol, formic acid and benzoic acid [14] as reaction products.

The kinetic curves of products formation during cumene ozonolysis are shown in Fig. 14.

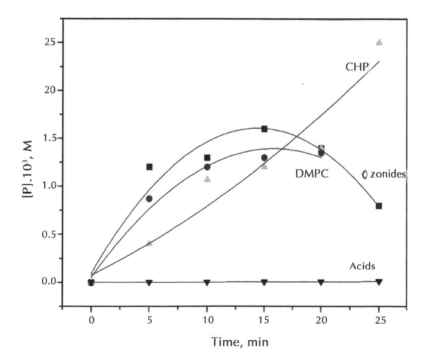

FIGURE 14   Kinetics of products formation during ozonolysis of cumene [RH] =7.17 M; $W_{O3} = 8.6 \times 10^{-8}$ $M.s^{-1}$, 20°C.

The reaction products include ozonides (OZ), DMPC, CHP, DCHP, H_2O and acids.

The linear character and the absence of induction periods in the curves of OZ, CHP and DMPC formation are an indication that these products are obtained in parallel reactions. The pattern of the kinetic curves is typical for the formation of intermediate products. In the case of DMPC the curve reaches a plateau in 10 min, and that of OZ, which at the beginning is identical to the latter after 15 min begins to go down. This fact is apparently related to the ozonides decomposition and polymerization and from this part of the curve its rate of decomposition was calculated as 3×10^{-3} s^{-1}.

The dependencies of the initial rates of DMPC and OZ formation on W_{O3} and of CHP on $(W_{O3})^{1/2}$ are depicted in Figs. 15a and 15b.

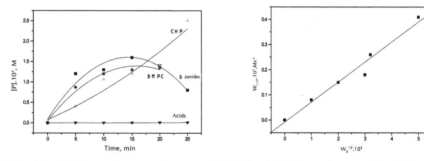

FIG. 15 (a) Dependence on the rates of DMPC and OZ formation ver. W_{O3}, 20°C; (b) Dependence on the rate of CHP formation on $(W_{O3})^{1/2}$, 20°C.

It is seen that while the rates of DMPC and OZ formation depend linearly on W_{O3}, the rate of CHP formation is proportional to $(W_{O3})^{1/2}$. These results suggest the occurrence of parallel steps of the formation of the reaction products as well as of radical stages in the reaction mechanism.

The chemiluminiscent kinetic curve shown in Fig. 16, resulting from the $RO_2{}^{•}$ – radicals recombination [55, 105] becomes a straight line in coordinates $\ln[(I_0^{1/2}+I^{1/2})/(I_0^{1/2}-I^{1/2})]/t.$, where I_0 is the intensity of the chemiluminiscent signal, I – the current intensity. The tangent of the curves slope is equal to $0.5.(W_i.k_6)^{-1/2}$, where W_i is the rate of the initiation equal to $\alpha.W_{O3}$ ($\alpha<1$); k_{6-} the rate constant of the square termination.

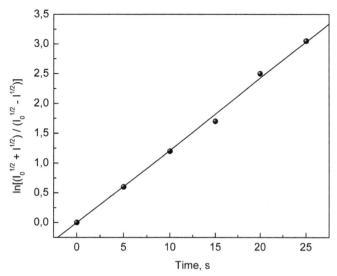

FIGURE 16 Dependence of the chemiluminiscence inetnsity on time at $W_o = 4.2.10^{-5}$ M.s^{-1}.

The results obtained could be described by the following mechanism:

SCHEME 5

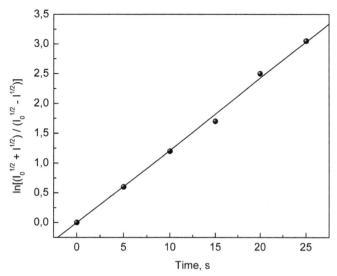

$$RH + HO^{\bullet} = R^{\bullet} + H_2O \tag{1}$$

$$ROO^{\bullet} + RH = ROOH + R^{\bullet} \tag{2}$$

$$2ROO^{\bullet} = 2RO^{\bullet} + O_2 \tag{6}$$

$$RO^{\bullet} + RH = ROH + R^{\bullet} \tag{5}$$

At quasi-stationary approximation [14, 28, 104] the rates of products formation will be presented as Eqs. (44)–(46):

$$W_{DMPC} = (1 + 3\alpha).\beta.W_0, \tag{44}$$

$$W_{CHP} = k_2 (2\alpha\beta W_0/k_6)^{1/2} [RH] \tag{45}$$

$$W_0 = W_{DMPC} + W^I_{DMPC} + W_{CHP} + W_{O3}, \tag{46}$$

where, W'_{DMPC} is the rate of DMPC formation at square termination.

The values of α and β have been determined from the slopes of the kinetic curves in Figs. 15a and 15b and on the basis of the literature values for the rate constants k_2 and k_6 [95] and they are 0.04 and 0.23, respectively. It follows form the tangent of the angle on the OZ curve (Fig. 15a) that 75% of the effective collisions between ozone and cumene are with the benzene ring and only 25% are with the alkyl chain of cumene molecule, which fits very well the calculated value of β from Fig. 16.

From Fig. 15a we have estimated that the value of W_{DMPC} comprises only 7% from that of W_{DMPC}, hence the basic part of carbinol is formed via the first reaction in the kinetic cage.

From the slope of the curve in Fig. 16 and using the literature values of k_6 we have found out that 0.3% of W_{O3} and 1/6 of RO_2-radicals yield products which chemiluminescence. These results correlate well with the aforementioned EPR data.

12.5.1   OZONOLYSIS IN THE PRESENCE OF TRANSITION METAL COMPOUNDS

The data on the ozonolysis of cumene in the presence of transition metals are very scant [106, 107]. This reaction has been studied mainly in sulfuric acid medium and low cumene concentrations. Having this in mind we have carried out the ozonolysis reaction both in pure and 50% solutions of acetic acid [57].

As it has been shown the main products of cumene ozonolysis are OZ, DMPC and CHP. In acetic acid, the product composition remains the same, but the amount of the products is decreased by about 1.6 times. The rates of CHP and OZ formation are lower by about 8.8 and 1.6 fold, respectively, and the rate of DMC accumulation grows 1.6 times. The ratio of the products is found to be 2.1:1.9:1 = OZ : DMPC : CHP. The rate constant (k) characterizing the ozone-acetic interaction at ambient temperature amounts to 1.5×10^{-5} M.s^{-1} that is by about 3.5×10^{-4} smaller than that found for the ozone reaction with cumene (k = 0.35 M^{-1}.s^{-1}). It is known that AcOH deactivates ozone, inhibits RO$_2$·-radicals and protects the benzene ring [9–101]. The deactivation of ozone in AcOH is also confirmed by the change of the molar coefficient in the UV-spectrum as compared with that in CCl$_4$.

The catalytic properties of Co^{2+} in the hydrocarbon oxidation have been the subject of intensive investigations [107]. It has been established that during the cumene-AcOH ozonolysis in 1:1 (v:v) in the presence of Co(AcO)$_2$ the oxidation reaction is accelerated (Fig. 17).In contrast to the noncatalysed process in the catalyzed by transition metal salts the ozonolysis is characterized by: 1) absence of ozonides formation that is indicative of the absence of ozone interaction with the phenyl ring and 2) the main product is DMPC, the accumulation rate of which proportional to the concentration of Co^{2+} after the 10 min. The initial rates of CHP formation do not vary with the changes in Co^{2+} but after the 15 min the rates increase with [Co^{2+}]. It can be seen from Table 9 that if we assume the ozonolysis of pure cumene as a reference then the addition of AcOH results in autoretardation of the oxidation rate and to reduction of the products yield. The ratio [ΣP]/[O$_3$} reaches value of 6.9.

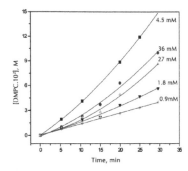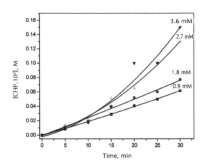

FIGURE 17   Kinetics of DMPC and CHP formation on [Co(AcO)$_2$], 20°C.

TABLE 9   Formation of the sum of products (ΣP) and their ratio per absorbed ozone [ΣP]/ [O$_3$] in non-catalyzed and catalyzed reaction in the presence of AcOH (cumene:AcOH = 1:1) at ambient temperature.

	Time, min	Cumene	Cumene+ AcOH	[Co^{2+}]= 1.10^{3}M	[Co^{2+}]= 2.10^{3}M	[Co^{2+}]= 3.10^{3}M	[Co^{2+}]= 4.10^{3}M	[Co^{2+}]= 5.10^{3}M
[ΣP]×10^3 M	5	0.3	0.2	0.8	0.8	0.9	1.0	1.1
[ΣP]/[O$_3$]		1	0.7	2.7	2.7	3.0	3.3	3.7
[ΣP]×10^3 M	10	0.7	0.4	1.4	1.4	1.8	2.2	3.3
[ΣP]/[O$_3$]		1	0.6	2.0	2.0	2.6	3.1	4.7
[ΣP]×10^3 M	15	1.0	0.6	2.1	2.1	2.9	4.0	5.9
[ΣP]/[O$_3$]		1	0.6	2.1	2.1	2.9	4.0	5.9
[ΣP]×10^3 M	20	1.4	0.8	2.8	2.8	4.5	6.1	8.7
[ΣP]/[O$_3$]		1	0.6	2.0	2.0	3.2	4.3	6.2
[ΣP]×10^3 M	25	1.7	1.0	3.5	3.5	6.2	8.1	11.7
[ΣP]/[O$_3$].		1	0.6	2.1	2.1	3.6	4.8	6.9
[ΣP]×10^3 M	30	2.0	–	4.2	4.2	8.0	10.1	–
[ΣP]/[O$_3$]		1	–	2.1	2.1	4.0	5.1	–

These data indicate the participation of oxygen accompanying ozone in the formation of the reaction products. The course of the kinetic curves of DMPC formation on the Co(AcO)$_2$ concentration is connected with its formation in reactions with participation of Co(AcO)$_2$ (Fig. 17).

The catalytic properties of the metal ions under other equal conditions depend on their rate constant with ozone [102]. For that reason we have determined the values of k in the presence of different transition metals in cumene-AcOH medium. In Table 10 are presented the rate constants of ozone reaction with some transition metals ions at 20°C in cumene: AcOH solution (1:1, v:v).

TABLE 10   Rate constants of the zone reaction with some transition metal ions.

Reagent	$[Me^{+n}]$, mM	k, $M^{-1}.s^{-1}$
Co^{2+}	1	500
Mn^{2+}	1	1500
Cr^{3+}	0.5	64
Cumene	$6.52.10^3$	0.032

Upon ozonolysis of cumene in AcOH the value of W_{O3} is 2.10^{-6} $M.s^{-1}$ and at addition of $Co(AcO)_2$ in concentrations of $(0.9, 1.8, 2.7$ and $4.5).10^{-3}$ its value is 0.6, 1.1, 1.7 and 2.9 $M.s^{-1}$, respectively. This clearly shows that first of all ozone reacts with the Co^{2+} ions.

In conformity with the data obtained we have proposed the following scheme of ozonolysis;

SCHEME 6

$$M^{n+}L + O_3 \rightarrow M^{n+1}L(O_3^-) \tag{0}$$
$$M^{n+1}L(O_3^-) + RH \rightarrow R^\bullet + M^{n+1}L\,(\hat{I}H) + O_2 \tag{1}$$
$$M^{n+1}L(OH) + RH \rightarrow M^{n+}L + R^\bullet + H_2O \tag{1'}$$
$$R^\bullet + O_2 \rightarrow RO_2^\bullet \tag{2'}$$
$$RO_2^\bullet + RH \rightarrow ROOH + R^\bullet \tag{2}$$
$$R^\bullet + O_3 \rightarrow RO^\bullet + O_2 \tag{3}$$
$$\rightarrow (\alpha)\ 2\ RO^\bullet + O_2$$
$$RO_2^\bullet + RO_2^\bullet \,| \tag{6}$$

$$\rightarrow (1 - \alpha)\ ROOR + O_2$$

$$RO^\bullet + RH \rightarrow ROH + R^\bullet \qquad\qquad (4)$$

Applying the quasi-state approximation approach in relation to WDMPC we will obtain:

$$W_{DMPC} = 2.[\alpha/(1-\alpha)].k_0\ .[Me^{n+}L].[O_3].[RH] \qquad (47)$$

The last Eq. (47) has been derived without considering Eqs. (2) and (3) because of their insignificant contribution. It follows from Eq. (47) that W_{DMPC} is proportional to the concentration of the catalyst applied.

By applying literature values for k_2 and k_6 in cumene oxidation [14, 103–105] and $[RO_2]_{st}$ as determined from Fig. 17 the value α is found to be 0.63 that is in a good agreement with that reported by other authors [14]. The analysis of Fig. 18, Table .10 and considering the concentrations of ozone gives the same value of α. The good correlation of the kinetic results with the proposed Eq. (16) testifies its validity.

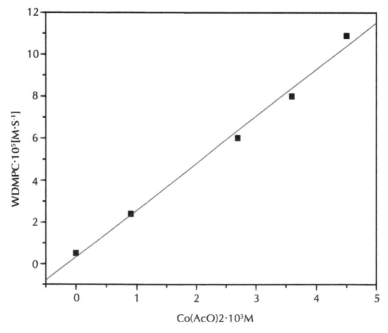

FIGURE 18   Dependence of W_{DMPC} on $[Co(AcO)_2]$

The addition of Co and Mn acetates to the reaction mixture changes the general features of the products formation kinetics (1.19). Thus W_{OZ} increases and acetophenone has been identified (AcP). The sum of the products exceeds by 7.3-fold that at cumene ozonolysis. OZ is obtained at the interaction of ozone with the benzene ring, and AcP is provided by the monomolecular decomposition reaction of $RO^{\bullet}$-radicals. Most likely the main role of Mn is in accelerating of these two reactions. The catalytic properties of the metal salts studied are confirmed by the ratio of the amount of the products formed in the catalyzed and noncatalysed processes per unit of absorbed ozone. At Co:Mn=5:1 this ratio becomes equal to 7.3 and is greater by about 6% than that in the absence of Mn. The cumene conversion is increased but the selectivity of the process is reduced. The contribution of OZ and ApC to the total sum of the products formed is 27%. The synergism of the simultaneous action of the both salts (Fig. 19) can be associated with the occurrence of the following reaction:

$$Mn^{3+}(AcO)_2(O_3^-) + Co(AcO)_2 = Co^{3+}(AcO)_2(O_3^-) + Mn(AcO)_2$$

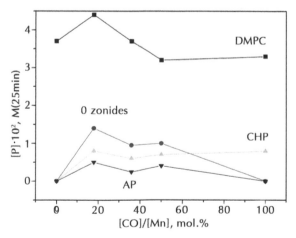

FIGURE 19   Reaction products formation depending on the ratio Co/Mn.

The authors of Ref. [108] stated that with increasing the Mn concentration the catalytic effect should decrease mainly on account of its non-effective interaction with the oxy radicals. The data of Fig. 19, however, show that such a trend is not observed as the variation in the concentrations of DMPC, CHP and AcP are insignificant.

The profiles of the curves are explained by the oxy-reduction interactions of the catalysts and with the radical intermediates.

12.5.2 OZONOLYSIS IN THE PRESENCE OF NIO

The ozonolysis of cumene in the presence of heterogeneous additives, which are effective in ozone decomposition [14], is of interest from two points of view: (a) for elaboration of the mechanism of the ozone interaction with the heterogeneous surface; and (b) to increase the oxidation selectivity.

At bubbling of ozone (0.1 l/min) through a powder NiO (33 m²/g) depending on the oxide concentration a partial or complete ozone decomposition is observed (Fig. 20). The ozone concentration becomes zero at NiO concentrations, 140 mg.

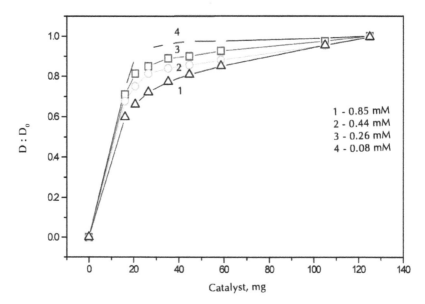

FIGURE 20 Dependence of the ozone decomposed quantity $\Delta D/D_{ot}$ on the NiO amount. $\Delta D = D_o - D_t$; D_o – ozone concentration at the reactor inlet at flow rate of 6 l/h.

In the EPR spectrum a bad-resolved singlet is registered which, however, disappears in the course of the reaction. This signal was assigned to O_3^- anion radicals. The UV spectra have a maximum at 320 nm, which is attributed to the electron transfers in Ni^2. The intensity of the latter goes down in the process of ozone decomposition (the gray color of the fresh NiO turns black during the reaction proceeding). We have estimated that the catalytic coefficient, i.e., the number of decomposed ozone molecules per one NiO molecule is about 20.

By special experiments it has been established that on one Ni^{2+} of the gas phase consequently are absorbed about 133 ozone molecules, but the probability of Ni^{2+} to be transformed to Ni^{3+} is 1/7, i.e. only one from seven ions is oxidized. These results indicate that the reaction of ozone with Ni^{2+} proceeds probably in two ways:

$$NiO + 2O_3 \rightarrow NiO + 3O_2 + Q$$

$$NiO + O_3 \rightarrow ONi^{3+}O_3^-$$

With the formation of surface compounds of $ONiO_3^+O_3^-$ type. They can further react with a new ozone molecule from a neighboring center, with NiO or with the oxides:

$$ONi^{3+}O_3^- + O_3 \rightarrow NiO_2 + 2.5O_2$$

$$ONi^{3+}O_3^- + ONi^{3+}O_3^- \rightarrow Ni_2O_3 + 2O_2$$

$$ONi^{3+}O_3^- + NiO \rightarrow Ni_2O_3 + O_2$$

$$NiO_2 + NiO \rightarrow Ni_2O_3$$

If a hydrocarbon is added to this system a proceeding of oxidation with participation of $ONiO_3^+O_3^-$ might be expected. For this purpose the cumene ozonolysis was carried out in the presence of NiO.

Figure 21 illustrates that OZ, DMPC and CHP are the main products of the cumene ozonolysis in the presence of NiO.

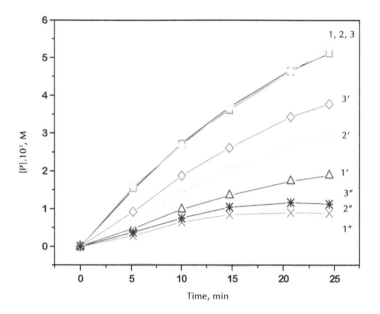

FIGURE 21   Kinetics of products formation of the ozone reaction with cumene in the presence of NiO at 20°C. 1, 1', 1" – DMPC; 2, 2', 2" – CHP and 3, 3', 3" – OZ;. 1, 2 and 3 – 0.635; 1', 2' and 3' – 0.325; 1", 2" and 3" –0.05wt.% NiO.

The rates of OZ, DMPC and CHP formation are increase with [NiO]. The ozonides are obtained from the ozone interaction with C=C bonds in the benzene ring, and the ozonolysis of the hydrocarbon part takes place via a hydrogen atom abstraction mechanism. No ozonides formation on the surface and ozonation of the alkyl part dominates in homogeneous and heterogeneous part of the reaction. The ozone is activated on the NiO surface via the formation of the surface compound Ni^{3+}O$_3^-$according to the scheme oxidizes the alkyl part of cumene:

$$Ni^{3+}O_3^- \rightarrow Ni^{3+} + O_3^-$$

$$O_3^- + RH \rightarrow R^\bullet + HO^- + O_2$$

$$Ni^{3+} + RH \rightarrow R^\bullet + Ni^{3+} + H^+$$

The contribution of these reactions to the formation of the reaction products, is however small, as the total rate of products formation is only from 5 to 17% higher as compared to the rate of ozone consumption.

12.5.3   OZONOLYSIS IN THE PRESENCE OF MO AND V OXIDES

We have studied ozonolyzis of cumene catalyzed by Mo–V catalysts. The reactions have been performed at ambient temperature in a glass reactor with a centrifugal pump with 33 ml cumene in the presence of 0.35wt.% mixtures of MoO_2 and V_2O_5 on SiO_2 (Fig. 22) at various ozone concentrations.

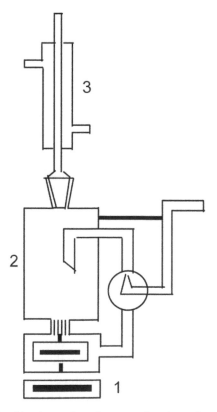

FIGURE 22   Reactor for kinetic studies; 1- magnetic stirrer; 2-reactor; 3-cooler; 4 – centrifugal pump with rotation of 100–1100 r/min.

The catalysts presented in Table 11 were prepared by soaking of SiO_2 with water solutions of ammonium paramolibdate and ammonium vanadate with subsequent treatment at 400°C.

TABLE 11   Catalysts sample at total amount of oxides 10%.

No.	$MoO_2 : V_2O_5$	S, m²/g
1.	0 : 1	13
2.	1 : 99	10
3.	1 : 9	11
4.	1 : 3	5
5.	1 : 1	4
6.	9 : 1	5
7.	1 : 0	10
8.	0 : 0	14

Figure 23 A–C illustrates the kinetic curves of major products formation – CHP, DMPC and OZ . The rate of CHP formation for sample 5 with Mo:V = 1:1 ratio amounts to $1.6.10^{-5}$ M.s⁻¹. The values of W_{CHP} for samples Nos 7 and 8 are similar to that value. In case of sample No1 the initial rate of CHP formation is to $0.4.10^{-5}$ M.s⁻¹ and is the same as that for the noncatalysed reaction (curve K). The untreated SiO_2 (sample 8) exerts also some catalytic properties. The integral areas below (under) the kinetic curves corresponding to the amount of the products formed vary in the presence and in the absence of the catalyst. For example, the area below the kinetic curve of CHP formation in the catalyzed reaction is greater as compared to that of the noncatalyzed one. The stationary concentration of CHP depends on the type of the catalyst and in relation to their catalytic activity the sample shows the following order 5>8>7>1>K. Curve K is characterized by a S-shape form whereby three periods are distinguished: (1) the start of the product accumulation; (2) stationary regime in relation to the product; (3) autocatalytic process of product formation (Fig. 23, A).

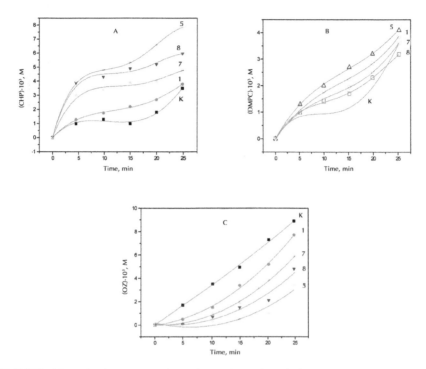

FIGURE 23 Kinetic curves of products formation during the catalyzed cumene ozonolysis. A-CHP, B-DMPC, C-OZ; 1, **5**, 7 and 8 – the number of catalysts according to Table 11; K- in the absence of catalyst, $W_{O3} = 1 \times 10^{-5}$ M.s^{-1}.

Upon DMPC formation (Fig. 23, B) one can also notice some difference in the course of the kinetic curves in the presence (curves 1, 5, 7 and 8) and in the absence (curve K) of the catalyst. It has been found that the rate of DMPC formation is one and the same for all samples studied and it is 4.2×10^{-6} M.s^{-1}. After 10 min curve K becomes constant and after 15 min it again rises, i.e., its profile is typical for an autocatalytic process. The curves of DMPC formation in the presence of catalyst (1, 5, 7 and 8) are characterized by two sections – an initial one with a bigger slope and a second one with a smaller slope and the transition point is in the interval 5–10 min. Obviously, W_{DMPC} is slowed down in the advanced reaction because of the occurrence of secondary reactions associated to the DMPC depletion or are due to the catalyst deactivation. With respect to their activity the samples have the following sequence: 5>1>7>8>K.

The kinetics of OZ formation is entirely different from those discussed before (Fig. 23, C). The initial section of the curves is characterized by the appearance of induction period up to 15 min. These results unambiguously point out that the catalyst direct selectively the process to the oxidation of the isopropyl substituent in the cumene molecule and only after the accumulation of a definite amount of products, which probably block the catalytic surface, the ozone attacs the benzene ring. In this case the order of activity of the samples is just opposite to that of CHP formation, i.e. 5<8<7<1<K.

The effect of the catalyst type on the amounts of products formed (20 min) is demonstrated in Fig. 24.

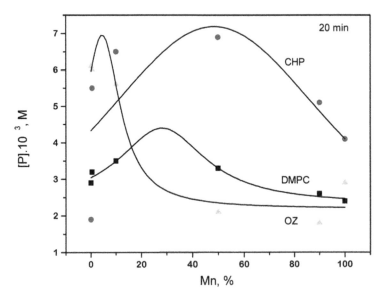

FIGURE 24   Kinetic curves of products formation depending on the Mo:V ratio.

The increase of the Mo content in the catalyst leads to changes in the amounts of DMPC and CHP. The latter rise and reach maximum value at Mo content of 28.2 and 48.5%, respectively. Simultaneously, the OZ yield is reduced with the rise of Mo content.

The catalytic effect of the samples studied is also demonstrated by the increase of the initial rate of oxidation that results in rise of CHP yield. This increase in time can be presented by the following equation:

$$[CHP] = a.t^2 + b.W_i^{1/2}.t \tag{48}$$

For example, the CHP amount formed in the presence of catalyst is 2.4-fold higher as compared with that of the noncatalysed process. The selectivity of the catalysts studied consists in directing the process to oxidation of the isopropyl substituent in the cumene molecule.

12.5.4 CUMENEHYDROPEROXIDE

In order to assess the effect of secondary reactions occurring during cumene ozonolysis we have investigated its reaction in the presence of CHP. The dependence of W_{O3} and $[O_3]$ is studied by means of stop-flow techniques at mixing time of 0.5 s.

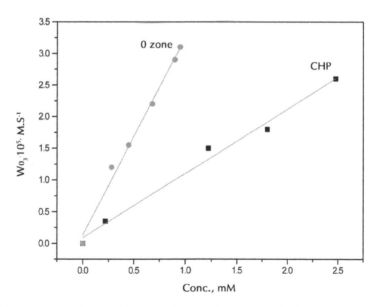

FIGURE 25   Dependence of the rate of ozone absorption on $[O_3]$ ($[CHP] = 2.5$ M) and $[CHP]$ ($[O_3] = 0.68$ mM), 20°C, CCl_4.

As seen the rate depends linearly on the concentration of the both reagents, thus, showing a first order kinetics with respect to each of them.

The rate constant k was determined from the slope and it amounts to to 14.5 ± 0.2 $M^{-1} \cdot s^{-1}$. The main reaction products were found to be DMPC and OZ, while AcP was obtained in much smaller amounts (Fig. 26).

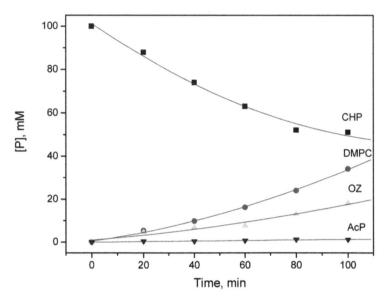

FIGURE 26   Kinetics of products formation during ozonolysis of CHP in CCl_4, $[O_3]_0 = 0.36$ mM.

Ozone reacts simultaneously with both the alkyl and phenyl moiety in the CHP molecule yielding DMPC, AcP and OZ. In the former case the attack on the OOH-group is much more kinetically beneficial resulting in the formation of cumeneperoxide and hydroxy peroxide radical. These radicals according to the classical concepts of oxidation further undergo transformation giving rise to DMPC and AcP. Upon the attack on the phenyl ring, ozone is incorporated into the double bond forming mono-, di- and triozonides. The formation of monoozonide is the limiting step as that moment the aromatic character of the ring is destroyed while the ozone interaction with dienes and olefins occur with high rates and low activation energy [1, 25]. In view of this we have proposed the operation of the following mechanism for this reaction:

SCHEME 7

$$ROOH + O_3 \rightarrow \alpha\,(OZ) + \beta\,(RO_2{}^{\bullet} + HO^{\bullet} + O_2) + \gamma\,(RO^{\bullet} + HO_2{}^{\bullet}) \qquad (1)$$

$$RO_2{}^{\bullet} + RO_2{}^{\bullet} \rightarrow 2\delta\,(RO^{\bullet}) + (1{-}\delta)\,(ROOR) \qquad (6)$$

$$RO^{\bullet} + ROOH \rightarrow DMPC + RO_2{}^{\bullet} \qquad (4)$$

$$RO^{\bullet} \rightarrow AcP + CH_3{}^{\bullet} \qquad (5)$$

On the basis of the above pointed scheme and the available value of k for Eqs. (4) and (5) the theoretical ratio $W_{DMPC}{:}W_{AcP}$ is 25/30:1. The experimentally observed value is 25:1 which is in a very good fit with the theoretical one. Another interesting ratio has been obtained for $W_{DMPC}{:}W_{OZ} = 2.5{:}1$ (Fig.16) and the values of $\alpha{=}0.4$ and $(\beta{+}\gamma){=}0.6$ have been found. The good correlation between the theoretical and experimental values of the $W_{DMPC}{:}W_{AcP}$ ratio suggests that $\gamma{<}1$, i.e. the free-radicals formation is insignificant. The rate constant of ozone reaction with DMPC at ambient temperature has been determined in a separate kinetic experiment and $k{=}3.4{\pm}0.1\ M^{-1}\ s^{-1}$. At this value of k the rate of end products formation becomes commensurable at [CHP]:[DMPC]=1:4.3. The linear profile of the kinetic curves of DMPC, OZ and AcP formation depending on time duration and the absence of induction periods can be regarded as a kinetic evidence for the independent formation of the products in parallel reactions (Fig. 26). Upon ozonolysis of cumene 1 mol absorbed ozone yields 1 mol products, hence the reaction is not a chain one.

The main products in acetic acid medium (Fig. 27) are OZ, DMPC and AcP. The curve of CHP consumption is found to decrease exponentially (it becomes a straight line in semilogarithmic coordinates), the curve of OZ formation passes through a maximum, DMPC is accumulated linearly with the time and that of AcP formation is strongly accelerated after 50 min. The ratio of the initial rates of DMPC:AcP and DMPC:OZ formation are 1:1 and 1:5, respectively. Phenol (PhOH), which have been identified in the oxidate could be obtained from the well-known heterogeneous decomposition of CHP in acid medium giving rise to phenol and acetone.

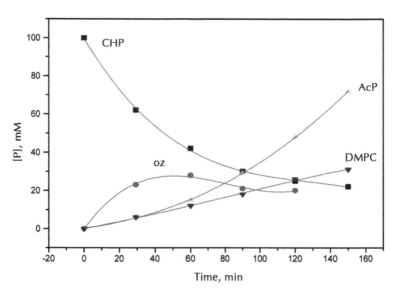

FIGURE 27   Kinetics of products formation during CHP ozonolysis in AcOH, $[O_3]$=0.8 mM.

SCHEME 8

$$Ph-C(Me)_2-OOH + H^+ \Leftrightarrow Ph-C(Me)_2-O-^+OH_2$$

$$Ph-C(Me)_2-O-^+OH_2 \rightarrow Ph-C(Me)_2-O^+ + H_2O$$

$$Ph-C(Me)_2-O^+ \rightarrow Ph-^+O=C(Me)_2$$

$$Ph-^+O=C(Me)_2 + H_2O \rightarrow Ph-O-C(Me)_2-^+OH_2$$

$$Ph-O-C(Me)_2-^+OH_2 \Leftrightarrow Ph-O-C(Me)_2-OH + H^+$$

$$Ph-O-C(Me)_2-OH \; (H^+) \rightarrow PhOH + (Me)_2C=O$$

The composition of products has been found to be different in ozonolysis in $CHCl_3$ as compared with that in pure cumene and in AcOH solution (Fig. 28).

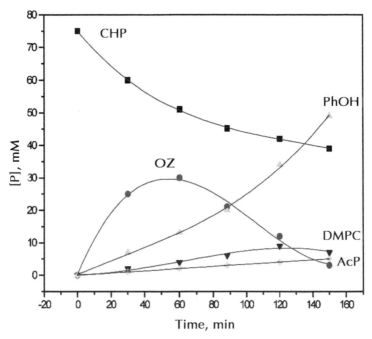

FIGURE 28   Kinetics of products formation at CHP ozonolysis in $CHCl_3$, $[O_3]_0$=0.8 mM.

It has been established that the main products are again OZ, DMPC and AcP but higher amounts of phenol are identified. In addition, the ozone reaction with $CHCl_3$ yields inorganic chlorocontaining acids and the pH of the medium approaches values of 1–2. These acids (probably HCl and HOCl) are efficient catalysts for the CHP decomposition to phenol. The ratios of the rates are $W_{DMPC}:W_{AcP}$ = 2.5:1 and $W_{DMPC}:W_{OZ}$=1:14.

The kinetic curve of ozonides formation is indicative of its formation in consecutive reactions.

The rates of AcP and PhOH formation rise just at the moment of the drop on the kinetic curve of OZ formation (Fig.27 and 28) thus suggesting the participation of OZ in their formation.

The dependence of the initial rate of DMPC, OZ and AcP on W_{O3} in the cumene ozonolysis in CCl_4 is presented in Fig.29. The values of $\alpha = 0.4$, $\beta = 0.6$, $\gamma < 1$ and $\delta = 0.04$ calculated from this figure are in a good agreement with those obtained from the analysis of Scheme 7 and Fig. 26.

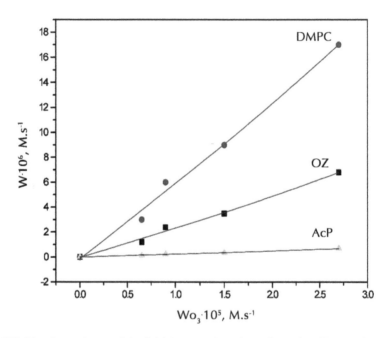

FIGURE 29   Dependence of the initial rates of products formation (W) on the rate of ozone absorption (W_{O3}) in CCl_4, 20°C.

12.6   POLYETHYLENE AND POLYPROPYLENE

We have extended our investigations on ozonation of the C-H bonds in polymer analogs of paraffins – polyethylene, polypropylene, their mixtures and polystyrene, with the aim of elaborating the effect of the polymer structure on this reaction.

The ozonolysis of polyolefins leads to change of the structural-physical and dynamic parameters of the polymer matrix such as: the degree of crystallinity (κ); density of the amorphous phase; orientation factor and segmental mobility of the macromolecules [109, 110]. The variations of these parameters in polymer mixtures have been scarcely studied. This provokes us to study the effect of ozone reaction with oriented polyethylene high density (PE), oriented isotactic polypropylene (PP) and their mixtures on the melting point of crystallites (T_t), the relative melting heat, and thus on the degree of crystallinity and the zise distribution in the crystallites of the two components. The polymer blends were prepared from noninhibited PP samples with average weight mass (M_w) of $2.86.10^5$, average number mass (M_n) of $6.23.10^4$ and the ratio between them (M_w/M_n) of 4.6 and for PE samples with the corresponding values of $4.15.10^4$, $2.71.10^4$ and 1.53. The isotropic films were prepared by pressing of granules at 463 K and under pressure of 150 MPa on a cellophane support and tempered with water at 373 K. The oriented films of 12 ± 2 µm thickness and stretching degree ($\lambda=9$) were obtained by local heating up to 373-383 K and 393–403 K. The melting point and the degree of crystallinity were evaluated by means of DSC method (Fig. 30).

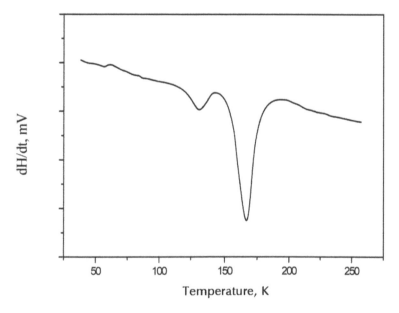

FIGURE 30   DSC curves of the PE:PP = 90:10 blend at $\lambda=9$.

The values of (κ) for PP and PE were estimated using the following values of the relative heats of crystallites melting, i.e., 284.2 and 135.9 kJ/g, respectively. The polydispersity of the crystallites was characterized by the half width of its peak of melting Δh. The oriented samples were subjected to ozonolysis in a thermostated ozone chamber. The reaction kinetics was followed by monitoring the variations of the IR spectral intensity at $v = 1715$ cm^{-1}, characteristic of C=O. As an internal reference the frequency at 1455 cm^{-1}, which remains constant during the course of the experiment, was used.

The dependencies of the oxidation level per unit thickness of the sample on the mixture composition at various oxidation times are presented in Fig. 31.

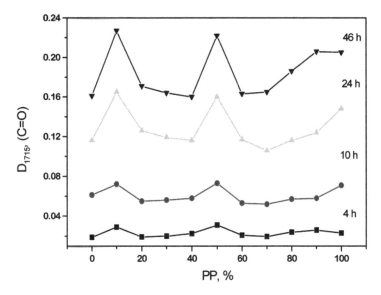

FIGURE 31   Kinetics of C=O accumulation in oriented films form PE/PP mixtures with $\lambda = 9$, $[O_3]_0 = 1$ mM.

After 4 h in the IR spectra of the samples containing 10 and 50% PP are registered two weak maxima. The appearance of the latter could be associated to the more porous structure leading to the better dissolving of ozone in them. This also suggests that the rate of oxidation should increase with increasing the time of ineteraction. We will discuss further the changes of T_T and κ of PE and PP in the ozonized and fresh samples in dependence on the mixture composition. The relationship between T_T of the crystallites in the initial and those ozonized for 45 h at ambient temperature are shown in Fig. 32.

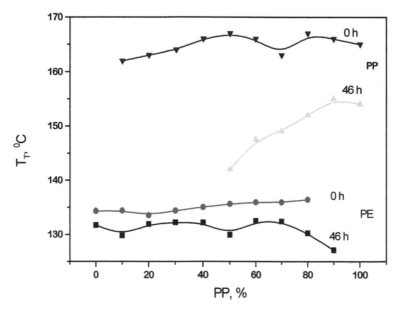

FIGURE 32   Variation of the melting points of PE and PP crystallites in mixtures.

As seen T_T goes down for the PE and PP samples irrespective of the composition and duration of treatment. For PE this decrease is 2–6°C depending on the composition of the mixture and in the case of PP it is considerably higher – 12°C for the initial PP and 24°C for the 1:1 mixture. As for the ozonized samples of PP content within 0–30%, PP crystallites are not observed and for those with PP 40 wt.% their amount is within the limits of accuracy of the DSC method.

The lower melting point of the crystallites films points to a reduced ordering of the crystal structure of PP and PE or for increase of the relative free surface energy of the crystallites sides (σ_e) as compared with that of the nonozonized ones. The disturbance of the crystallites ordering might be ascribed to the penetration of ozone in them. At PE ozonation the ozone penetration in the crystallites depth is much weaker since T_T in the ozonized and nonozonized samples is similar (Fig. 32) or it reacts only with the folded surface of the crystallites [115, 116]. The data in Ref.117 support the latter supposition for it has been found that the treatment by 100% nitric acid of the oriented PE samples of stretching degree $\lambda=10$ leads to 5–8°C reduction of T_T of the crystallites. This decrease is related to the increase of σ_e of the crystallites walls under this treatment. A similar decrease of T_T (2–6°C) has been also established upon ozonation of oriented PE with $\lambda=9$ (Fig. 32) which gives us ground to assume that similar to nitric acid, ozone reduces T_T also on account of σ_e rise.

The ozone ability to penetrate the PP crystallites is accelerated with the decrease of its content from 90 to 50% in the mixtures (Fig. 32).

The marked decrease of m.p. (12–24°C) upon PP ozonolysis could not be explained by the increase of σ_e of the crystallites as they simply disappear. More probably the amorphization is connected to destroy of crystallites as a result of the ozone action. Actually, in the DSC curves of the ozonized mixtures with 0–40% PP content crystallites are not observed while in the IR spectra the amount of the CH_3 groups is not altered. Ozone reacts with the tertiary hydrogen bong in the amorphous and crystal part. The former leads to increase of the number of defects which facilitates the subsequent ozone access. Thus the crystallites are destroyed and further get amorphized.

The melting point appears to be a very sensitive parameter in investigating the behavior of the ozonized samples. From Figs. 31 and 32, it is observed that an antibat dependence between the curves illustrating the dependence of the oxidation level on composition and T_T of the crystallites in blends with various PE content. The periods of acceleration, retardation and of maximum occurrence correspond to decrease, increase and minimum appearance depending on T_T of the PE crystallites in the mixture composition.

Figure 33 demonstrates the effect of mixtures composition on the degree of crystallinity (κ).

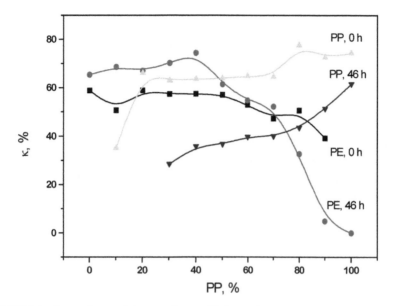

FIGURE 33   Dependence of degree of crystallinity on the polymer blends composition.

First of all we have to note the decrease in the values of κ for PE and PP with the decrease of their content in the mixture both for the nonozonized and in the ozonized samples. However, in the case of the nonozonized samples the profile of κ is characterized by a range whereby it is constant.

The value of κ for PE in mixtures with PE content varying from 0 to 50% amounts to 60% and in those of PP content of 50–70% is 64%. The observation of the decrease of crystallinity degree of the components in the system PE/PP is also confirmed by other authors [117].

In the case of PE the values of κ for the ozonized samples are found to be higher than those of the nonozonized ones. An exception are the samples containing 10 and 20% PE for which κ falls down abruptly. The increase, probably could be explained by the fact that ozone does not penetrate into the PE crystallites but rather reacts with PE in the transition zone on their surface. This reaction leads to break of the macromolecules in the transition zone and thus the crystallinity is enhanced on account of the section that provided by the cut-off molecules in the transition zone. The authors of Ref.118 have also observed a similar increase of κ for PE during its ozonation. These changes of κ are also demonstrated by the curves in Fig. 33.

It has been found that the values of κ in the PP crystallites of ozonized samples are always higher as compared to that in the nonozonized ones (Fig. 33). This observation could be related to the fact that ozone penetrates into the crystallites and reacts further with tert C-H bonds. Thus they are gradually crushed and transformed in an amorphous state. This phenomenon is characterized by a decrease of T_T and κ in the ozonized samples whereby the initial content of PP was lower than 30% and the crystal structure is completely destroyed.

It is well known that the polydispersity of crystallites can be characterized by the halfwidth of the melting peaks provided by the DSC method. The changes of this parameter for the two components are illustrated in Fig. 34.

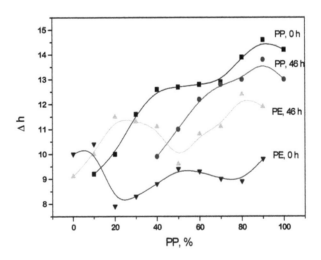

FIGURE 34   Dependence of half width of melting peaks on mixture composition.

For PE, Δh in the ozonized samples is higher than that of the nonozonized ones in the whole range of the mixtures composition with an exception of the pure PE and the mixture with 1 : 9 content. The comparison of the relationship between Δh and κ allow us to assume that the increase in polydispersity of PE crystallites can be explained by physical reasons stipulated by the mixture composition. Thus, in mixtures with high PE content ranging from 30 to 100% is observed a simultaneous increase of Δh and κ. The polydispersity of the crystallites is resulting from the rebuilding of the destructed parts of the macromolecules from the transition zones. In mixtures with lower PE content, below 30%, the rise of polydispersity is accompanied by a drop of the crystallinity degree.

It has been found that the polydispersity of PP crystallites monotonously decrease with the decrease of PP content in the mixtures (Fig. 34). In each case the polydispersity and κ in the ozonized samples are always lower than those in the nonozonized ones in the whole range of mixtures composition. Apparently, the decrease in the PP polydispersity is associated with its initial reaction with the smallest crystallites.

Upon more prolonged ozonation of PE and PP their amorphous phases get considerably altered as a result of the accumulation of oxygencontaining compounds of various molecular mass, which can also destroy them.

12.7 POLYSTYRENE

When bubbles ozone through a polystyrene (PS) solution in CCl_4 it is absorbed [58].

The rate constant at ambient temperature k has been determined according to Eq. (14) and it is 0.37 $M^{-1}.s^{-1}$. The rate of ozone absorption (W_{O3}) remains constant up to the uptake of 3.2 mol ozone per a monomer unit. Further, at consumption of 4.5 mol ozone it has been reduced 10 times. W_{O3} does not depend on the ozone flow rate in it but increases linearly with the increase of ozone and PS concentration (Table 12).

TABLE 12   Dependence of the rate constant on ozone and PS concentration and on the ozone flow rate (v) at 25°C.

$v \times 10^3$, l/s	0.6	0.8	0.8	1.0	1.0	1.0	1.0	1.0	1.0	1.4	2.1	2.8
[PS], M	0.01	0.027	0.054	0.027	0.027	0.027	0.027	0.027	0.027	0.068	0.12	0.072
$[O_3] \times 10^4$, M	0.75	0.75	0.81	0.09	0.23	0.75	1.7	2.9	7.1	0.72	0.78	0.71
$W \times 10^7$, $M.s^{-1}$	4.2	8.4	34	1.2	3.9	11.6	39	35.2	131	35	80	30.4
k, $M^{-1}.s^{-1}$	0.31	0.23	0.45	0.27	0.35	0.32	0.47	0.25	0.38	0.40	0.48	0.33

The analysis of the IR spectra of the ozonized samples (Table 13) shows that the content of t-CH_2-groups practically does not change during ozonolysis as the intensity of the bands at 700 and 2930 cm^{-1} do not change essentially.

TABLE 13   The intensity of the IR-absorption bands of the products of PS ozonolysis: [PS]=27.4 mM; [O_3] = 0.88 mM; ambient temperature.

Time, min	700 cm^{-1}, %	1500 cm^{-1}, %	2930 cm^{-1}, %	$-lg(I_t/I_0)$ at 1500 cm^{-1}
0	36.1	23.0	36.0	0
10	35.4	21.2	35.8	0.035
15	36.0	20.1	35.9	0.059
25	34.5	19.3	35.1	0.076
30	34.9	18.0	35.8	0.106

From the other hand the intensity at 1500 cm^{-1}, which is assigned to the aromatic ring vibrations decreases in the course of the reaction. This fact confirms the ozone interaction with the benzene ring giving rise to ozonides. The latter have been identified by the occurrence of C-O vibrations in the range of 1100–1200 cm^{-1}, as well as by the iodometric titration of the oxidates. It is known that the stoichiometry of the ozone reaction with benzene is 1:3. The kinetic curves presented in Fig. 35 demonstrate a linear kinetics of ozonides formation.

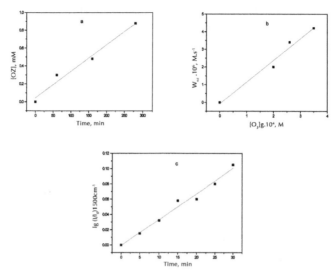

FIGURE 35 Kinetics of ozonides formation (a)–25°C, [PS] = 24.7 mM; $[O_3]_g$=1.2.10^{-5} M; (b) – dependence of the rate   of ozonides formation on ozone concentration at the reactor outlet, 25°C; [PS] = 27.4 mM; (c) variation of the band intensity at 1500 cm^{-1}, 25°C; [PS] = 27.4 mM; $[O_3]_g$ =8.8.10^{-4} M.

The rate of the ozonides formation ate 25°C has a magnitude of 0.29±0.01 M^{-1}.s^{-1}. This value correlates well with the rate constant of benzene ring consumption for cumene which is 0.28±0.02 M^{-1}.s^{-1}.

The kinetics of hydroperoxides accumulation is illustrated by the curves in Fig. 36. Its profile suggests the intermediate character of the hydroperoxides formed in this reaction.

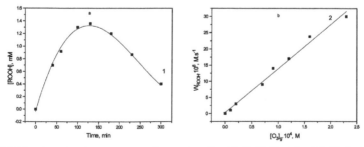

FIGURE 36   (a) Kinetics of ROOH formation (1), 25°C; [PS]$_0$=60 mM; $[O_3]_g$=1.4×10^{-4} M and (b) Dependence of the initial rate of ROOH formation on the ozone concentration at reactor outlet (2), 25°C; [PS]$_0$=27.4 mM.

It is seen that at the initial stage of the reaction [ROOH] increases and W_{ROOH} depends linearly on $[O_3]$ (Fig. 36, curve 2). The ROOH content at 40% conversion is about 6% irrespective of [PS] which is in the range 0.1–1.2%.

The kinetic curves of the variation of PS molecular weight at [PS]=1.5% passes through a maximum (Fig. 37, 1).

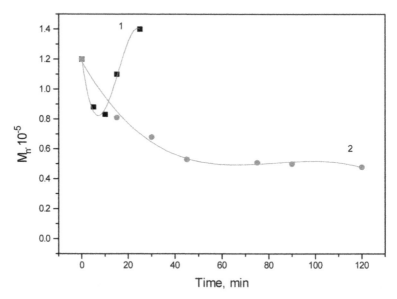

FIGURE 37   Kinetics of PS molecular weight variation with average number molecular mass 1.2×10^5 upon ozonolysis (1) [PS]=3.5%; (2) [PS] =0.3%, $W_{O3} = 2 \times 10^{-4}$ mol/(kg.s); 25°C.

It has been established that carbonyloxides take part in cross-linking processes [1].

For this reason we have studied the PS degradation in solutions with concentrations below 1.5%. Upon ozonolysis of dilutes PS solutions with concentrations below 0.5% we have observed a decrease of the average molecular weight (Fig. 37, 2). The rate of degradation is constant up to a conversion degree of 30%.

$$W_d = (1/M - 1/M_0).t^{-1}.[PS]_0^{-1} \qquad (49)$$

It is linearly dependent on the ozone uptake rate and temperature (Fig. 38).

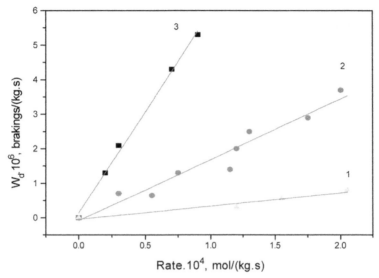

FIGURE 38   Dependence of the rate of PS degradation on the rate of ozone consumption. 1: 0°C; 2: 25°C; 3: 60°C; [PS]<1%.

The number of chain breaks in diluted solutions related to 1 mol absorbed ozone $n=W_d/W_i$ does not depend on the polymer concentration but rises with temperature (Table 14).

TABLE 14   Dependence of the rate of PS degradation on the solution composition and temperature.

t, °C	0	0	25	25	25	25	25	25	25	60	60
[PS],%	1.5	0.7	0.19	0.30	0.42	0.71	1.5	6.5	10.4	1.5	0.7
$W_d\times10^6$	0.88	1.6	3.4	3.8	5.0	1.4	5.8	1.4	0.4	6.0	7.7
$W\times10^4$	2.2	3.1	1.2	2.0	2.2	1.2	3.4	1.7	2.1	1.0	1.2
$n\times10^2$	0.4	0.54	2.8	1.9	2.3	1.2	1.7	0.8	0.2	6.0	6.4
$n_{av}\times10^2$	0.47	±0.07		1.8	±0.6			0.9±0.8		6.2	±0.2

Under these conditions the reactions leading to increase of the molecular mass do not exhibit any essential influence. This is also confirmed by the profiles of the molecular mass distribution (MMD) of low and high molecular components (Fig. 39).

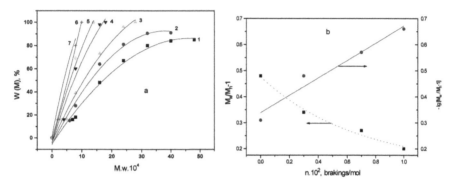

FIGURE 39   (a) Integral curves of MMD at 25°C, $W_{O3}=6.6.10^{-5}$ mol/(kg.s), [PS] = 1% and ozonation time of: 1–0 min, 2 – 15 min, 3 – 30 min, 4 – 45 min, 5– 60 min, 6 – 90 min and 7 – 120 min; (b) MMD in the course of ozonation.

The decrease of the molecular mass and the rise of the number of chain breaking (n) is accompanied by narrowing of the MMD (Fig. 39a) and the ratio M_w/M_n approaches a value equal to 1 (Fig. 39b).

By means of ESR it has been established that the ozonolysis of PS in solid phase generates peroxide radicals [126], which arise not by the direct ozone reaction with PS but are also secondary products. The kinetic scheme of the reaction, proposed by us, is based on the experimentally obtained data and can be presented as follows:

SCHEME 9

$$(-CH_2-(Ph)CH-)_m + O_3 \rightarrow \alpha P_1 + (1-\alpha)(P_2 + P_3 + ...) \qquad (1)$$

$$P_1 + O_3 \rightarrow RO_2^{\bullet} \qquad (2)$$

$$(-CH_2-(Ph)CH-)_m + RO_2^{\bullet} + O_2 \rightarrow ROOH + RO_2^{\bullet} \qquad (3)$$

$$RO_2^{\bullet} + RO_2^{\bullet} \rightarrow \beta(2RO^{\bullet}) \ (6') + (1-\beta) \ ROOH + O_2, \qquad (6'')$$

where; P_1, P_2 and P_3 are intermediate products.

The reactions describing the ozonides formation and the ozone interaction with ROOH are included in the scheme pointed above:

$$(-R-)_m Ph + O_3 \rightarrow monoozonide \ (slow)$$

$$monoozonide + O_3 \rightarrow diozonide$$

$$diozonide + O_3 \rightarrow triozonide \qquad (4)$$

$$\rightarrow \gamma ROH + 2O_2 \qquad (5')$$

$$ROOH + O_3 \dashv$$

$$\rightarrow (1-\gamma) \ (RO_2^{\bullet} + HO^{\bullet} + O_2^{\bullet}) \qquad (5'')$$

The linear dependence of the rate of ozone absorption on its concentration as well as the stoichiometric coefficient of the reaction in the range of 4.2–4.5 can be regarded as evidences for the absence of ozone absorption in a chain reaction. According to Eq. (9) the reaction of ozone consumption will be presented as:

$$W = (k_1 + 3k_4)[PS][O_3] + k_2[P_1][O_3] \qquad (50)$$

and

$$[P_1] = \alpha.[PS]_0.k_1.(k_2 - k_1)^{-1}.\{exp \ [-k_1[O_3].t - exp[-k_2[O_3].t\} \qquad (51)$$

It is apparent that reaction 2 has an insignificant effect in the total balance of ozone consumption since only 6% of ROOH are obtained at the absorption of 1 mol ozone. The value of the observed rate constant k_{obs} is equal to $k_1 + k_4$ and then $k_1 = 0.13$ $M^{-1}.s^{-1}$.

The insignificant amounts of ROOH testify that actually the oxidation of tert.-H bonds does not take place via a chain reaction. At the initial stage of ozonation $[P_1] = \alpha.k_1.[PS]_o[O_3].t$, since $[RO_2] \sim O_3$ and $W_{ROOH} = k_3[RO_2].[PS] \sim O_3$. This is also observed experimentally (Fig.36, 2). The results suggest that the RO_2 radicals formation in PS is not dependent on the phase state of the reaction occurrence. The recombination of RO_2–radicals according to Eq. (6)" gives rise to RO-radicals. The macromolecule destruction takes place via a monomolecular decomposition of RO-radicals:

$$RO^{\bullet} \rightarrow r^{\bullet} + (R')_2 C=O \text{ (degradation)} \qquad (7)$$

In case the degradation process takes place via reaction (7), then $W_d - W_{O3}$. In order to elaborate the mechanism of degradation process we have carried out an initiated oxidation of PS {[AIBN]=2%, 60°C, $W_i = 1 \times 10^{-5}$ mol/(kg.s)}, where AIBN – azoisobutyronitrile. For the inhibited oxidation W_d has been found as 9×10^{-7} breaks/(kg.s) and upon ozonation a similar value for W_d has been obtained at $W_{O3} = 1.5 \times 10^{-5}$ mol/(kg.s) (Fig. 38, 3). If we assume that the initiated oxidation and the ozonolysis differ only in the pathway of RO_2–radicals generation, then the mechanism of destruction and the kinetic relationships should be similar. The experimentally found values of the ratio W_i/W_{O3} accounting for the occurrence of radical steps in the ozonolysis are: 0.05 at 0°C, 0.2–25°C and 0.69 – 60°C (Fig. 38, 1–3). The activation energy of degradation E_a calculated on the basis of Fig. 38 amounts to 7.6 kcal/mol and correlates well with the activation energy of RO-radical izomerisation E_a [120]. The ratio W_i/W could be estimated using the values of k_1 and k_4 and the quasistationary conditions in relation to P_1:

$$W_i/W = 0.43\alpha\beta \qquad (52)$$

From the experimentally measured ratio $W_i/W = 0.2$ at 25°C it follows that $\alpha\beta = 0.47$. Because of lack of any data for the magnitude of β for PS we have used the known value of β for cumene oxidation (low molecular analog of PS) that is 0.86 and then we have obtained 0.55 for α. The latter demonstrates that the major part of RO-radicals have been generated not via reaction 6" but as a result of the direct ozone attack on the PS macromolecule.

Thus it can be reasonably expected that the degradation will occur via C=C bonds, for which is known [1, 25] the rate of their reaction with ozone, i.e., $10^5 M^{-1}.s^{-1}$ and similar to polydienes they would be responsible for the PS degradation. However the amount of C=C bonds in PS is of order of $(3\pm0.5).10^{-2}$ mol/kg and at average number mass of PS – 10–4 mol/(kg.s) which means that one macromolecule comprises 1–2 double bonds. Upon statistical distribution of C=C bonds along the macromolecule length, degradation at $W_{03}\times10^{-4}$ mol/(kg.s) should finish in 5–10 min. However we failed to observe a marked decrease of the molecular weight within this time. Further, W_d should begin to go down. Actually the rate of degradation remains constant even after 150 min when the conversion is already 30%. Apparently, the C=C bonds in PS are not effective in this, mainly because they are terminal double bonds. This supposition has been confirmed by measuring the concentration of C=C bonds which during ozonation decreases with the simultaneous rise of PS molecular mass.

If we take into account that the degradation proceeds via the random law then the equilibrium width of MMD should approach value of 2 only after 2–3 breaks. However this is in contrast to the experimentally obtained data illustrated in Fig. 39b. It is seen that MMD depends on the conversion degree and approaches to 1. This effect could be explainded by the variation of the ratio of the values of the elementary constants, which characterize the breaking of C-C bonds in the macromolecule. A decrease of W_d has been registered at the transition to lower molecular polymers, as is the case with fractionated PS and with polymers with broad MMD. This might be the explanation of the decrease of the effective W_d upon ozonolysis and the narrowing of the MMD [71–73, 108, 111–114].

KEYWORDS

- **benzenering**
- **cumenehydroperoxide**
- **Franck-Condon principle**
- **Hammond empirical rule**
- *mopac6*

REFERENCES

1. Razumovskii, S. D.; Rakovski, S. K.; Shopov, D. M.; Zaikov, G. E. Ozone and Its Reactions with Organic Compounds, Publ. House of Bulgarian Academy of Sciences, Sofia, **1983**; Rakovsky, S.; Zaikov, G., "Kinetics and Mechanism of Ozone Reactions with Organic and Polymeric Compounds in Liquid Phase", Nova Sci. Publ., Inc., Commack, New York, **1998**, 1–345; Rakovsky, S. K.; Zaikov, G. E. "Kinetic and Mechanisms of Ozone Reactions with Organic and Polymeric Compounds in Liquid Phase", (second edition), Nova Sci. Publ., Inc. New York, 2007, 1–340; Zaikov, G. and Rakovsky, S. "Ozonation of Organic and Polymer Copmounds", Smithers Rapra Publ., *Shawbury, Shewsbary, Shropshire*, SY4 4NR, UK, **2009**, 412 pp.
2. Klejmanov, N. A.; Nalbandjan, A. B. Dokl. Akad. *Nauk SSSR*, **1958**, *122(103)*, Ishiguro, K.; Nojima, T.; Sawaki, Y. Novel aspects of carbonyl oxide chemistry, *J. Phys. Org. Chem.* **1997**, *10*, 787–79.
3. Schubert, C. C.; Pease, R. N. The Oxidation of Lower Paraffin Hydrocarbons. I. Room Temperature Reaction of Methane, Propane, n-Butane and Isobutane with Ozonized Oxygen. *J. Am. Chem. Soc.* **1956**, *78*, 2044–2048.
4. Schubert, C. C.; Pease, R. N. Reaction of Paraffin Hydrocarbons with Ozonized Oxygen: Possible Role of Ozone in Normal Combastion. *J. Chem. Phys.* **1956**, *2(4)*, 919–920.
5. Dillemuth, F. J.; Skidmore, D. R.; Schubert, C. C. The Reaction of Ozone with Methane. *J. Phys. Chem.* **1960,** *64*, 1496–1499.
6. Morrissey, R. J.; Schubert, C. .C. The reaction of Ozone with Propane and Ethane. *Comb. Flame 1,* **1963**, 263–268.
7. Dardin, V. J.; Albright, L. N. Partial Oxidation of Propane Initiated by Ozone. Ind. Eng. *Chem. Proc. Des. Dev.* **1965**, *4(61),;* Pullabhotla, V. S. R. R.; Southway, C.; Jonnalagadda, S. B. Oxidation of n-hexadecane with uranyl loaded/anchored microporous zeolites and ozone. *Catalysis Communications* **2008**, *9*, 1902–1912
8. Grewer, T. H. Chapter IV. 4.3. Oxidation of Hydrocarbons by Ozone in: Low Temperature Oxidation (Ed. W. Jost) **1964**, *1*, 186–188
9. Hamilton, G. A.; Ribner, B. S.; Hellman, T. M. On The Mechanism of Alkane Oxidation by Ozone. Oxidation of Organic Compounds. *Adv. Chem. Ser.* **1969**, *77(15)*.

10. Shljapintoh, V. Ya.; Kefeli, A. A.; Gol'denberg, V. I.; Razumovskii, S. D., *Dokl. Akad. Nauk SSSR.* **1969**, *168*, 1132.
11. Razumovskii, S. D.; Kefeli, A. A.; Zaikov, G. E. *Zurnal Org. Khim.* **1971**, *7*, 2045.
12. Durland, J. R.; Adkins, H. *J. Am. Chem. Soc.* **1939**, *61*, 429.
13. Geiseler, F.; Asinger, H. *Wien. Chem. Ber.* **1959**, *92*, 958; Suarez-Bertoa, R.; Saliu, F.; Bruschi, M.; Rindone, B. Reaction products and mechanism of the regioselective oxidation of N-phenylmorpholine by ozone. *Tetrahedron*, **2012**, *68*, 8267–8275.
14. Emanuel', N. M.; Denisov, E. T.; Z. K. Majzus., Hudrocarbon Chain Oxidation Reactions in Liquid Phase. *Nauka*, **1965**, p. 202.
15. Whiting, M. C.;, A. Bolt, J. N.; Parish, J. H. The Reaction between Ozone and Saturated Compounds. Int. Symp. Oxidation Preprints, San Francisco 1967, *2*, 267–287; Giamalva, D. H.; Church, D. F.; Pryor, W. A. Kinetics of ozonation. 5. Reactions of ozone with carbon-hydrogen bonds, *Journal of the American Chemical Society*, **1986**, *108(24)*, 7678–7681.
16. Tjutjunnikov, N.; Ljashenko, A. I.; Shiman, A. M.; Suhoterin, I. S.; Moskwina, G. I.; Drozdov, A. S.; Sanchenko, G. F. *Khim. Promish.* **1971**, 653; Rezaei, E.; Soltan, J.; Chen, N. Catalytic oxidation of toluene by ozone over alumina supported manganese oxides: Effect of catalyst loading. *Applied Catalysis B: Environmental*, **2013**, 136–137 239–247.
17. Williamson, D. G.; Cvetanovic, R. J. Rates of Ozone Reactions in Carbon Tetrachloride Solution. *J. Am. Chem. Soc.* **2949**, 1970, 92
18. Razumovskii, S. D.; Kefeli, A. A.; Trubnikov, G. R.; Zaikov, G. E.; *Dokl. Akad. Nauk SSSR*, **1970**, *192*, 1313.
19. Kang, E. T.; Neoh, K. G.; Zhang, X.; Tan, K. L.; Liaw, D. J. Surface modification of electroactive polymer films by ozone treatment. *Surface and Interface Analysis*, **1996**, *24(1)*, 51–58; Mastan, E.; Wu, J.; Doan, H. An investigation into surface modification of polyethylene films for hydrophilicity enhancement by catalytic ozonation. *Journal of Applied Polymer Science* **2013**,*128*, 828–835.
20. Basiuk, V. A. Organic reactions on silica surface: Synthetic applications. *Uspekhi Khimii*, **1995**, *64(11)*,, 1073–1090; Cataldo, F.; Rosati, A.; Lilla, E.; Ursini, O. On the action of ozone at high concentration on various grades of polyethylene and certain straight chain paraffins. *Polymer Degradation and Stability*, **2011**, *96(5)*, 955–964.
21. Horwath, M.; Vitzky, L.; Huttner, G. Ozone. Budapest Academy Kiadoi **1985**,; Patel, D.; Wu, J. , Chan, P. Upreti, S. Turcotte, G. Y. T. Surface modification of low density polyethylene films by homogeneous catalytic ozonation. *Chemical Engineering Research and Design* **2012**, *90* 1800–1806.
22. Razumovskii, S. D.; Zaikov, G. E. Izv. Akad. *NAuk SSSR. ser. khim.* **1973**, 2244.
23. Tedder, M. Quart. Revs. **1960**, *14* 336; Nakamura, Y.;. Sawada, T.; Komatsu, A.;. Ethanol production from raw starch by a recombinant yeast having saccharification and fermentation activities. *Journal of Chemical Technology and Biotechnology* **2002**, *77* 1101–1106.
24. Herron, T.; Huie, R. E. *J. Phys. Chem.* **1969**, *73* 3327; Hureiki, L.; Croué, J. P.; Legube, B., Doré, M. Ozonation of amino acids: Ozone demand and aldehyde formation. *Ozone: Science and Engineering 1998, 20 381–402.*
25. Bailey, P. S. "Ozonattion in Organic Chemistry", Academic Press, New York, **1982**, v.*I*, 1978, v.II, Adv. Chem. Ser.-A, (Ed. Wasserman, H. H.), **1982**, 39 I, 39-II. 1978.

26. Denisov, E. T.; Denisova, T. G. Reaction Ability of Ozone in the Reactions with C-H. Bonds in Hydrocarbons, Alcohols, and Ketones: Analyst in the Frame of Parabolic Model. Kinet. Katal. **1996**, *37(51)*.

27. Nangia, P. S.;Benson, S. W. Thermochemistry and Kinetics of Ozonation Reactions. *J. Am. Chem. Soc.* **1980**, *102* 3105–3115.

28. Benson, S. W. The Foundations of Chemical Kinetics. Moscow, Mir **1964.**

29. Atkinson, R. Gas-phase tropospheric chemistry of volatile organic compounds. 1. Alkanes and alkenes. J. Phys. Chem. Ref. Data **1997**, *26(20)*, 215–290; Hisahiro Einaga, Shigeru Futamura. Oxidation behavior of cyclohexane on alumina-supported manganese oxides with ozone. Applied Catalysis B: Environmental **2005**, *60* 49–55.

30. Reiser, O. Oxidation of Weakly Activated C-H. Bonds. Angewandte Chemie-International Edition in English **1994**, *33(1)*, 69–72.

31. Razumovskii, S. D. Zurnal Priklod. *Khim.* **1969**, *42*, 2118; Oliveira, R. C. D. M.; Bauerfeldt, G. F. hermochemical analysis and kinetics aspects for a chemical model for camphene ozonolysis. *Journal of Chemical Chem. Physics Phys.* 137 Article number (134306).

32. Razumovskii, SD. *Khim. Promishl.* **1967,** 491.

33. Hatakeyama, S.; Lai, H.; Gao, S.; Murano, K. Production of Hydrogen Peroxide and Organic Hydroperoxides in the Reaction of Ozone with Natural Hydrocarbons in Air. *Chemistry Letters* **1993**, *8*, 1287–1290.

34. Hudson, A. Applications of Electron Spin Resonance to Environmental Problems. Research on Chemical Intermediates **1993**, *19(7)*, 635–641; Ghigo, G.; Maranzana, A.; Causà, M.; Tonachini, G. Theoretical mechanistic studies on oxidation reactions of some saturated and unsaturated organic molecules. *Theoretical Chemistry* Accounts **2007**, *117*, 699–707.

35. Kulik, E. A.; Ivanchenko, M. I.; Kato, K.; Sano, S.; Ikada, Y. Peroxide generation and decomposition on polymer surface. Journal of Polymer Science, Part A-Polymer Chemistry **1995**, *33(2)*, 323–330; Kwamena, N. O. A.; Earp, M. E.; Young, C. J.; Abbatt, J. P. D. Kinetic and product yield study of the heterogeneous gas-surface reaction of anthracene and ozone. Journal of Physical Chemistry **2006**, *A. 110*, 3638–364.

36. Jacob, P.; Wehling, B.; Hill, W.; Klockow, D. Feasibility study of Raman spectroscopy as a tool to investigate the liquid-phase chemistry of aliphatic organic peroxides. *Appl. Spectrosc* **1997**, *51(1)*, 74–80.

37. Gross, Z.; Simkhovich, L. Ozone as primary oxidant in iron (III). porphyrin catalyzed hydroxylation of hydrocarbons. *J. Mol. Catal. A-Chem.* **1997**, *117(1-3)*, 243–248.

38. Atkinson, R. Gas-phase tropospheric chemistry of volatile organic compounds. 1. Alkanes and alkenes. *J. Phys. Chem. Ref. Data* **1997**, *26(2)*, 215–290.

39. J. Cerkovnik.; Plesnicar, B. Characterization and Reactivity of Hydrogen Trioxide, (HOOOH). A. Reactive Intermediate Formed in the Low- Temperature Ozonation of 2-Ethylanthrahydroquinone. *J. Am. Chem. Soc.* **1993**, *115(25)*, 12169–12170; Syroezhko, A. M.; Begak, O. Yu. Proskuryakov, V. A. Mechanism of Methylcyclohexane Ozonolysis. *Russian Journal of Applied Chemistry* **2003**, *76(5)*, 785–790.

40. Marshall, J. A.; Garofalo, A. W. Oxidative Cleavage of Monosubstituted, Disubstituted, and Trisubstituted Olefins to Methyl Esters Through Ozonolysis in Methanolic NaOH. *J.Org.Chem.* **1993**, *58(14)*, 3675–3680; Kahan, T. F.; Kwamena, N. -O. A.;

Donaldson, D. J. Heterogeneous ozonation kinetics of polycyclic aromatic hydrocarbons on organic films. *Atmospheric Environment* **2006**, *40*, 3448–3459.

41. Dunkin, I. R.; Mccluskey, A. Tetrachlorocyclopentadienone O-Oxide, a Facile Oxygen Atom Transfer Reagent-The Oxidation of Cyclohexane. *Journal of Photochemistry and Photobiology A-Chemistry* 1993, *74(2-3), 159*–164.

42. Dhandapani, B.; Oyama, S. T. Gas phase ozone decomposition catalysts. *Appl Catal B-Environ.* **1997**, *11(2)*, 129–166.

43. Encinar, J. M.; Beltran, F. J.; Frades, J. M. Liquid Phase Ozonation of Cyclohexanol Using Acetic Acid as Solvent. Anales de Quimica 1993, 89(7–8), 682–690; Marcq, O.; Barbe, J. M.; Trichet, A.; Guilard, R. Reaction pathways of glucose oxidation by ozone under acidic conditions. *Carbohydrate Research* **2009**, *344* 1303–1310.

44. Galstyan, G. A.; Galstyan, T. M.; Mikulenko, L. I. The Kinetics and Mechanism of the Catalytic Reaction of Ozone with Methylbenzenes in Acetic Acid. *Kinetics and Catalysis* **1994**, *35(2)*, 231–236.

45. Rakovsky, S. K.; Cherneva, D. R.; Deneva, M. Ozone Reactions with Aliphatic Ethers in CCl4-Kinetics and Mechanism. *International Journal of Chemical Kinetics* **1995**, *27(2)*, 153–165.

46. Deninno, M. P. ''Anomalous" ozonolysis of cyclic allylic alcohols: Mechanism and synthetic utility. *Journal of the American Chemical Society* **1995**, *117(39)*, 9927–9928.

47. Razumovskii, S. D.; Rakovski, S. K.; Zaikov, G. E. Influence of the Paraffine Hydrocarbon Structure on its Reaction Rate with Ozone. *Izv. Akad. Nauk, ser. khim.* **1975**, *9*, 1963–1967.

48. Rakovsky, S. K.; Popov, A. A.; Razumovskii, S. D.; Shopov, D. M.; Zaikov, G. E. On the Mechanism of the reaction of Cycloparaffins with Ozone. Comm. *Dep. Chem-Bulgarian Academy of Sciences* **1975**, *8(3)*, 544–557.

49. Popov, A. A.; Rakovski, S. K.; Razumovskii, S. D.; Shopov, D. M.; Zaikov, G. E. Steric Strain Effects on the Kinetic of Chemical Reactions of Cycloparaffins with Ozone. *Comm. Dep. Chem-Bulgarian Academy of Sciences* **1975**, *8 (3)*, 571–577.

50. Rakovsky, K. S.; Razumovskii, S. D.; Zaikov, G. E. Investigation of the Ozone Paraffine Reactions by ESR. *Izv. Akad. Nauk, ser. K`him.* **1976**, *3* 701– 403.

51. Popov, A. A.; Rakovski, S. K.; Shopov, D. M.; Ruban, L. V. Mechanism of the Ozone Reactions with Saturated Hydrocarbons. *Izv. Akad. Nauk, ser. Khim.* **1976**, *40(5)*, 982–990.

52. Ruban, L. V.; Rakovski, S. K.; Popov, A. A. Kinetics and Mechanism of Cyclohexane Oxidation by Ozone. *Izv. Akad. Nauk, ser. Khim.* **1976**, *5*, 1950–1956.

53. Ruban, L V.; Rakovski, S. K.; Razumovskii, S. D.; Zaikov, G. E.; On the Composition of Peroxide Products during Cyclohexane Ozonation. *Izv. Akad. Nauk, ser. Khim.* **1976**, *5*, 2104–2106.

54. Rakovski, S. K.; Cherneva, D. R.; Shopov, D. M.; Razumovskiy, S. D. Application of the Barbotage Method in Investigating the Kinetics of Ozone Reactions with Organic Compounds. *Comm. Dep. Chem-Bulgarian Academy of Sciences* **1976**, *9(4)*, 711–717.

55. Rakovski, S. K.; Cherneva, D. R.; Shopov, D. M.; Parfenov, V. M. Kinetics and Mechanism of Cumene Ozonation. *Comm. Dep. Chem-Bulgarian Academy of Sciences* **1978**, *11(1)*, 153–159.

56. Rakovski, S. K.;, Nenchev, L. K.; Cherneva, D. R.; Shopov, D. M. Mechanism of the Reaction of Ozone with Cumene in the Presence of Silica-Supported MoO3 and V2O5 Catalysts. *Comm. Dep. Chem-Bulgarian Academy of Sciences* **1979**, *12(2)*, 254–1257.
57. Rakovsky, SK.; Cherneva, D. R.; Shopov, D. M.; Cumene Oxidation by Ozone-Oxygen Mixture in the Presence of Transition Metal Salts. *Izv. Akad. Nauk, ser. Khim.* **1979**, *9*, 1991–1995.
58. Kefely, A. A.; Rakovski, S. K.; Shopov, D. M.; Razumovskii, S. D.; Rakovski, K. S.; Zaikov, G. E. Kinetic Relationships and Mechanism of Ozone Reaction with Polystyrene in CCl4 Solution. *J. Polymer Science: Polymer Chem.* Edition **1981**, *19(6)*, 2175–2184.
59. Razumovskii, S. D.; Rakovski, S. K.; Shopov, D. M.; Zaikov, G. E. Ozone and Its Reactions with Organic Compounds (monograph in Russian). *Publishing House of the Bulgarian Academy of Sciences* **1983**, *1*, 278.
60. Rakovski, S. K.; Cherneva, D. R.; Shopov, D. M. Ozonolysis of Cumene Hydroperoxide in Different Solvents. *Oxid. Commun.* **1984**, *7(1)*, 181 – 190.
61. Blinov, N. N.; Popov, A. A.; Rakovski, S. K.; Stoyanov, A. K.; Shopov, D. M.; Zaikiv, G. E. Change of the Melting Point, Polymolecularity and Polycristalinityof Components of High Density Polyethtylene-Polypropylene Blend under Deep Ozone Oxidaton. *Visoko Molekul. Soedin.* **1989**, *A. 31*, 2212–2218.
62. Rakovski, S. K.; Shopov, D. M.; Cherneva, D. R. The Investigation of Ozone Reaction with Cumene Hydroperoxide as a Model of Polyolefinic Degradation. 25-th Microsymposium "Processing and Lohg-Term Stabilities of Hydrocarbon Polymers", Prague, July 18–21, **1983**, *33*, 1–4.
63. Rakovsky, S. K.;, Nenchev, L. K.; Cherneva, D. R.; Shopov, D. M. Catalytic Degradation of Ozone over NiO. 7-th Intern. *Symp. Heter. Catal., Varna*, **1979**, *2*, p.231–236.
64. Rakovsky, S. K.; Cherneva, D. R.; Spozhakina, A. A.;Novakov, K. . Catalytic Decomposition of Ozone over Suported Metal Oxide Catalysts. 8-Th Intern. *Symp. Heter. Catal., Varana*, 5–9 October, **1996**, p. 407–413.
65. Martinez, R. I. The Mechanism of O3-Aldehyde Reactions. *Int. J. Chem.Kinet.* **1982**, *14* 237–249.
66. Komissarovva, I. N., Komissarov, V. D.; Denisov, E. T. *Izv. Akad. Nauk ser. khim.* 1978, *9*, 1991; Syroezhko, A. M.; Proskuryakov, V. A.; Begak, O. Yu. *Russian Journal of Applied Chemistry* **2002**, *75(8)*, 1296–1303.
67. Hellman, T. M.; Hamilton, G. A. *J. Am. Chem. Soc.* **1974**, *96* 1530.
68. Stary, F. E.; Emge, D. E.; Murray, R. W. Ozonation of Organic Substrates. Hydrotrioxide Formatin and Decompozition to Give Singlet Oxygen. *J. Am. Chem Soc.* **1976**, *98* 1880–1884.
69. Vedeneev, I.; Gurvich, L. V.; Kondrat'ev, V. N.; Medvedev, V. A. Frankevich, E. L. Energy Breaking of the Chemical Bonds , Ionization Potencials, and Electron Affinity. Publ. House Akad. Nauk SSSR, Moscow, 1962, p. 90; Gronert, S. An alternative interpretation of the C-H. bond strengths of alkanes. *Journal of Organic Chemistry* **2006**, *71(3)*, 1209–1219.
70. Voevodskii, V. V. Physics and Chemistry of Elementary Chemical Processes. Nauka, Moscow, **1969.**
71. Frank-Kamenetzkii, D. A.; Diffusion and Heat Transfer in Chemical Kinetics. Nauka, Moscow, **1987.**

72. Sherwood, T. K.; Pigford, R. L.; Wilke, C. R.; Mass Transfer. McGraw Hill, New York, **1975.**
73. Danckwerst, P. V. Gas-Liquid Reactions. McGraw Hill Book Comp., New York, **1970.**
74. Astarita, G. Mass Transfer with Chemical Reaction. Elsevier, Amsterdam, **1967.**
75. Hammond, G. S. *J. Am. Chem. Soc.* **1955,** *77,* 334.
76. Hisatsune, I. C.; Kolopajlo, L. H.; Heicklen, *J. J. Am. Chem. Soc.* **1977,** *99,* 3704.
77. Alcock, W. G.; Mile, B. *J. Chem. Soc. Chem. Comm.* **1976,** *5.*
78. Bukert, U.; Allinger, N. L. Molecular Mechanics. *ASC. Monograph* **1982,** *177,* Washington, D.C.
79. Chen, K. S.; Allinger, N. L.; A. Molecular Mechanics Study of Alkyl Peroxides. *Journal of Computational Chemistry* **1993,** *14(7),* 755–768.
80. Eliel, E. L.; Allingerí. N. L.; Angyal, S. J.; Morrison, G. A. Conformational Analysis, Moscow, Mir **1969.**
81. Eyring, H.; Lin, S. H.; Lin, S. M. Basic Chemical Kinetics. *J. Wiley,* New York **1980.**
82. Levine, I. N.; Physical Chemistry.McGraw Hill, New York, **1988.**
83. Benson, S. W.; The Foundations of Chemical Kinetics. Moscow, Mir **1964.**
84. Tyutyulkov, N.; Quantum Chemistry. Nauka i Izkustvo, Sofia **1978.**
85. Klark, T.; A. Handbook of Coputational Chemistry, *J. Wiley,* New York 1985,; Olesen, S. G.; Hammerum, S. Hydrogen bonding to alkanes: Computational evidence. *Journal of Physical Chemistry* **2009,** *A. 113(27),* 7940–7944.
86. Dobbs, K. D.; Dixon, D. A.; Komornicki, A. Ab initio Prediction of the Barrier Height for Abstraction of, H. from CH4 by OH. *J.Chem.Phys.* **1993,** *98(11),* 8852–8858.
87. Dewar, M. J. S.; Hwang, J. C.; Kuhn, D. R.; An AM1 Study of the reactions with Ozone with Ethylene and 2-Butene. J. Am. Chem. Soc. 1991, 113(3), 735–741; Saeys, M.; Reyniers, M. F.; Marin, G. B.; Van Speybroeck, V.; Waroquier, M.; Ab Initio Calculations for Hydrocarbons: Enthalpy of Formation, Transition State Geometry, and Activation Energy for Radical Reactions. *Journal of Phys.Chem.* **2003,** *A. 107(43),* 9147–9159.
88. Meredith, C.; Quelch, G. E.; Schaefeer, H. F.; III. Open-Chain and Cyclyc Protonated Ozone: The Ground-State Potencial-Energy Hypersurface. *J. Am. Chem. Soc.* **1991,** *113* 1186–1193.
89. Karadakov, P.; Rakovsky, S. K.; Computer Program.-Calculation of Pre-exponents According to the Activated Complex Theory and Collision Theory. *Inst. Org. Chem., Inst Catlal.* **1988.**
90. Herzberg, G.; Molecular Spectra and Molecular Structure (two volumes), New York, **1954/1957.**
91. Benson, S.; Thermochemical Kinetics. Moscow, Mir **1971.**
92. Pitzer, K. S. J. *Chem. Phys.* **1946,** *14(239).*
93. Bazilevskii, M. V.; Molecular Orbitals Method and Reaction Ability of Organic Molecules. Moscow, Khimia **1969;** Vreven, T.; Morokuma, K.; The accurate calculation and prediction of the bond dissociation energies in a series of hydrocarbons using the IMOMO. (integrated molecular orbital+molecular orbital) methods. *Journal of Chem. Phys.* **1999,** *111(19),* 8799–8803.
94. Bazilevskii, M. V.; Trosman, Z. T. *Uspehi Khimii* **1960,** *41(3);* Förner, W., Badawi, H. M. Theoretical vibrational spectra of cyclohexanecarboxaldehyde. *Journal of Molecular Modeling* **2001,** *7(8),* 288–305.

95. Radinovitch, E.; *Trans. Farady Soc.* **1937**, *33*, 1225.

96. OEvans, M. G.; Polany, M. *Trans. Farady Soc.* **1936**, *32*, 1333.

97. Galimova, L. G.; Komissarov, V. D.; Denisov, E. T. *Izv. Akad. Nauk SSSR, ser. khim.* **1973**, 307.

98. Dioxin, M.; Ebb, E. Ferments. Mir, Moscow, **1966**, p. 277; Cerkovnik, J.;, Eržen, E.; Koller, J.; Plesni?ar, B.; Evidence for HOOO. radicals in the formation of alkyl hydrotrioxides (ROOOH) and hydrogen trioxide (HOOOH) in the ozonation of C-H. bonds in hydrocarbons. *Journal of the American Chemical Society* **2002**, *124(3)*, 404-409.

99. Karnogitzkii, V.; Organic Peroxides. IL, Moscow, **1961**, p. 115.

100. Kritchfild, F.; Analysis of Basic Functional Groups in Organic Compounds. Mir **1965**, p. 188; Baj, S. Quantitative determination of organic peroxides. Fresenius' Journal of Analytical Chemistry **1994**, *350(3)*, 159-161.

101. Hawkins, E. J. E.; Organic Peroxides. Khimia, Moscow **1965**.

102. Johnson, R. M.; Siddiqi, J. W. The Determination of Organic Peroxides. Pergamon Press, New York **1970**, p. 15.

103. Denisov, E. T.; Rate Constants of the Homolytical Liquid-Phase Reactions. Nauka, Moscow **1971**.

104. Denisov, E. T.; Kinetic of Homogeneous Chemical Reactions. Visshaya Shkola **1988**.

105. Shlyapintoh, V. Ya.;, Karpuhin, O. N.; Postnikov, L. M.; Zaharov, I. V.; Vichutinskii, A. A.; Tzepalov, V. F. Chemiluminescent Methods of Investigation of the Slow Chemical Processes. Nauka, Moscow, **1966**.

106. Yakubi, V. A.; Ponomarev, B. A.; Tupalo, N. F.; Galstyan, G. A.; Petrenko, L. P.; Tyutyunnik, S. M.; Karpuhin, P. P. Theory and Practis of Liquid Phase Oxidation. Ed. Emanuel' N. M.; Nauka, Moscow **1974**, p. 303; Ghanbari, B.; Ferdosi, S. R.; Tafazolian, H.; Solvent-free oxidation of cumene by molecular oxygen catalyzed by cobalt salen-type complexes. *Research on Chemical Intermediates* **2012**, *38(3-5)*, 871–883.

107. Galstyan, G. A. Pross. II-nd Ozone Conference, MGU, Moscow, **1977**, p.40; Potapenko, E. V.; Andreev, P. Yu.; Catalytic oxidation of alkylbenzenes with ozone in acetic acid in the presence of strong acids. *Petroleum Chemistry* **2012**, *52(2)*, 113–118.

108. Nenchev, L. K.; Shopov, D. M. Neftekhimia **1978**, *18(2)*, 256.

109. Karpova, S. G.; Popov, A. A.; Chvalun, S. N.; Zubov Yu, A.; Zaikov, G. E. *Visoko Molekul. Soed.* **1985**, *B. 27(9)*, 686.

110. Karpova, S. G.; Popov, A. A.; Chvalun, S. N.; Zubov Yu, A.; Zaikov, G. E. *Visoko Molekul. Soed.* **1986**, *A. 28(7)*, 1404.

111. Popov, A. A.; Rusak, A. V.; Ledneva, O. A.; Zaikov, G. E. *Visoko Molekul. Soed.* **1986**, *A. 28(9)*, 1836.

112. Blinov, N. N.; Popov, A. A.; Popova, E. S.; Butusov, A. V.; Neverov, A. N.; Zaikov, G. E.; Irinyi, G. *Polym. Deg. Stab.* **1988**, *22(7)*.

113. Cipitoin, A.; Constantinesku, A.; Dobresku, V. *IUPAC. Macro'83, Budapest, Sec.* **1983**, *4* P.727.

114. Patel, K. H. H. J. *Polym. Sci. Polym. Phys.* **1975**, *13(2)*, 303.

115. Popov, A. A.; Blinov, N. N.; Zaikov, G. E. *Dokl. Akad. Nauk SSSR.* **1986**, *286(4)*, 922.

116. Selihova, V. I.; Ozerina, L. A.; Ozerin, L. N.; Bakeev, N. F. *Visoko Molekul. Soed.* **1986**, *A. 28(2)*, 342; Jean Jacques Robin, Overview of the Use of Ozone in the Synthesis of New Polymers and the Modification of Polymers. *Advances in Polymer Science* **2004**, *167(3)*, 5–80.

117. Priest, D. J. *J. Polym. Sci.* **1971**, *A. 29(10), 1777.*
118. Karpova, S. G. PhD. Theses. *Inst. Chem. Phys. Moscow*; Robert, W. Keller, Oxidation and Ozonation of Rubber. *Rubber Chemistry and Technology* **1985,** *58(3),* 637–652.
119. Gaponova, I. S.; Gol'dberg, V. M.; Zaikov. G. E.; Kefely, A. A.; Pariiskii, G. B.; Razumovskii, S. D.; Toptigin, V. Ya. *Visoko Molekul. Soed.* **1978,** *A. 20,* 2037.
120. Howard, J. A.; Ingold, K. V. *Can. J. Chem.,* **1969,** *47,* 3797.

CHAPTER 13

ADAPTOGENS DECREASE THE GENERATION OF REACTIVE OXYGEN SPECIES BY MITOCHONDRIA

V. ZHIGACHEVA and E. B. BURLAKOVA

CONTENTS

13.1   INTRODUCTION

In this chapter, we studied the influence of spatially-hindered phenols and plant growth regulators (PGR) on the intensity of lipid peroxidation (LPO) in membranes of mitochondria. Under stress conditions, the intensity of fluorescence of LPO products in membranes of these organelles increases 3 to 4 times. Spatially-hindered phenols and PGRs decrease the LPO intensity to the control values and maintain thereby a high functional activity of mitochondria. Prevention of dysfunction of mitochondria is associated with enhancement of the resistance of plant and animal organisms to the action of stressors.

Various stresses lead to disfunction of bioenergetic functions of mitochondria and an excess produce of ROS that underlie the development of pathological processes [1]. One of the most important tasks is the search for preparations and methods for protecting the organism from an oxidative stress caused by the action of adverse environmental impacts. We assumed that the main property of adaptogens is the reduction of the of ROS level in cells. Since mitochondria are the main source of ROS under stress conditions, we proposed the hypothesis that the basic mechanism of action of preparations-adaptogens is the reduction of generation of ROS.

The aim of this chapter was to verify the hypothesis using antioxidants and plant growth regulators.

Mitochondria, as an energy metabolism regulator, play one of the basic roles in the organism response to the action of stressors. About 1 to 3% of oxygen consumed by mitochondria form, as a result of 1-electron reduction, reactive oxygen species (ROS), which participate in the cellular redox-signaling.

Normally, the bound level of ROS in organs and tissues is rather low (on the order of $10^{-10}-10^{-11}$ M) due to the enzymatic and non-enzymatic systems of regulation of the accumulation and elimination of ROS. A shift in the antioxidant–prooxidant relationship towards increasing the ROS generation is a result of the action of stressors and leads to the development of pathological states. In this connection, the problem of search for new preparations having adaptogenic properties is very urgent.

Under stress conditions, the main sources of ROS are mitochondria [1]. Consequently, it was possible to put forward the assumption that these agents should affect primarily the generation of ROS by these organelles.

The main candidates for this role are antioxidants. Since the respiratory chain of mitochondria of plants and animals has a common pattern of organization, and the principal differences relate to the CN-resistant electron transfer and the structure of the NADH-dehydrogenase region of the respiratory chain [2], the basic mechanisms of adaptogens functioning were studied in both animal and plant mitochondria.

It is well known that spatially-hindered phenols have antioxidant properties [3, 4]. Therefore, the test subjects used were preparations that are spatially-hindered phenols: potassium phenosan (2,6-ditert-butyl-4-hydroxyphenyl-propionic acid potassium salt) and anphen sodium (1-N-(acetylamido)-1-(3,5-di-tert-butyl-4-hydroxybenzyl)-methyl-malonate).

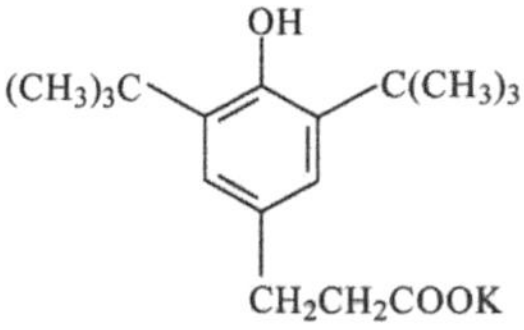

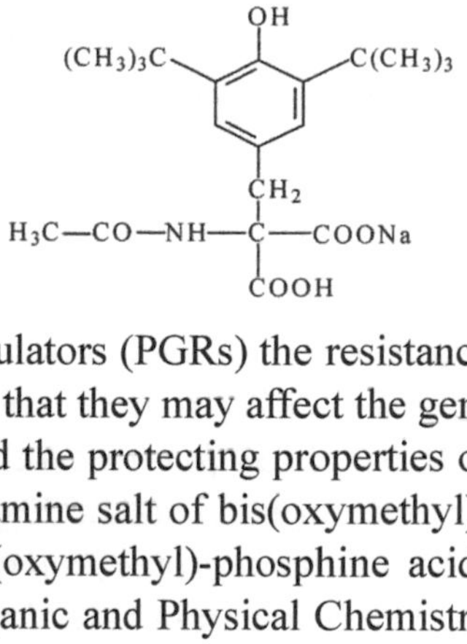

Since plant growth and development regulators (PGRs) the resistance of plants to stresses [5], it could be assumed that they may affect the generation of ROS by mitochondria. We studied the protecting properties of PGRs using as test subjects melaphen (melamine salt of bis(oxymethyl)-phosphine acid) and pyraphen (salt of bis(oxymethyl)-phosphine acid) (synthesized in the Arbuzov Institute of Organic and Physical Chemistry of the Russian Academy of Sciences, Kazan Science Center):

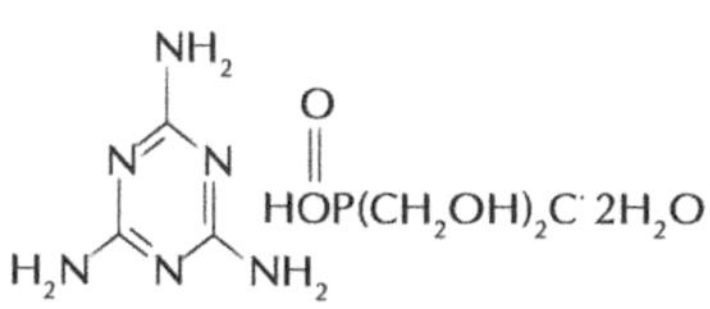

Melaphen

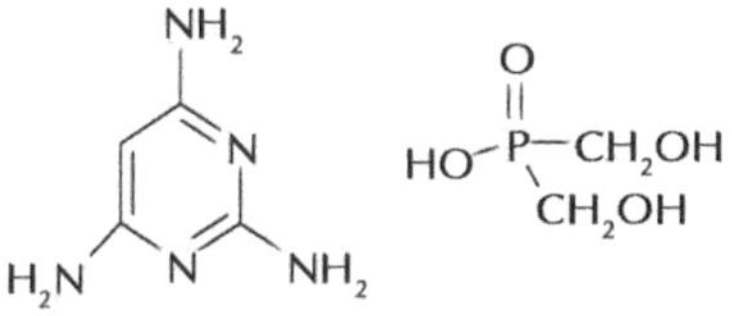

Pyraphen

The aim of the chapter was also to study the bioenergetic characteristics of mitochondria under stress conditions and the effect of the preparations on these characteristics

13.2 MATERIALS AND METHODS

The study was carried out in Wistar line rat liver mitochondria, etiolated pea germs mitochondria, and mitochondria isolated from sugar beetroot storage parenchyma.

The isolation of liver mitochondria was carried out by a method of differential centrifugation [6]. The first centrifugation was carried out at 600 g for 10 min.; the second, at 9000 g for 10 min. The precipitate was resuspended in the isolation medium. The tissue: medium ratio was 1:0.25. The isolation medium was: 0.25 M sucrose, 10 mM HEPES, pH 7.4.

The isolation of mitochondria from 6-day pea germ epicotyls *(Pisum sativum)* or sugar beet roots *(Beta vulgaris L.)* performed by a method of [7] of our modification. Pea epicolyls having a length of 1.5 to 5 cm (20–25 g) or 25 to 30 g of the sugar beet storage parenchyma were homogenized with 100 ml of the isolation medium containing 0.4 M sucrose, 5 mM EDTA, 20 mM KH_2PO_4 (pH 8.0), 10 mM KCl, 2 mM dithioerythrite, and 0.1% BSA (free of fatty acids). The homogenate was centrifuged at 25,000 g for 5 min. The second centrifugation was carried out at 3000 g for 3 min. *The supernatant was centrifuged for 10 min at 11,000 g for mitochondria sedimentation. The sediment was re-suspended in 2–3 ml of solution contained: 0.4 M sucrose, 20 mM KH_2PO_4 (pH 7.4), 0.1 % BSA (without fatty acids) and mitochondria were precipitated by centrifugation at 11,000 g for 10 min. The suspension of mitochondria (about 6 mg of protein/ml) was stored in ice.*

The respiration rate of mitochondria from rat liver, sugar beet roots, and pea germs were recorded using a Clark type electrode on an LP-7 polarograph (Czechia). The incubation medium of liver mitochondria contained 0.25 M *sucrose*, 10 mM tris-HCl, 2 mM $MgSO_4$, 2 mM KH_2PO_4, 10 mM KCl (pH 7.5) (28°C). The incubation medium of sugar beetroots mitochondria and pea germs mitochondria contained 0.4 M *sucrose*, 20 mM HEPES-tris-buffer (pH 7.2), 5 mM KH_2PO_4, 4 mM $MgCl_2$, 0.1% BSA (28°C).

Protein was measured with a biuretic method.

The level of LPO was measured with a fluorescence method [8]. Lipids were extracted by the mixture of chloroform and methanol (2 : 1). Lipids of mitochondrial membranes (3–5 mg of protein) were extracted in the glass homogenizer for 1 min at 10°C. Thereafter, equal volume of distilled water was added to the homogenate, and after rapid mixing the homogenate was transferred into 12mL centrifuge tubes. Samples were centrifuged at 600 g for 5 min. The aliquot (3 ml) of the chloroform (lower) layer was taken, 0.3 ml of methanol was added, and fluorescence was recorded in 10 mm quartz cuvette with a spectrofluorometer (FluoroMax-HoribaYvon, Germany). Background fluorescence was recorded using a mixture of 3 ml chloroform and 0.3 ml methanol. The excitation wavelength was 360 nm, the emission wavelength was 420–470 nm. The results were expressed in arbitrary units per mg protein.

The protecting activity of the preparations was studied using the following models:

1. **Model of acute hypobaric hypoxia.** The rarefaction corresponding to the altitude of nine thousand meters was created in the low-pressure chamber. The "ascent" to five thousand meters was performed during the first minute and to one thousand meters during every succeeding minute. (The residence time for rats at the altitude of nine thousand meters was 5 min).

2. **Model of acute alcoholic intoxication.** Ethanol at a dose of 140 mg was introduced to mice subcutaneously.

To all test animals the preparation was introduced intraperitoneally at a chosen dose 45 min before the event.

The study of stress impacts and plant growth regulators was carried out in pea germs using a model of insufficient watering. Pea (Pisum sativum L., cv. Alpha) seeds were washed with soapy water and 0.01% KMnO4. Control seeds were then soaked in water, experiment seeds – in 2×10^{-12} M melaphen for 1 h. Thereafter, seeds were transferred into covered trays on moistened filter chapter in darkness for a day. In a day, half of control and half of melaphen-treated seeds were transferred in the open trays on dry filter chapter (water deficit treatment). Another half of control plants were retained in closed trays on wet filter chapter, where they were kept for 5 days. After two days of water deficit treatment, seeds were transferred to covered trays on wet filter chapter, where they were kept for next two days. On the fifth day, mitochondria were isolated from seedling epicotyls.

Reagents: methanol, chloroform (Merck, Germany), saccharose, Tris, FCCP (carbonylcyanide-*p*-trifluoromethoxyphenylhydrazone), malate, glutamate, succinate (Sigma, USA), BSA (Sigma, USA), HEPES (MP Biomedicals, Germany).

13.3   RESULTS AND DISCUSSION

Since mitochondria are the main source of ROS under stress conditions, it was necessary to develop a stress-simulating model, i.e., to find conditions under which the generation of ROS by mitochondria will be increased and, consequently, LPO will be activated. We solved the problem by having developed a model of "aging" (incubation of mitochondria in a salt medium at room temperature). To activate LPO, mitochondria were placed for 15 min in a 0.5 ml medium containing 70mM KCl, 10 mM HEPES and 1mM KH_2PO_4, pH 7.4. The incubation of mitochondria in the hypotonic KCl solution resulted in a weak swelling of mitochondria and an increase in the ROS generation, which was expressed in a 3 to 4-fold increase in the intensity of fluorescence of LPO products. The introduction of 10^{-6} and 10^{-13}M anphen or 10^{-8}–10^{-16}M potassium phenosan into the incubation medium caused a decrease in the LPO intensity to control values (Fig. 1).

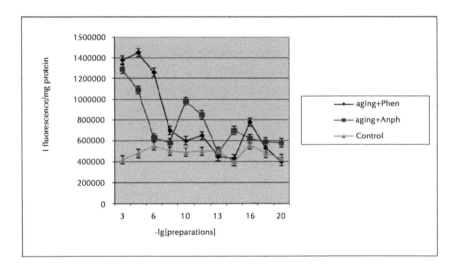

FIGURE 1   The effect of "aging", potassium phenosan (Phen), and anphen (Anph) on the intensity of fluorescence of LPO products in membranes of liver mitochondria. *Y-line: the fluorescence intensity in arbitrary units/mg of protein.*

The study of the influence of the preparations on lipid peroxidation was performed in intact animals subjected to a stress. For a stress impact we used models of acute hypobaric hypoxia (AHH), acute alcoholic intoxication, and a low-temperature stress. These stresses resulted in the LPO activation and dysfunction of liver mitochondria of rats. The acute hypobaric hypoxia and acute alcoholic intoxication resulted in a 1.5 to 3-fold increase in the intensity of fluorescence of LPO products in membranes of liver mitochondria of rats. The introduction of 10^{-14} M potassium phenosan or 10^{-6}M and 10^{-13}M anphen to rats 45 min before the impact prevented the activation of LPO (Figs. *2* and *3*).

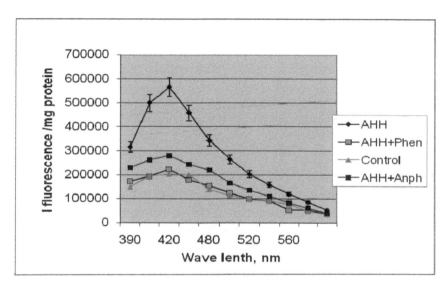

FIGURE 2   The effect of acute hypobaric hypoxia (AHH), potassium phenosan (Phen), and anphen (Anph) on fluorescence spectra of LPO products. *X-line: Wave length, nm; Y-line: Fluorescence intensity in artbitrary units/mg of protein.*

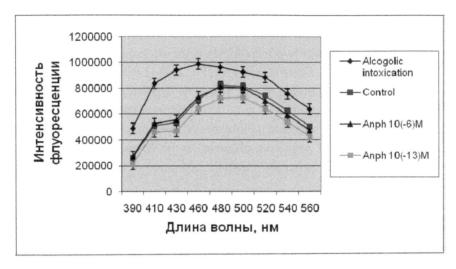

FIGURE 3 Fluorescence spectra of LPO products in membranes of liver mitochondria of rats under acute alcoholic intoxication and on the introduction of anphen (Anph) to rats under alcoholic intoxication. *Y-line: Fluorescence intensity in arbitrary units/mg of protein; X-line: Wave length, nm*

Under conditions of AHH, there was observed a decrease in the highest rates of NAD-dependent substrate oxidation: the rates of malate +glutamate oxidation in the presence of ADP or uncoupler (FCCP) were by 24.3–30% lower than in control mitochondria (Table 1).

TABLE 1 The effect of hypobaric hypoxia, potassium phenosan, and anphen on rates of oxidation of NAD-dependent substrates (Oxidation rates are expressed in ng moles of O_2/ mg.protein min, number of experiments is 10).

Group	Vo	V_3	V_4	V_3/V_4	FCCP
Control	6.5 ±1.4	28.1±1.1	8.0±0.4	3.51±0.04	27.5±1.0
AHH	7.3±1.2	21.2±1.6	9.3±0.2	2.27±0.03	19.4±2.0
AHH+ potassium phenosan	8.3±1.6	28.4±1.3	7.8±0.7	3.62±0.07	32.4±2.5
AHH+ Anphen	7.0±2.0	27.6±1.4	7.8±0.9	3.54±0.2	28.4±1.3

The incubation medium contained: 0.25 M *sucrose*, 10 mM tris-HCl, 2 mM KH_2PO_4, 5 mM $MgSO_4$, 10 mM KCl, pH 7.5. Additives: 200 μM ADP, 10^{-6}M FCCP carbonylcyanide-*p*-trifluoromethoxyphenylhydrazone), 4 mM glutamate, 1 mM malate.

The introduction of potassium phenosan or anphen to rats prevented changes in the functional characteristics of liver mitochondria.

We noted that the studied preparations used at concentrations of 10^{-12}–10^{-16}M increased 3.5 to 4.5 times the lifespan and increased the survival of mice by 20 to 30% under condition of different types hypoxia, low-temperature stress, and acute alcoholic intoxication (Table 2).

TABLE 2 Protective action of potassium phenosan and anphen (Results of 10 experiments).

Impact	Parameter	Potassium phenosan, M		
		0	**10^{-5}**	**10^{-14}**
Ascent to the altitude of 11.5 thousand m	Lifespan, min., survival %	4.0±1.1 20%	6.7 ±1.2 30%	15. 0 ±2.8 50%

Impact	Parameter	Potassium phenosan, M		
		0	**10^{-5}**	**10^{-14}**
Injection of sodium azide 20 mg/kg	Lifespan, min., survival %	3.0±0.6 0%	8.5±2.6 40%	11.0±1.1 50%
Injection of sodium nitrite 250 mg/kg	Lifespan, min., survival %	15.1±2.5 20%	33.6±1.8 20%	53.5±3.0 60%
Swimming with load at 2˚C	Lifespan, min., survival %	3.2±0.6 0%	7.2±0.6 0%	14.3±2.4 0%

Anphen, M				
0 10^{-13} 10^{-6}				
Injection of sodium azide 20 mg/kg	Lifespan, min., survival %	2.3±0.5 50%	10.5±0.8 30%	5.2±1.2 40%
Injection of sodium nitrite 250 mg/kg	Lifespan, min., survival %	20.5±3.1 0%	38.6±5.3 20%	35.7±6.4 15%
Injection of ethanol 140 mg	Lifespan, min., survival %	35.4±6.1 0%	100.4±26 20%	137.4±31.1 12%

The hypothesis that the main property of the preparations-adaptogens is decreasing the excessive generation of ROS and, consequently, the reduction of the intensity of lipid peroxidation in biological membranes and, first of all, in membranes of mitochondria was verified using the preparations that influence the growth and development of plants, i.e., plant growth regulators. The studies were carried out in mitochondria isolated from the sugar beetroot storage parenchyma.

NOTES: 1. In accordance with the model of hypobaric hypoxia, white rats (males, 25 g) were placed in a glass low-pressure chamber connected with a vacuum pump; the rarefaction corresponding to the altitude of 11.5 thousand meters was created in the chamber. The ascent rate was 100 m/sec. **2.** In the model of cytotoxic hypoxia, sodium azide at a dose of 20 mg/kg was injected mice intraperitoneally. **3.** In the model of hemic hypoxia, sodium nitrite at a dose of 250 mg/kg was injected mice intraperitoneally. **4.** In the model of a combined action of a low-temperature stress and a muscular exercise, 500 mg loads were attached to mice placed to a 20°C bath. **5.** In the model of acute alcohol intoxication, ethanol at a dose of 140 mg was injected mice subcutaneously.

To all test animals potassium phenosan or anphen at a corresponding dose was introduced intraperitoneally 30 min before the event.

It appeared that PGRs under study affect the intensity of fluorescence of LPO products depending on the functional state of mitochondria and concentration of the preparations in the incubation medium. Melaphen at concentrations of 2×10^{-7}, 2×10^{-12} and $2\times10^{-18} - 2\times10^{-22}$ M decreased the intensity of fluorescence of LPO products in membranes of "aged" mitochondria (Fig. 4).

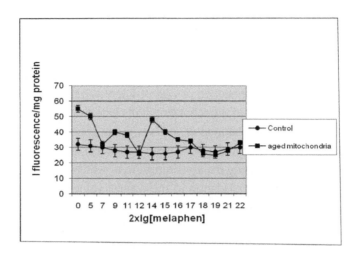

FIGURE 4   The intensity of fluorescence of LPO products on the introduction of different concentrations of melaphen to the incubation medium of mitochondria isolated from the sugar beetroot storage parenchyma. *Y-line: the fluorescence intensity in arbitrary units/mg of protein; X-line: melaphen concentration, M.*

A pyrimidine analogue of melaphen – pyraphen used at concentrations of 10^{-5}, 10^{-7}, and 10^{-13}–10^{-18} M decreased the content of LPO products in membranes of mitochondria to the control values (Fig. 5).

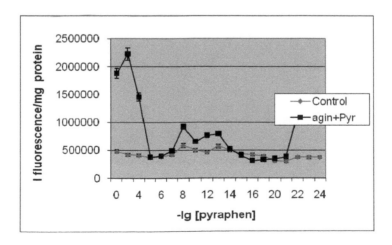

FIGURE 5   The effect of different concentrations of pyraphen (Pyr) on the intensity of fluorescence of LPO products in membranes of mitochondria. *Y-line: the fluorescence intensity in arbitrary units/mg of protein; X-line: pyraphen concentration, M.*

Our hypothesis that the protective properties of the plant growth regulators are due to the effect of these preparations on the functional state of mitochondria was verified using the model of insufficient watering. It is known that a water deficit decreases the functional activity of both chloroplasts and mitochondria [9]. In this connection, it was of interest to find out whether the bioenergetic characteristics of mitochondria will change under conditions of a water deficit and after the treatment of seeds with melaphen or pyraphen. The protective properties of the preparations were checked at the concentrations of melaphen and pyraphen, using the concentration in which these plant-growth regulators reduced lipid peroxidation to the control values. The test subjects were mitochondria isolated from 5-day etiolated pea germs. The insufficient watering resulted in activation of the free radical oxidation in membranes of pea germs mitochondria; this was evident from a 3-fold increase in the intensity of fluorescence of products of lipid peroxidation (LPO). The treatment of the seeds with melaphen or pyraphen decreased the content of LPO products. Melaphen appeared the most efficient preparation that decreased the intensity of fluorescence of LPO products to the control values (Fig. 6).

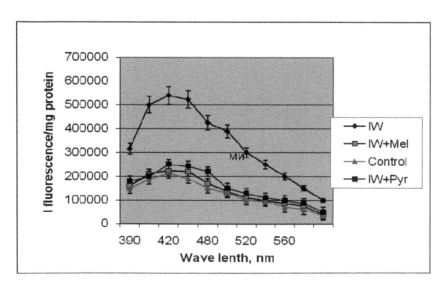

FIGURE 6 Fluorescence spectra of LPO products in membranes of pea germs mitochondria under conditions of insufficient watering (IW) and on the treatment of seeds with melaphen (Mel) or pyraphen (Pyr). *Y-line: the fluorescence intensity/mg of protein; X-line: wave length, nm.*

The insufficient watering resulted in a decrease in the maximum rates of oxidation of NAD-dependent substrates (Table 3).

TABLE 3   The effect of insufficient watering and the treatment of pea seeds with melaphen or pyraphen at a rate of oxidation of a glutamate+malate pair by mitochondria of 5-day pea germs (Oxidation rates are expressed in ng atoms of O_2/mg.protein min; number of experiments is 10).

Group	V_0	V_3	V_4	V_3/V_4	V_{FCCP}	V_{KCN}
Control	20.0±1.5	68.0±4.1	30.0±2.0	2.27±0.01	70.0±4.6	20.1±1.5
Insufficient watering	12.0±2.0	48.6±3.0	40.2±1.0	1.70±0.02	48.9±3.2	18.3±2.0
Insufficient watering+treatment of seeds with melaphen	19.8±3.0	66.0±2.4	27.5±1.3	2.40± 0.02	75.3±5.2	18.9±1.2
Insufficient watering+treatment of seeds with pyraphen	18.5±2.4 (10)	50.0±2.1	26.3±1.1	1.90±0.02	52.0±3.4	17.6±1.8

The incubation medium: 0.4 M *sucrose*, 20 mM HEPES-Tris buffer (pH 7,2), 5 mM KH_2PO_4, 4 mM $MgCl_2$, and 0.1% BSA, 10 mM malate, 10 mM glutamate. Additives: 200 μM ADP, 10^{-6}M FCCP (carbonylcyanide-*p*-trifluoromethoxyphenylhydrazone).

The oxidation rates of glutamate+malate in the presence of FCCP decreased from 70.0±4.6 to 48.9±3.2 ng of oxygen atom/mg of protein □ min and the respiratory control rate (RCR) decreased from 2.27±0.01 to 1.70±0.02. The treatment of seeds with melaphen prevented changes in the efficiency of oxidative phosphorylation caused by the water deficit. Also, such treatment restored the oxidation rates of NAD-dependent substrates in the presence of ADP or FCCP to the control values. It was found out that the treatment with pyraphen, which affects the LPO intensity to a lesser degree, nearly does not affect the maximum rates of oxidation of NAD-dependent substrates. The maximum rates of oxidation of NAD-dependent substrates were measurable with the corresponding values of oxidation of these substrates by mitochondria of germs under conditions of insufficient watering and untreated with pyraphen. However, the efficiency of the oxidative phophorylation increased: the respiratory control rate increased from 1.70±0.02 to 1.90±0.02. It can be assumed that maintaining a high functional activity of mitochondria is the base for the resistance of plants to stress impacts, in particular, to insufficient watering.

The antistress properties of the preparations showed themselves in the physiological parameters (Fig. 7). Insufficient watering inhibited the growth of roots and sprouts of pea germs. The treatment of pea seeds with melaphen or pyraphen stimulated the growth of roots under conditions of insufficient watering 5 and 1.75 times, respectively. Pyraphen, which affects the LPO intensity to a lesser degree, nearly does not affect the growth of sprouts, whereas melaphen stimulates the growth of sprouts 3.5 times. We emphasize that the discovered stimulation of growth of roots of sprouts under the drought condition is of great importance for adaptation.

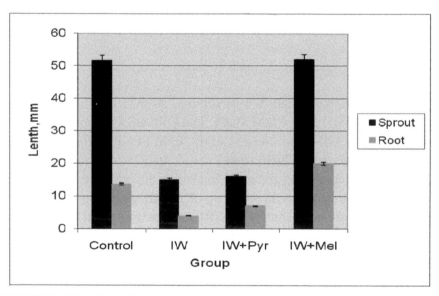

FIGURE 7   The effect of insufficient watering (IW) and the treatment of pea seeds with melaphen (Mel) or pyraphen (Pyr) on the length of sprouts and roots of 5-day germs.

However, the efficiency of melaphen in the protection of plants from a water stress is somewhat higher than that of pyraphen; this is due to a more effective protection of membranes from lipid peroxidation

Thus, the studied biologically active compounds and plant growth regulators and antioxidants reduce the intensity of lipid peroxidation in membranes of mitochondria, which is the basis for maintaining a high functional activity of mitochondria and determines the protective activity of these compounds.

KEYWORDS

- aging
- lipid peroxidation
- plant growth regulators
- reactive oxygen species
- spatially-hindered phenols

REFERENCES

1. Plotnikov, E.; Chupyrkina, A.; Vasileva, A.; Kazachenko, A.; Zorov D. The role of reactive oxygen and nitrogen species in the pathogenesis of acute renal failure. **2008**, *BBA 1777*, S58–S59.
2. Shugaev, A. G. Some aspects of structural organization and oxidative activity of the respiratory chain of plant mitichondria. *Uspekhi sovremennoi biologii*, **1991**, *111(2)*, 178–191
3. Ershov, V. V.; Volodkin, A. A.; Prokofiev, A. I.; Solodovnikov, S. P. Reactions *of spatially hindered phenols and derivatives thereof with one electron transfer. Uspekhi khimii*, **1973**, *42(9)*, 1622–1649
4. Volodkin, A. A.; Zaikov, G. E. Potassium and sodium 2,6-di-tert-butylphenolates and properties thereof. *Izvestiya Akademii Nauk, Seriya Khimicheskaya.* **2006**, *12*, 2138–2143.
5. Chalova, L. I.; Ozeretskovskaya, O. L. *Biological inductors of protective reactions of plants and their possible practical use. Biochemistry of immunity, rest, and aging of plants.* Ed. I. V. Berezin, M. "Nauka", **1984**, 41–57.
6. Mokhova, E. N.; Skulachev, V. P.; Zhigacheva, I. V. Activation of external pathway of NADH oxidation in liver mitochondria of cold-adapted rats. *BBA* **1977**, *501*, 415–423.
7. Popov, V. N.; Ruge, E. K.; Starkov, A. A. Effect of electron transport inhibitors on the generation of reactive oxygen species in the oxidation of succinate by pea mitochondria *Biokhimiya*, **2003**, *68(7)*, 910–916.

8. Fletcher, B. I.; Dillard, C. D.; Tappel, A. L. Measurement of fluorescent lipid peroxidation products in biological systems and tissues. *Anal. Biochem.* **1973**, *52*, 1–9.
9. Shugaeva, N. A.; Vyskrebentseva, E. I.; Orekhova, S. O.; Shugaev, A. G. Effect of water deficit on the respiration of the conducting bundle of sugar beetroot chard. *Fiziologiya rastenii,* **2007**, *54(3)*, 373–380.

INDEX